Advances in
ORGANOMETALLIC CHEMISTRY

VOLUME 31

Advances in Organometallic Chemistry

EDITED BY

F. G. A. STONE

DEPARTMENT OF INORGANIC CHEMISTRY

THE UNIVERSITY

BRISTOL, ENGLAND

ROBERT WEST

DEPARTMENT OF CHEMISTRY

UNIVERSITY OF WISCONSIN

MADISON, WISCONSIN

VOLUME 31

ACADEMIC PRESS, INC.

Harcourt Brace Jovanovich, Publishers

San Diego   New York   Boston
London   Sydney   Tokyo   Toronto

ACADEMIC PRESS, INC.
San Diego, California 92101

United Kingdom Edition published by
ACADEMIC PRESS LIMITED
24-28 Oval Road, London NW1 7DX

LIBRARY OF CONGRESS CATALOG CARD NUMBER: 64-16030

ISBN   0-12-031131-3   (alk. paper)

PRINTED IN THE UNITED STATES OF AMERICA
90 91 92 93   10 9 8 7 6 5 4 3 2 1

Contents

Highly Reduced Metal Carbonyl Anions: Synthesis, Characterization, and Chemical Properties

JOHN E. ELLIS

I.	Introduction	1
II.	Tetracarbonylmetallates (3−) of Manganese and Rhenium	2
III.	Pentacarbonylmetallates (3−) of Vanadium, Niobium, and Tantalum	15
IV.	Tricarbonylmetallates (3−) of Cobalt, Rhodium, and Iridium	31
V.	Tetracarbonylmetallates (4−) of Chromium, Molybdenum, and Tungsten and Related Materials	39
VI.	Concluding Remarks	47
	References	48

The Interplay of Alkylidyne and Carbaborane Ligands in the Synthesis of Electronically Unsaturated Mixed-Metal Complexes

F. GORDON A. STONE

I.	Introduction	53
II.	Synthesis of Salts Containing the Anions $[W(\equiv CR)(CO)_2(\eta^5\text{-}C_2B_9H_9R'_2]^-$ (R = alkyl or aryl; R′ = H or Me), $[Mo(\equiv CC_6H_4Me\text{-}4)(CO)\{P(OMe)_3\}$-$(\eta^5\text{-}C_2B_9H_9Me_2)]^-$, and $[W(\equiv CR)(CO)_2(\eta^6\text{-}C_2B_{10}H_{10}Me_2)]^-$ (R = aryl)	56
III.	The Carbaborane Group as a Spectator Ligand	59
IV.	Expolyhedral B–H→Metal Bonding in Di- and Trimetal Compounds	63
V.	Formation of Exopolyhedral σ Bonds between Cage Boron Atoms and Transition Elements	70
VI.	Reactions Leading to Transfer of Alkylidyne Groups to the Carbaborane Cage	77
VII.	Protonation of the Anionic Complexes $[W(\equiv CR)(CO)_2(\eta^5\text{-}C_2B_9H_9R'_2)]$ (R = alkyl or aryl, R′ = H or Me)	84
VIII.	Conclusion	87
	References	88

Transition Metal Complexes Incorporating Atoms of the Heavier Main Group Elements

NEVILLE A. COMPTON, R. JOHN ERRINGTON, and NICHOLAS C. NORMAN

I.	Introduction	91
II.	Gallium, Indium, and Thallium	92
III.	Germanium, Tin, and Lead	108
IV.	Antimony and Bismuth	130
V.	Tellurium	154
	References	176
	Note Added in Proof	182

Organo-Transition Metal Compounds Containing
Perfluorinated Ligands

RUSSELL P. HUGHES

I.	Introduction	183
II.	Fluorocarbons vs. Hydrocarbons	183
III.	Organometallic Compounds with Perfluorinated Ligands	184
IV.	Octafluorocycloocta-1,3,5,7-Tetraene and Its Valence Isomers	202
V.	Unsuccessful Attempts to Prepare Transition Metal Complexes of OFCOT	204
VI.	Manganese Complexes of OFCOT and Perfluorotricyclo[4.2.0.0^{2,5}]Octa-3,7-Diene	205
VII.	Iron Complexes of OFCOT and Perfluorotricyclo[4.2.0.0^{2,5}]Octa-3,7-Diene	211
VIII.	Cobalt and Rhodium Complexes Derived from OFCOT and Perfluorobicyclo[4.2.0]Octa-2,4,7-Triene	217
IX.	Nickel, Palladium, and Platinum Complexes Derived from OFCOT	240
X.	Metallation of OFCOT and Its Valence Isomers	245
XI.	Vicinal Defluorination of OFCOT to Give μ-Hexafluorocyclooctatrieneyne Complexes of Cobalt	255
XII.	Concluding Remarks	261
	References	261

Alkyl- and Aryl-Substituted Main-Group Metal Amides

M. VEITH

I.	Introduction	269
II.	Structural Aspects of Alkyl and Aryl Metal Amides	270
III.	Organometallic Derivatives of Special Silazanes	276
IV.	Conclusion	298
	References	299

Heteronuclear Clusters Containing Platinum and the Metals of the
Iron, Cobalt, and Nickel Triads

LOUIS J. FARRUGIA

I.	Introduction	301
II.	Synthesis	302
III.	Bonding and Electron Counting	306
IV.	Fluxional Behavior and Isomerization Reactions	312
V.	Structural and Reactivity Aspects of Hetero-Pt Clusters	328
VI.	Catalysis by Hetero-Pt Clusters	383
VII.	Addendum	384
	References	384

INDEX	393
CUMULATIVE LIST OF CONTRIBUTORS	405

ADVANCES IN ORGANOMETALLIC CHEMISTRY, VOL. 31

Highly Reduced Metal Carbonyl Anions: Synthesis, Characterization, and Chemical Properties

JOHN E. ELLIS

Department of Chemistry
University of Minnesota
Minneapolis, Minnesota 55455

I

INTRODUCTION

Pioneering work by Hieber (*1*), Hock and Stuhlmann (*2*), Feigl and Krumholz (*3*), Behrens (*4*), and Hein (*5*) and later studies by Fischer (*6*), Wilkinson (*7*), Calderazzo (*8*), King (*9*), Stone (*10*), and others (*11*) established the existence of numerous binary and substituted carbonylmetallate mono- and dianions and demonstrated these to be quite useful reagents in chemical synthesis. In the course of preparing a review on metal carbonyl anion chemistry in 1973, which emphasized parallels in the reactivity patterns of these materials and formally analogous main-group anions, it became evident that several derivatives of carbonylmetallate trianions existed, but the parent compounds were unknown (*12*). The existence of electronically equivalent (or isolobal) main-group analogs such as P^{3-} and As^{3-} gave us hope that carbonylmetallates($3-$) could be prepared, especially in the presence of strongly interacting cations. For these and related reasons, the reduction of $[Mn(CO)_5]^-$ was examined with the hope of preparing $[Mn(CO)_4]^{3-}$, the manganese analog of $[Fe(CO)_4]^{2-}$, which was the first characterized carbonylmetallate (*3,13*).

In the unusually effective reducing medium of sodium-hexamethylphosphoramide (HMPA), $[Mn(CO)_5]^-$ underwent facile reduction to form a golden yellow to yellow brown solution. On the basis mainly of derivative chemistry and infrared spectra, the major soluble component was formulated as $Na_3[Mn(CO)_4]$ (*14*). Once this material was in hand it was evident that similar binary carbonyl trianions should be possible for other *d*-block transition elements of odd atomic number. Since 1976, syntheses of mononuclear binary carbonyl trianions of V, Nb, Ta, Re, Co, Rh, and Ir have also been reported from this laboratory (vide infra). Because these substances contain transition metals in their lowest known oxidation states, they have often been referred to as "superreduced" species (*15*).

1

TABLE I

KᴺOWN AʟᴋAʟɪ MᴇᴛAʟ CARBONYLMETALLATES(3− AND 4−)

		Group		
5	6	7	8	9
$Cs_3[V(CO)_5]^a$	$Na_4[Cr(CO)_4]$	$Na_3[Mn(CO)_4]$	—	$Na_3[Co(CO)_3]$
$Cs_3[Nb(CO)_5]^a$	$Na_4[Mo(CO)_4]$	—	—	$Na_3[Rh(CO)_3]$
$Cs_3[Ta(CO)_5]^a$	$Na_4[W(CO)_4]$	$Na_3[Re(CO)_4]$	—	$Na_3[Ir(CO)_3]$

[a] Also known in the form of quite thermally unstable trisodium compounds.

Soon after the first examples of mononuclear carbonyl trianions had been synthesized, it was of interest to determine whether even more highly reduced carbonyl anions, i.e., "tetraanions," could be prepared. Attempts to reduce $Na_2[M(CO)_5]$ (M = Cr, Mo, W) to the previously unknown tetrasodium salts, $Na_4[M(CO)_4]$, were unsuccessful either because the disodium salts were inert toward further reduction (Cr) or did not provide carbonylmetallates (Mo, W). These failures led to the development of a *reductive labilization method* for the synthesis of highly reduced organometallics. This method is based on the useful generalization that reduction of a noncluster metal compound containing both good and poor π-acceptor groups often causes preferential loss of the weaker or nonacceptor groups, provided the latter are not susceptible to reduction. Thus, in the Na−NH_3 reduction of $M(CO)_4$-(TMED) (M = Cr, Mo, and W; TMED = N,N,N',N'-tetramethylethylenediamine), two amine groups were effectively replaced by four sodium atoms to provide high yields of $Na_4[M(CO)_4]$ (*16*). In Table I, the known carbonylmetallate tri- and tetraanions are listed. The first detailed review on the syntheses and properties of these materials will now be presented.

II

TETRACARBONYLMETALLATES (3−) OF MANGANESE AND RHENIUM

A. *Syntheses and Characterizations*

A variety of strong reducing agents, including solutions of alkali metals in liquid ammonia, sodium solubilized by crown ethers or cryptands in tetrahydrofuran (THF), and alkali metal naphthalenides in THF, have been found to reduce $M_2(CO)_{10}$ and/or $[M(CO)_5]^-$ (M = Mn and Re) to the respective $[M(CO)_4]^{3-}$; however, $[Re(CO)_5]^-$ has often been observed to be more resistant to reduction than is the manganese analog. For example, sodium–benzophenone slowly converted $Na[Mn(CO)_5]$ to $Na_3[Mn(CO)_4]$ in refluxing dioxane, but under these conditions $Na[Re(CO)_5]$ appeared to be unreactive. This $Na–Ph_2CO$ route is entirely analogous to that developed by the Collman group for the conversion of $Fe(CO)_5$ to $Na_2[Fe(CO)_4]·\frac{3}{2}$ dioxane [Eq. (1)] (17). Although the $Na_3[Mn(CO)_4]$ obtained from this

$$Na[Mn(CO)_5] \xrightarrow[\substack{dioxane \\ \Delta}]{NaPh_2CO} Na_3[Mn(CO)_4] + \cdots \qquad (1)$$

method was generally dark brown and obviously impure, one reaction provided an attractive homogeneous light tan solid. Its Nujol mull infrared spectrum in the $v(CO)$ region was nearly superimposable on that of pure $Na_3[Mn(CO)_4]$ (vide infra), but attempts to obtain satisfactory analyses for any composition of the type $Na_3[Mn(CO)_4] \cdot (dioxane)_x$ were unsuccessful. Both $[Mn(CO)_5]^-$ and $[Re(CO)_5]^-$ are reduced by $Na–NH_3$, or more rapidly by $K–NH_3$, to provide insoluble and impure (especially for Re) tan to brown solids containing the trianions. Similar insoluble materials were obtained by the rapid reduction of $M_2(CO)_{10}$ with deep blue solutions of sodium dicyclohexyl-18-crown-6 in THF, which very likely contain Na^- (18), an extremely potent reducing agent. Unfortunately, however, these materials generally reacted with various electrophiles to provide lower yields of products, which were usually less pure and consequently more difficult to crystallize than those obtained from corresponding reactions with solutions of $Na_3[M(CO)_4]$ in hexamethylphosphoric triamide (HMPA).

Sodium naphthalenide in THF has also been shown to reduce $Na[M(CO)_5]$ (M = Mn, Re). Initially it was believed that solutions of $Na_3[M(CO)_4]$ were obtained, but a more careful examination established that only very fine suspensions of insoluble and impure brown trisodium salts had formed. For example, treatment of $Re_2(CO)_{10}$ with 10 equivalents of $NaC_{10}H_8$ in THF provided a complete conversion to finely divided brown insoluble $Na_3[Re(CO)_4]$ within an hour at room temperature (19). Although $Na_3[Re(CO)_4]$ prepared with excess $Na[C_{10}H_8]$ was impure, we discovered that these HMPA-free suspensions were useful in reactions in which removal of HMPA was very difficult. Attempts to find practical routes to relatively pure salts of $[M(CO)_4]^{3-}$, which do not involve the use of the toxic and high-boiling HMPA, have not been entirely successful to date. A summary of routes to insoluble (and impure) forms of $A_3[M(CO)_4]$, where A = Na and K and M = Mn and Re, is shown in Eqs. (2) and (3).

$$\begin{matrix} \frac{1}{2}[M_2(CO)_{10}] \\ or \\ A[M(CO)_5] \end{matrix} + \begin{Bmatrix} excess \ A-naphthalene, \\ or \ 3A, \ crown \ ether \end{Bmatrix} \xrightarrow[THF]{20°C} A_3[M(CO)_4] + \cdots \qquad (2)$$

$$A = Na, K; M = Mn, Re$$

$$\begin{array}{c} \frac{1}{2}[Re_2(CO)_{10}] \\ or \\ A[Mn(CO)_5] \end{array} + 3A \xrightarrow[\ NH_3\ ]{-78\ to\ -33°C} A_3[M(CO)_4] + \frac{1}{2}A_2C_2O_2 \qquad (3)$$

$$A = Na\ (Mn\ only),\ K\ (Mn\ and\ Re)$$

Indeed, the Na–HMPA route consistently provided the cleanest products and has been the only synthesis to provide *solutions* of $Na_3[M(CO)_4]$. It is often important to use solutions rather than slurries of trianion salts to minimize the formation of side products during the reactions of these materials with electrophiles. Until recently, product separation from the viscous and high-boiling HMPA has always been a problem (and remains so in some cases). For example, addition of excess THF to solutions of $Na_3[M(CO)_4]$ in HMPA invariably resulted in the formation of sticky solids that contained HMPA and did not analyze satisfactorily (14). But recently, it was discovered that addition of these HMPA solutions to excess liquid ammonia resulted in practically quantitative precipitation of tan to pale yellow brown solids, which provided satisfactory elemental analyses of unsolvated $Na_3[M(CO)_4]$ (M = Mn, Re). Virtually all impurities remained in the HMPA–NH_3 filtrate [Eqs. (4) and (5)].

$$Na[Mn(CO)_5] + 3\ Na \xrightarrow{\ i,\ ii\ } Na_3[Mn(CO)_4]\downarrow + \cdots \qquad (4)$$

$$(95\text{--}100\%)$$

$$\frac{1}{2}Re_2(CO)_{10} + 4\ Na \xrightarrow{\ i,\ ii\ } Na_3[Re(CO)_4]\downarrow + \cdots \qquad (5)$$

$$(85\text{--}90\%)$$

(i) HMPA, 12 hr, 20°C; (ii) Excess liq. NH_3, $-78°C$

Unlike the related $Na_3[M'(CO)_5]$, where M' = V, Nb, and Ta, which undergoes thermolysis below 0°C (vide infra), these materials possess remarkable thermal stabilities for metal carbonyls and briefly survive without melting at temperatures as high as 300°C. By comparison, $K_2[Fe(CO)_4]$, another metal carbonyl salt of high thermal stability, has been reported to melt at 270–273°C with decomposition (20). The related $K[Co(CO)_4]$ melts at about 203°C with decomposition (21).

Lower yields of $Na_3[Mn(CO)_4]$ were obtained from the direct reduction of $Mn_2(CO)_{10}$ in Na–HMPA, because the neutral dimer underwent slow disproportionation in this medium to form $[Mn(HMPA)_x][Mn(CO)_5]_2$, in contrast to $Re_2(CO)_{10}$, which showed no tendency to react with HMPA at room temperature. Of all neutral binary carbonyl dimers known, $Re_2(CO)_{10}$ appears to be the most resistant toward Lewis base-promoted disproportionation reactions. The slightly lower yields of $Na_3[Re(CO)_4]$, compared to those of $Na_3[Mn(CO)_4]$, may have arisen from the fact that $Re_2(CO)_{10}$ does not cleanly reduce to $[Re(CO)_5]^-$ in HMPA or other solvents (22). It should also be noted that more than two equivalents of sodium are required for the complete reduction of $[M(CO)_5]^-$, undoubtedly because some of the liberated CO undergoes reduction by Na–HMPA. Although we have firm infrared spectral evidence for the formation of disodium acetylenediolate, $Na_2C_2O_2$ (23), during the reduction of $Na[Mn(CO)_5]$ by sodium in liquid ammonia, the nature of possible CO reduction products in HMPA remains unknown.

This new procedure for the isolation of $Na_3[M(CO)_4]$ from HMPA is straightforward and promises to be of general importance for the purification and/or isolation of other highly reactive anions from HMPA, provided they form reasonably insoluble salts in and do not react with liquid ammonia. This method has also provided pure $Na_3[Ir(CO)_3]$ (vide infra). Evidently, ammonia interacts strongly enough with HMPA, perhaps by hydrogen bonding, to effectively cause full dissociation of $Na(HMPA)_x^+$ and subsequent precipitation of unsolvated or weakly ammoniated $Na_3[M(CO)_4]$. The only other solvents we have found to be effective in this regard include water, methanol, and ethanol, which are incompatible with the trianions and many of their derivatives. It is interesting to note that solutions of $Na_3[M(CO)_4]$ in HMPA were stable for at least a week at room temperature under strict anaerobic conditions and did not deposit any solid during this time. By comparison, after unsolvated $Na_3[Mn(CO)_4]$ or $Na_3[Re(CO)_4]$ precipitated from the HMPA–NH_3 mixture, they showed no tendency to redissolve— either in HMPA (at room temperature) or any other unreactive solvent.

Infrared spectra of $Na_3[M(CO)_4]$ as solutions in HMPA or as mineral oil mulls of the solid are quite similar in terms of peak positions. For this reason, it seems likely that the $[M(CO)_4]^{3-}$ units in solution or in the solid state are not dramatically different in nature. However, as is usually the case for mull spectra of unsolvated alkali metal salts of carbonylmetallates, the absorptions in the carbonyl stretching frequency are considerably broader than those in solution (Fig. 1); i.e., more complex cation–anion interactions are likely to be present in the solid state, which in turn lead to a greater spread in the $v(CO)$ values for a given band compared to those in solution. However, even in a solvent like HMPA, which is normally very effective in minimizing tight ion-pair formation (24), the most intense $v(CO)$ bands (slightly incorrect positions were reported in Ref. 19) of $Na_3[Mn(CO)_4]$ are substantially broader than corresponding bands of $Na_2[M'(CO)_4]$, $M' = Fe$, Ru, and Os, or especially $Na[M''(CO)_4]$, $M'' = Co$, Rh, and Ir in HMPA. [A superb review by M. Darensbourg on ion-pairing effects in carbonylmetallates, which discusses these and similar phenomena, has been published (25).] For example, in Fig. 2 infrared spectra of $Na_3[Mn(CO)_4]$ and $Na_2[Ru(CO)_4]$ in HMPA at approximately the same concentrations are illustrated and show that the most intense band at half-height of the manganese compound is about twice

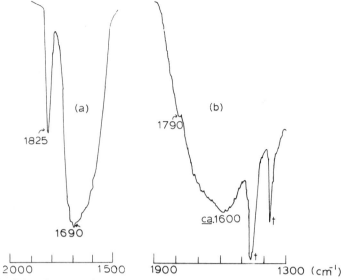

FIG. 1.   Infrared spectra of $Na_3[Re(CO)_4]$ in the $\nu(CO)$ region: (a), in hexamethylphosphoramide (HMPA)—1825, 1690 cm^{-1}; (b), mulled suspension of unsolvated compound in mineral oil—1790, ~1600 cm^{-1}. Absorptions marked with daggers are due to mineral oil.

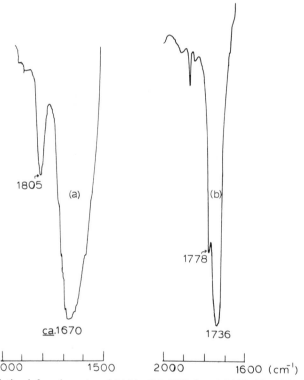

FIG. 2.   Solution infrared spectra of (a) $Na_3[Mn(CO)_4]$ and (b) $Na_2[Ru(CO)_4]$ in HMPA and in the $\nu(CO)$ region. Position of bands: (a), 1805, 1670 cm^{-1}; (b), 1778 w, 1736 vs cm^{-1}; weak spike at 1880 cm^{-1} is due to an oxidation product.

TABLE II

INFRARED SOLUTION SPECTRA OF $M(CO)_4^Z$ IN THE
CARBONYL STRETCHING FREQUENCY REGION[a]

Compound[b]	Medium	$\nu(CO)$, cm^{-1}
$Ni(CO)_4$	THF	2040 vs[c]
$[Co(CO)_4]^-$	HMPA	1890 vs[d]
$[Fe(CO)_4]^{2-}$	HMPA	1771 sh, 1729 vs[d]
$[Mn(CO)_4]^{3-}$	HMPA	1805 w, 1670 vs[d]
$[Cr(CO)_4]^{4-}$	Nujol mull	1657 w, 1462 vs[e]

[a] $Z = 0$ to -4, M = Ni-Cr.
[b] All anions contain Na$^+$ as the counterion.
[c] W. Beck and R. E. Nitzschmann, Z. Naturforsch. B 17, 577 (1962).
[d] All values from our research.
[e] Insoluble in all known unreactive solvents.

as broad as that of the tetracarbonylruthenate($2-$). Presumably the latter ion is less perturbed by ion pairing from its ideal tetrahedral symmetry in HMPA than is $[Mn(CO)_4]^{3-}$ (vide infra).

Band positions for the known first-row metal tetracarbonyls, $[M(CO)_4]^Z$, are collected in Table II. Interestingly, the most intense $\nu(CO)$ values are found to be nearly a linear function of Z, for $Z = 0$, -1, and -2. For these tetracarbonyls, each additional negative charge causes the corresponding $\nu(CO)$ value to drop by about 150–160 cm^{-1}. It is noteworthy that the observed IR-active $\nu(CO)$ value for crystalline $[Na(crypt.2.2.2)]_2[Fe(CO)_4]$ is very similar to that for dilute solutions of $Na_2[Fe(CO)_4]$ in HMPA. The cryptand salt has been established to contain an essentially unperturbed tetrahedral $[Fe(CO)_4]^{2-}$ unit (26), which is consistent with the rather sharp and symmetrical $\nu(CO)$ band observed for this species as a mineral oil mull IR spectrum (27). However, $Na_2[Fe(CO)_4]$ in HMPA also shows a weak shoulder at 1770 cm^{-1}. The latter peak is not present in the mull or solution spectrum of the Na(2.2.2)$^+$ salt and may be attributed to ion pairing involving a direct Na$^+$-iron interaction. This type of perturbation has been established to be present in crystalline $Na_2[Fe(CO)_4] \cdot \frac{3}{2}$ dioxane (28) and is known to cause an increase in the energy of a $\nu(CO)$ band relative to that of the free ion (25). Edgell first reported that the IR spectrum of $Na_2[Fe(CO)_4]$ in tetrahydrofuran consisted of a rather broad and intense band centered at about 1780 cm^{-1} (29). Direct sodium ion–iron interactions appear to be important in influencing this value, which is shifted about 50 cm^{-1} to higher energy than that of unperturbed $[Fe(CO)_4]^{2-}$. On the basis of the position and impressive breadth of the corresponding $\nu(CO)$ band of $Na_3[Mn(CO)_4]$, we believe that the tetracarbonyl–manganate unit is also involved extensively in direct sodium ion–manganese interactions in HMPA. If the most intense IR-active $\nu(CO)$ value for unperturbed $[Mn(CO)_4]^{3-}$ were a linear function of the molecular charge, as it is for the less reduced tetracarbonyls (vide supra), a reasonable estimate would be 1580 cm^{-1}, which is about 90 cm^{-1} lower than the observed maximum in HMPA. It would be useful to corroborate our proposal that the tetracarbonylmetallates(3−) are extensively involved in tight ion pairing with conductivity studies in HMPA, but so far we have been unable to obtain reliable data for these quite air-sensitive materials. We have also been unsuccessful in observing ^{13}C resonances for $Na_3[M(CO)_4]$ M = Mn or Re) in HMPA. Although tight ion pairing could conceivably cause the ^{13}C resonance signals of these trisodium salts to be broadened, the lifetimes of any such interactions would have to be at least on the order of the time scale of the NMR experiment, $\sim 10^{-2}$ sec, to have any influence in this regard (25).

When these studies involving the synthesis of carbonylmetallates in HMPA were initiated in the mid-seventies (30), it was obvious that solutions of sodium in HMPA were qualitatively more effective for many organometallic reductions than, for example, sodium amalgam, sodium sand in THF, sodium in liquid ammonia, or sodium naphthalenide, but only recently has it been established that sodium dissolves in HMPA to provide appreciable amounts of Na$^-$ (31). The presence of this potent reducing agent in Na–HMPA may well explain the unusually effective reducing ability of this medium. In retrospect, these initial studies represented some of the first investigations of the reactions of the sodide (or natride) ion with organometallics!

B. Protonations of $[M(CO)_4]^{3-}$

Protonation reactions of $[M(CO)_4]^{3-}$ have been examined in some detail (19). Corresponding studies on the pentacarbonylmetallates(3−) of V, Nb, and Ta and their conjugate acids, $[HM(CO)_5]^{2-}$ (vide infra), suggested that a similar examination of the trianions of manganese and rhenium would provide valuable information on the nature of these species. Proton NMR spectra of approximately 0.1 M HMPA solutions of the freshly prepared trianions at room temperature often showed very weak sharp singlets at $\delta - 7.8$ and -9.4 ppm for manganese and rhenium, respectively. These experiments suggested, but did not prove, that most of the manganese and rhenium species in solution were nonhydridic in nature. However, addition of one equivalent of ethanol to these solutions caused IR absorptions in the $\nu(CO)$ region attributed to the trianions to entirely disappear and new bands at higher energies to appear. Also, the aforementioned very weak singlets in the ^{1}H NMR spectra of these solutions grew dramatically on addition of the ethanol. Although the monoprotonated products have not been isolated as pure substances, their ^{1}H NMR and IR spectra are consistent with the presence of mononuclear dianions of C_{3v} symmetry of the general formula $[HM(CO)_4]^{2-}$. Prior to the deliberate addition of ethanol, water, or other proton source to freshly prepared solutions of $[M(CO)_4]^{3-}$ in HMPA, no IR bands due to the respective conjugate acids, $[HM(CO)_4]^{2-}$, were observed. This was an important indication that the highly reduced material in solution was essentially all in the form of the unprotonated $Na_3[M(CO)_4]$.

Perhaps the most persuasive chemical evidence for the existence of the monohydrides of manganese and rhenium is their high-yield conversion to corresponding dihydrides, $[H_2M(CO)_4]^-$, which can be isolated as relatively stable crystalline solids. Treatment of solutions of the trianions in HMPA with 1.6 equivalents of ethanol provided ^{1}H NMR spectra that showed monohydrides as well as new singlets at $\delta - 8.6$ and -7.2 ppm due to the dihydrides of manganese and rhenium, respectively. Two equivalents of ethanol or water caused complete conversion of $[M(CO)_4]^{3-}$ to the corresponding $[H_2M(CO)_4]^-$. These processes are represented by Eq. (6).

$$[M(CO)_4]^{3-} \xrightarrow[\text{HMPA}]{\text{EtOH}} [HM(CO)_4]^{2-} \xrightarrow[\text{HMPA}]{\text{EtOH}} [H_2M(CO)_4]^- \qquad (6)$$

$$(M = Mn, Re)$$

Addition of the resulting solutions to ice-cold aqueous tetraphenylarsonium chloride caused rapid and nearly quantitative precipitation of colorless to tan solids, as shown by Eq. (7). These materials provided satisfactory elemental analyses for the compositions $[Ph_4As][H_2M(CO)_4]$ $(M = Mn, Re)$.

$$[M(CO)_4]^{3-} + 2H_2O \xrightarrow[0°C]{\text{HMPA}} \xrightarrow[0°C]{\text{aqueous} \atop Ph_4AsCl} [Ph_4As][H_2M(CO)_4]\!\downarrow \qquad (7)$$

$$(70-80\%; M = Mn, Re)$$

The previously unknown dihydridomanganate, $[H_2Mn(CO)_4]^-$, was found to be more thermally stable than its isoelectronic neighbor, $H_2Fe(CO)_4$, the first known carbonyl hydride (32). However, $[H_2Mn(CO)_4]^-$ still proved to be a rather fragile molecule. As a Ph_4As^+ salt, it decomposed within days at room temperature as a crystalline solid or within minutes at room temperature in THF. Of considerable greater thermal stability is the rhenium analog, $[H_2Re(CO)_4]^-$, which appeared to survive indefinitely at room temperature under an inert atmosphere as crystalline Ph_4As^+, Ph_4P^+, or $(Ph_3P)_2N^+$ salts. Ciani and co-workers first isolated $[H_2Re(CO)_4]^-$ as the less stable Et_4N^+ salt in about 10% yields from the reaction of $Re_2(CO)_{10}$ with refluxing methanolic KOH, followed by metathesis. X-ray crystallography established [cis-$H_2Re(CO)_4$]$^-$ units to be present in [Et_4N][$H_2Re(CO)_4$] (33). Infrared spectra of [Ph_4As][$H_2Mn(CO)_4$] and [Ph_4As][$H_2Re(CO)_4$] are very similar to that of the Ciani salt and indicate the same structural units are also present in the anions of these salts. No attempts to further protonate [$H_2M(CO)_4$]$^-$ to form the neutral and presently unknown trihydrides, $H_3M(CO)_4$, have been reported. However, the existence of the formally analogous triphenylphosphine gold complexes, $(Ph_3PAu)_3M(CO)_4$ (vide infra), suggests that such materials may be accessible.

C. Reactions of [M(CO)$_4$]$^{3-}$ with Main-Group Electrophiles, Including Alkylating Agents

A variety of main-group electrophiles react with $Na_3[M(CO)_4]$ to provide octahedral complexes of the type [$E_2M(CO)_4$]$^-$. These include E = Ph_3Sn, Ph_3Pb, Ph_3Ge, Me_3Sn, and Me_3Ge. For the triphenylstannyl group, stepwise addition of Ph_3SnCl to the trianions has been shown to proceed via the intermediate formation of the very reactive dianions, [$Ph_3SnM(CO)_4$]$^{2-}$. On the basis of their infrared spectra in the $v(CO)$ region, these dianions have been proposed to have the same basic C_{3v} geometry as the structurally character-ized anion, [$Ph_3PMn(CO)_4$]$^-$ (34). As shown in Eq. (8), these intermediates may be protonated or reacted with an additional equivalent of Ph_3SnCl to provide appropriately substituted monoanions, which have infrared spectra consistent with the presence of cis-disubstituted tetracarbonylmetallates($1-$).

$$[M(CO)_4]^{3-} + Ph_3SnCl \xrightarrow{\text{HMPA}} [Ph_3SnM(CO)_4]^{2-} \begin{cases} \xrightarrow{Ph_3SnCl} [(Ph_3Sn)_2M(CO)_4]^- \\ \xrightarrow{HOAc} [H(Ph_3Sn)M(CO)_4]^- \end{cases}$$

$$(M = Mn, Re)$$

$$(8)$$

For $Na[(Ph_3Sn)_2Re(CO)_4]$, there is good spectroscopic evidence that addition of a third Ph_3Sn unit occurs in diethyl ether to provide the colorless seven-coordinate $(Ph_3Sn)_3Re(CO)_4$ [$v(CO)$ in heptane: 2108 (w), 2004 (s) cm^{-1}]. It undergoes substantial heterolysis in THF to provide $Ph_3Sn(THF)_x^+$ and [$(Ph_3Sn)_2Re(CO)_4$]$^-$ and has infrared $v(CO)$ bands of very similar relative intensities but of higher energies, compared to those of the related anion [$(Ph_3Sn)_3Cr(CO)_4$]$^-$, whose molecular structure shows the presence of seven-coordinate chromium (vide infra).

Interactions of [$HMn(CO)_4$]$^{2-}$ with a variety of methylating agents such as MeI or methyl tosylate in HMPA gave only small amounts of [$Mn(CO)_5$]$^-$

and uncharacterized products, but corresponding reactions with $[HRe(CO)_4]^{2-}$ provided isolable salts containing $[cis\text{-}H(CH_3)Re(CO)_4]^-$, which is analogous to the known $H(CH_3)Os(CO)_4$ (35). The rhenium anion was prepared by two independent routes [Eq. (9)] in 70–80% isolated yields.

$$Na_3[Re(CO)_4] \begin{array}{c} \xrightarrow[\text{(a)}]{\text{ETOH–HMPA}} Na_2[HRe(CO)_4] \xrightarrow{\text{MeOTs}} \\ \\ \xrightarrow[\text{MeOTs–HMPA}]{\text{(b)}} Na_2[MeRe(CO)_4] \xrightarrow{\text{H}_2\text{O}} \end{array} \xrightarrow[0°\text{C}]{[Ph_4E]Cl–H_2O}$$

$$[Ph_4E][H(CH_3)Re(CO)_4]\downarrow \quad (9)$$
$$(E = P, As)$$

Bergman and co-workers (36) previously reported that $[H(CH_3)Re(CO)_4]^-$ was the likely product formed in the reaction of [PPN][CpV(CO)$_3$H] with $CH_3Re(CO)_5$ [Eq. (10)].

$$[PPN][CpV(CO)_3H] + CH_3Re(CO)_5 \xrightarrow[\text{THF}]{50°\text{C}} [PPN][H(CH_3)Re(CO)_4] + CpV(CO)_4$$
$$(10)$$

Although they were unable to isolate it from the reaction mixture and no infrared spectrum was reported, its ^1H NMR spectrum in THF [$\delta - 0.65$ (d; $J = 3$ Hz, 3H], -5.56 (br; 1 H)] was essentially identical to those of the compounds prepared from $[Re(CO)_4]^{3-}$. On this basis, there can be little doubt that Bergman's complex is identical to the $[cis\text{-}H(CH_3)Re(CO)_4]^-$ derived from $[Re(CO)_4]^{3-}$.

Alkylation reactions of $[Re(CO)_4]^{3-}$ have provided the first alkyl derivatives of carbonylmetallates(3$-$). Although the reactions of alkylating agents with $Na_3[Mn(CO)_4]$ are not yet well understood, treatment of a solution of $Na_3[Re(CO)_4]$ in HMPA with two equivalents of methyl tosylate gave the first dialkyl derivative of a carbonyl trianion, as a thermally stable sodium salt. Addition of this solution to aqueous $[Ph_4E]Cl$ (E = P or As) caused nearly quantitative precipitation of product as shown in Eq. (11). High

$$[Re(CO)_4]^{3-} + 2CH_3OTs \xrightarrow[25°\text{C}]{\text{HMPA}} \xrightarrow[\text{Ph}_4\text{ECl}]{\text{H}_2\text{O}} [Ph_4E][(CH_3)_2Re(CO)_4]\downarrow \quad (11)$$
$$(E = P, As)$$

yields (80–90%) of nearly colorless solids were obtained. These products have spectroscopic properties consistent with the presence of discrete $[cis\text{-}(CH_3)_2Re(CO)_4]^-$ units. Both the Ph_4As^+ and Ph_4P^+ salts were of substantial and comparable thermal stability (dec $\geq 140°C$) and were indefinitely stable under nitrogen at room temperature. Infrared spectra of these compounds in the $v(CO)$ region are shown in Table III and have essentially the same features as those observed for $cis\text{-}(CH_3)_2Os(CO)_4$ (37), with the expected

TABLE III

INFRARED DATA FOR TETRACARBONYLMETALLATES(3−) OF MANGANESE AND
RHENIUM AND SELECTED DERIVATIVES

Compound	Medium[a]	$v(CO)$, cm^{-1}	Reference
$Na_3[Mn(CO)_4]$	HMPA	1805 m, 1670 vs br	45
$Na_3[Re(CO)_4]$	HMPA	1825 m, 1690 vs br	45
$[Ph_4As][H_2Mn(CO)_4]$	THF	2005 w, 1905 s, 1879 m	19
$[Ph_4As][H_2Re(CO)_4]$	THF	2024 w, 1919 s, 1892m	19
$Na_2[(Ph_3Sn)Mn(CO)_4]$	HMPA	1880 m, 1756 s	19
$Na_2[(Ph_3Sn)Re(CO)_4]$	HMPA	1900 m, 1780 sh, 1765 s	19
$[Et_4N][(Ph_3Sn)_2Mn(CO)_4]$	THF	2000 m, 1922 sh, 1913 vs	45
$[Et_4N][(Ph_3Sn)_2Re(CO)_4]$	THF	2032 m, 1968 m, 1945 sh, 1920 vs	45
$(Ph_3PAu)_3Mn(CO)_4$	THF	1954 m, 1882 s, 1867 sh	45
$(Ph_3PAu)_3Re(CO)_4$	THF	1999 m, 1922 sh, 1912 s, 1892 sh	45
$[Ph_4P][(CH_3)_2Re(CO)_4]$	THF	2024 w, 1923 vs, 1906 s, 1855 s	19

[a] HMPA, Hexamethylphosphoramide; THF tetrahydrofuran.

shifts in band positions to lower energy due to the anionic nature of the rhenium complex.

Attempts to isolate thermally stable salts containing $[cis\text{-}(C_2H_5)_2\text{-}Re(CO)_4]^-$ by the reaction of $Na_3[Re(CO)_4]$ and ethyl tosylate in HMPA have not been successful to date. White precipitates were obtained on treatment of these HMPA solutions with cold aqueous Et_4N^+, Ph_4P^+, Ph_4As^+, or $(Ph_3P)_2N^+$, but the products quickly decomposed into as-yet uncharacterized orange oils when the temperature rose from 0°C to room temperature. However, similar reactions with 1,4-butaneditosylate provided a fairly stable substance that proved to be the first example of an anionic metallacyclopentane species [Eq. (12)]. Although this substance appeared to be moderately

$$Na_3[Re(CO)_4] + TsOCH_2CH_2CH_2CH_2OTs \xrightarrow[Ph_4PCl]{HMPA \text{ or } THF} [Ph_4P]\left[(OC)_4Re\!\!\begin{array}{c}\bigcirc\end{array}\right] \quad (12)$$

stable in solution at room temperature, in contrast to the recently reported Group 8 analogues, $cyclo\text{-}(CH_2)_4M(CO)_4$ [M = Fe (38), Ru (39)], attempts to isolate the pure substance were unsuccessful owing to its poor thermal stability at room temperature. However, essentially pure solutions (by IR and NMR) of $[Ph_4P][cyclo\text{-}(CH_2)_4Re(CO)_4]$ were obtained. Its IR spectrum in the $v(CO)$ region closely matched that of $[cis\text{-}(CH_3)_2Re(CO)_4]^-$ and was qualitatively similar in terms of relative band intensities to that reported for the neutral ruthenium analogue (39). The synthesis of the rhenacyclopentane from $[Re(CO)_4]^{3-}$ was quite analogous to methods used by Lindner and co-workers [Eq. (13)] for the synthesis of corresponding neutral metallacyclopentanes of iron and ruthenium.

$$[M(CO)_3L]^{2-} + (F_3CSO_3CH_2CH_2)_2 \longrightarrow (OC)_3\underset{L}{M} \quad \bigcirc \quad (13)$$

[M = Fe, L = CO, PPh$_3$ (Ref. 38); Ru, L = CO (Ref. 39)]

Acylations of $Na_3[Mn(CO)_4]$ or $Na_3[Re(CO)_4]$ have not been examined to date. Although a variety of diacyltetracarbonylmetallates(1−) of these metals have been obtained by treatment of an appropriate acylpentacarbonyl metal with an organolithium reagent [Eq. (14)] (40–44), monoacylates of the general formula $[(RCO)M(CO)_4]^{2-}$ are presently unknown and should be available by monoacylation of the carbonyl trianions. Also, the alternative acylation route to diacylmetallates should be attractive when the corresponding organolithium species are unknown or thermally unstable.

$$RC(O)M(CO)_5 + R'Li \rightarrow [cis\text{-}(RCO)(R'CO)M(CO)_4]Li \quad (14)$$

Most of the known reactions of $Na_3[Re(CO)_4]$ are summarized in Scheme 1. Several of these also proceed as shown for $Na_3[Mn(CO)_4]$, but the products are often of lower thermal stability than are their rhenium analogs.

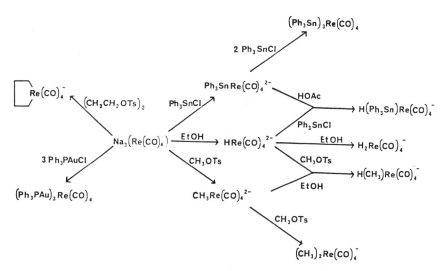

SCHEME 1.   Several reactions of $Na_3[Re(CO)_4]$.

D. *Reactions of [M(CO)₄]³⁻ with Transition Metal Electrophiles*

Least studied among reactions involving $Na_3[M(CO)_4]$ are those with transition metal electrophiles. Of particular interest are the reactions with Ph_3PAuCl that provide the formally seven-coordinate species $(Ph_3PAu)_3M$-$(CO)_4$ (*45*). A single-crystal X-ray study of the molecular structure of $(Ph_3PAu)_3Mn(CO)_4$ has been carried out and a view of the core atoms is shown in Fig. 3 along with selected interatomic distances and angles and other crystal data. Although there is considerable scatter in the positions of the light atoms in the unit cell and for this reason we doubt that the rather acute C(4)–Mn–Au(2) angle of 66° is of any significance, the structure of the $(PAu)_3Mn$ core is firmly established to be planar. Attempts made to refine the structure in centrosymmetric space groups were unsuccessful. Apparently, $[(Ph_3PAu)_3Ir(PPh_3)_2(O_2NO)]^+$ is the only other structurally characterized compound that contains a planar Au_3M core (*46*). More common appear to be tetrahedral Au_3M cluster cores such as those present in $(Ph_3PAu)_3V(CO)_5$ (vide infra) and $[(Ph_3PAu)_3(\mu\text{-}H)Rh(PPh_3)_2(CO)]^+$ (*47*). Although the latter vanadium complex may be considered to contain a symmetrical $(Ph_3PAu)_3^+$ unit (which is isolobal with H_3^+) coordinated to a low-spin d^6 octahedral $V(CO)_5^-$ fragment, the planar geometry of the manganese cluster may be rationalized on the basis that an open allylic-like $(Ph_3PAu)_3^-$ unit is bound to a

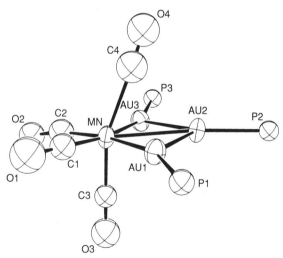

FIG. 3.   View of the core atoms of $(Ph_3PAu)_3Mn(CO)_4$. Selected interatomic distances (Å) and angles (degrees): Mn–Au(1), 2.720(4); Mn–Au(2), 2.655(4); Mn–Au(3), 2.601(3); Au(1)–Au(2), 2.695(1); Au(2)–Au(3), 2.866(1); Au(1)–Au(3), 4.746(1); Au–P (av), 2.23(6); Mn–C (av), 1.92(20); C–O (av), 1.14(12); Mn–Au–P (av), 168(6); Mn–C–O (av), 172(3). Crystal data: orthorhombic, space group P_{na21} (no. 33), $a = 20.446(6)$ Å, $b = 16.439(3)$ Å, $c = 17.636(7)$ Å, $V = 5928(5)$ Å³, $Z = 4$, $R = 0.056$, $R_w = 0.065$, GOF = 1.816

low-spin d^6 octahedral $Mn(CO)_4^+$ fragment. Of course, in terms of simple electron counting schemes the neutral $(Ph_3PAu)_3$ unit may be considered to be a three-electron donor, whether it binds with a metal to form a tetrahedral or planar Au_3M cluster. In view of the similarity of the infrared spectra (Table III) of $(Ph_3PAu)_3Mn(CO)_4$ and the rhenium analogue, it seems likely that the rhenium cluster also contains a planar Au_3M cluster core.

III

PENTACARBONYLMETALLATES(3−) OF VANADIUM, NIOBIUM, AND TANTALUM

A. Synthesis and Characterization of Na₃[V(CO)₅]

Soon after the reduction of $Mn(CO)_5^-$ by sodium metal in HMPA was examined and shown to provide new materials, corresponding reactions of $[V(CO)_6]^-$ were investigated. Whereas the reaction of Na–HMPA and $[Na(diglyme)_2][V(CO)_6]$ at room temperature provided a dark brown reaction mixture that showed the presence of only diminished amounts of $[V(CO)_6]^-$ and no new carbonyls of vanadium, the corresponding reduction of $[V(CO)_6]^-$ by sodium in liquid ammonia at $-78°C$ gave a beautiful deep red solution containing a thermally unstable substance. By analogy with the corresponding reduction of $Cr(CO)_6$ in Na–NH_3, which Helmut Behrens and associates had found to give $Na_2[Cr(CO)_5]$ in important research in the late fifties (4), we felt that this deep red substance could be $Na_3[V(CO)_5]$. These formally analogous reactions are shown in Eqs. (15) and (16)

$$Cr(CO)_6 + 3Na \rightarrow Na_2[Cr(CO)_5] + \tfrac{1}{2}Na_2[OCCO] \tag{15}$$

$$Na[V(CO)_6] + 3Na \rightarrow Na_3[V(CO)_5] + \tfrac{1}{2}Na_2[OCCO] \tag{16}$$

Following filtration at low temperature to remove small amounts of sodium amide and disodium acetylenediolate, and removal of the ammonia, a thermally unstable deep red solid could be obtained. However, this material decomposed quickly above $0°C$ to give a heterogeneous brown–black solid which contained only small amounts of $[V(CO)_6]^-$. The nature of the bulk of the decomposition material remains an interesting problem (vide infra). Treatment of the deep red solution with various electrophiles [Eq. (17)] gave

$$Na_3[V(CO)_5] \quad
\begin{matrix}
\xrightarrow{\ 2Ph_3ECl\ } & [(Ph_3E)_2V(CO)_5]^- \\
& (E = Sn, Pb) \\
\xrightarrow{\ 3Ph_3PAuCl\ } & (Ph_3PAu)_3V(CO)_5
\end{matrix}
\tag{17}$$

the first known derivatives of $[V(CO)_5]^{3-}$ (48). Later, about 90% yields of thermally stable (but potentially explosive) sparingly soluble trirubidium and tricesium salts of $[V(CO)_5]^{3-}$ were obtained by metathesis of the sodium salt with RbI and CsI in liquid ammonia as shown in Eq. (18).

$$Na_3[V(CO)_5] + 3A^+I^- \xrightarrow[-70°C]{NH_3} A_3[V(CO)_5]\downarrow + 3Na^+I^- \qquad (18)$$
$$A = K, Rb, Cs$$

The tricesium salt was obtained as an orange–brown powder and the corresponding trirubidium compound was isolated as a lovely golden to red–brown shiny microcrystalline substance. Both of these materials provided satisfactory elemental analyses for the unsolvated complexes (49) and have remained intact under nitrogen at room temperature for nearly 10 years. The tricesium compound proved to be of substantial thermal stability and survived for brief periods at about 180°C. When exposed to fairly moist air, however, it almost instantly bursts into a shower of bluish-white sparks, a rather dramatic display even to chemists who routinely handle pyrophoric and other extremely air-sensitive compounds. For this reason, the tricesium salt has been popular and useful in lecture demonstrations to give an idea of the sensitivity of the compounds we routinely handle. It is possible, however, that these demonstrations may have had the undesirable effect of intimidating individuals (including potential graduate students!) who might otherwise be interested in studying the chemistry of these interesting materials.

In practice, these exceptionally reactive materials are generally not isolated. In solution or as a slurry, they are not shock sensitive and are no more difficult to handle than any other very air-sensitive substance. In view of their sensitivity, we were gratified when excellent elemental analyses were obtained for these materials (on the first try!). In this manner we were reassured that the methods developed for the handling and isolation of such sensitive compounds were satisfactory. An account of some of the apparatus used in our laboratory for this research has been published (50). The isolated potassium salt, $K_3[V(CO)_5]$, was far less thermally stable than either the cesium or rubidium compounds. It resembled $Rb_3[V(CO)_5]$ when initially isolated and had essentially the same mineral oil mull infrared spectrum in the $v(CO)$ region (Fig. 4), but within hours at room temperature it had turned increasingly darker and became more sensitive to shock. Also, unlike the tricesium or trirubidium salt, $K_3[V(CO)_5]$ immediately inflamed in Fluorolube or perfluorokerosine. [In this respect, $K_3[V(CO)_5]$ behaved like the tetrasodium salts, $Na_4[M(CO)_4]$ (vide infra).] Within 24 hr it had changed to a jet-black form, while maintaining its microcrystalline appearance. However, infrared spectra and derivative studies showed that very little $[V(CO)_5]^{3-}$ was still present in

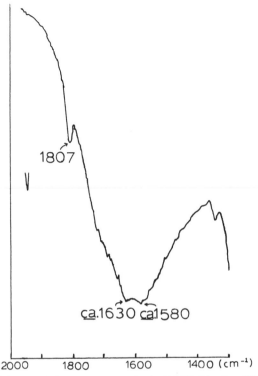

FIG. 4. Fluorolube mull infrared spectrum of $Rb_3[V(CO)_5]$ in the $\nu(CO)$ region: 1807 w, 1630 vs br, 1580 vs br cm^{-1}.

this black compound. Surprisingly, the latter material gave satisfactory elemental analyses for the composition $K_3VC_5O_5$! The nature of this black substance is still unknown.

There may be a connection between the thermal instability of unsolvated $K_3[V(CO)_5]$ and our inability to prepare unsolvated $K_2[Ti(CO)_6]$ (51). Alkali metal ions appear to promote the decomposition of highly reduced early transition metal carbonyl anions, but the details of these processes remain obscure. On the basis of work by Lippard and co-workers (52), however, it seems likely that the alkali metal ions form ion pairs with the carbonylmetallates and thereby promote some type of reductive coupling of coordinated carbon monoxide groups. The electropositive transition metal is thereby oxidized in this process. Although we originally proposed that the formation of acetylenediolate salts, i.e., $Na_2C_2O_2$ and $K_2C_2O_2$, during the reduction of $[V(CO)_6]^-$ in liquid ammonia occurred by the reduction of uncoordinated CO [formed, for example, by loss of CO from a putative 19-electron $[V(CO)_6]^{2-}$ species] (49), the very interesting possibility that the product containing $[C_2O_2]^{2-}$ arose via reductive coupling of coordinated carbon monoxide should be investigated.

B. Syntheses and Characterizations of $[Nb(CO)_5]^{3-}$ and $[Ta(CO)_5]^{3-}$: Thermal Instability and Shock Sensitivity of Unsolvated Alkali Metal Salts of Early Transition Metal Carbonyls

Syntheses of the corresponding trianions of niobium and tantalum had to await our development of a facile route to $[Nb(CO)_6]^-$ and $[Ta(CO)_6]^-$ (53). Once these anions were in hand as the $[Na(diglyme)_2]^+$ salts, their reduction by sodium in liquid ammonia was shown to provide deep red solutions containing the corresponding thermally unstable $Na_3[M(CO)_5]$. After filtration and cation exchange, deep red (M = Nb) or deep brown–red (M = Ta) slightly soluble and apparently amorphous solids were obtained in 40–50% yields, which provided satisfactory analyses for unsolvated $Cs_3[M(CO)_5]$ (54). These materials had infrared spectra that were very similar to those of $Cs_3[V(CO)_5]$ (49) (Fig. 5). Although they were only slightly less thermally stable than the vanadium analogue, as dry solids they proved to be much more shock sensitive. For this reason, our studies thus far have been largely limited to the reactions of $Na_3[M(CO)_5]$ formed *in situ* in liquid ammonia.

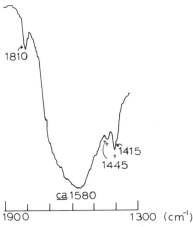

FIG. 5.   Silicone fluid mull infrared spectrum of $Cs_3[Nb(CO)_5]$ in the $v(CO)$ region: 1810 w, $\sim$1580 vs br cm^{-1}. Absorptions marked with daggers at 1445 and 1415 cm^{-1} are due to silicone fluid (113).

It may have been a minor miracle to have obtained any elemental analyses, let alone acceptable ones, on these materials, as they proved to be quite touchy substances. Garry Warnock, who first investigated the synthesis and properties of these materials, recorded the following observations in his laboratory notebook concerning $Cs_3[Nb(CO)_5]$:

The main sample (~ 0.9g) completely exploded when I attempted to powder the big lumps (in a nitrogen-filled dry box). An enormous BANG, louder than any I ever experienced before [occurred during this operation]. A minor explosion also resulted in an attempt to scrape the rest off the frit.

We have observed that the materials are especially prone to violent decomposition when they are even gently scraped off of a fritted glass disk with a metal spatula. However, in one instance an evacuated filtration apparatus containing about 2 g of dry $Cs_3[Ta(CO)_5]$ was almost instantly reduced to glass slivers by a violent explosion when nitrogen gas was rapidly introduced into the glassware in a dry box. There is no evidence that pure nitrogen gas reacts chemically with these trianions under any conditions, so we assume that the sudden impact of the tricesium salt against the glass frit on introduction of the nitrogen gas caused the explosion. After this event, which I personally experienced, these materials were handled with even more caution. The explosions resulted in the formation of a finely divided black soot which did not appear to react with air but was not further examined.

Although the related unsolvated $Na_2[Cr(CO)_5]$ (4) does not normally explode or decompose at room temperature under an inert atmosphere, we have observed that a finely divided black powder, similar in appearance to the aforementioned one, rapidly formed when small amounts of air were inadvertently introduced into an evacuated flask of unsolvated $Na_2[Cr(CO)_5]$. Under these conditions, the latter violently exploded, due to thermolysis of the pyrophoric product. Interestingly, unsolvated $Na[(C_5H_5)Ti(CO)_4]$ has also been observed to easily explode at room temperature on mild impact under an *argon* atmosphere. Thus, *unsolvated* alkali metal salts of all early transition metal carbonyl anions must be regarded as potentially explosive or shock-sensitive substances and manipulated carefully and handled with considerable care and respect.

As in the case of $Na_3[V(CO)_5]$, liquid ammonia solutions of the trisodium salts of $[Nb(CO)_5]^{3-}$ and $[Ta(CO)_5]^{3-}$ were also quite air and moisture sensitive but showed no tendency to decompose violently at or below $-33°C$. A summary of our results on the syntheses of the tricesium salts are shown in Eq. (19). Unlike $Cs_3[V(CO)_5]$, which is nearly insoluble in liquid ammonia, the niobium and tantalum analogs are appreciably soluble in this medium even at $-78°C$. Because the solid tricesium salts must be washed with liquid ammonia to remove all but traces of sodium iodide, the isolated pure

$$Na_3[M(CO)_5] \xrightarrow[NH_3]{3CsI} Cs_3[M(CO)_5]\downarrow + 3NaI \qquad (19)$$

85–94% isolated yield for M = V
40–50% isolated yields for M = Nb, Ta

compounds were obtained in substantially lower yields than were those of the vanadium compound. But from the good yields (often $\geq 80\%$) of derivatives obtained from unisolated $Na_3[Nb(CO)_5]$ and $Na_3[Ta(CO)_5]$ (54,55), it seems very likely that the pentacarbonylmetallates(3−) of niobium and tantalum were obtained in about the same yield ($\sim 90-95\%$) as was the vanadium analogue in these reductions. We have not explored the possible use of alkylamine solvents, such as $MeNH_2$ or Me_2NH, in separating NaI from $Cs_3[M(CO)_5]$. It seems quite possible that the tricesium salts, which are totally insoluble in ethers, would show substantially diminished solubility in alkyl amines compared to liquid ammonia, whereas the good solubility of NaI may not be as adversely affected.

C. Protonations of $[V(CO)_5]^{3-}$

Protonation reactions of $Na_3[V(CO)_5]$ in liquid ammonia have been examined in some detail. Treatment of the blood-red ammoniacal solutions of $[V(CO)_5]^{3-}$ with one equivalent of ethanol in THF at $-70°C$, followed by filtration and metathesis with $[Et_4N][BH_4]$ (which, along with $NaBH_4$, has good solubility in liquid ammonia), resulted in the rapid precipitation of an air-sensitive, microcrystalline yellow solid (55). Elemental analyses and infrared spectra of this material were consistent with the formulation $[Et_4N]_2[HV(CO)_5]$; however, it was necessary to obtain NMR spectra of the corresponding thermally unstable disodium salt in liquid ammonia to exclude an alternative formulation involving the still unknown tetraanion $[V_2(CO)_{10}]^{4-}$. The latter would have practically the same elemental analyses and possibly quite similar infrared spectra in the $v(CO)$ region as would those of $[HV(CO)_5]^{2-}$.

Proton and ^{51}V-NMR spectral studies on the protonation of $Na_3[V(CO)_5]$ in liquid ammonia proved to be particularly informative. Although a freshly prepared solution of $Na_3[V(CO)_5]$ showed no metal hydride signals in its ^1H-NMR spectrum, the ^{51}V-NMR spectrum of this substance exhibited a well-resolved singlet at -1962 ppm ($W_{1/2} = 18$ Hz). Treatment of $Na_3[V(CO)_5]$ with less than one equivalent of EtOH at $-50°C$ caused reduction in the intensity of the signal assigned to the trianion and a sharp doublet ($J_{V-H} = 27.6$ Hz, $W_{1/2} = 8$ Hz) grew in at -1985 ppm. The corresponding ^1H-NMR spectrum of this monoprotonated species (Fig. 6) showed a well-resolved octet ($J_{V-H} = 27.6$ Hz) centered at -4.76 ppm, due to the coupling of one vanadium (^{51}V is 99.75% abundant; $I = \frac{7}{2}$) to hydrogen. These spectra indicated that a

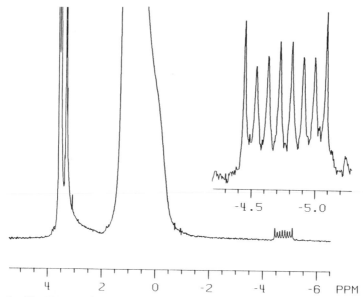

FIG. 6. The 300-MHz ^{1}H NMR spectrum of $Na_2[HV(CO)_5]$ in liquid ammonia at $-50°C$ showing blowup of metal hydride region; $\delta \sim 3.5$ (diethyleneglycol dimethyl ether), ~ 0.5 (ammonia), -4.76 [$HV(CO)_5^{2-}$] ppm.

mononuclear vanadium monohydride was present and were consistent with the formulation $[HV(CO)_5]^{2-}$ (55). Addition of more than one equivalent of ethanol to $Na_3[V(CO)_5]$ caused the singlet -1962 ppm to diminish and the doublet at -1986 ppm to become less intense. A new and rather broad singlet at -1604 ppm ($W_{1/2} = 70$ Hz) due to $[V(CO)_5NH_3]^-$ had grown in. Interestingly, except for small amounts of $[V(CO)_6]^-$, the only vanadium compounds observed during these protonation reactions were $[V(CO)_5]^{3-}$, $[HV(CO)_5]^{2-}$, and $[V(CO)_5NH_3]^-$ (Fig. 7). Indeed, in a separate experiment it was shown that $[V(CO)_5]^{3-}$ and $[V(CO)_5NH_3]^-$ did not react with one another or with $[V(CO)_6]^-$ in refluxing liquid ammonia. By comparison, under similar conditions, $Cr(CO)_6$ and $[Cr(CO)_5]^{2-}$ reacted rapidly, providing $[Cr_2(CO)_{10}]^{2-}$ (4), the chromium analogue of unknown $[V_2(CO)_{10}]^{4-}$. We were particularly encouraged that the ^{51}V-NMR spectrum of unenriched (i.e., natural abundance) ^{13}CO, $[V(CO)_5]^{3-}$, and $[HV(CO)_5]^{2-}$ showed sharp satellites due to $^{51}V-^{13}C$ coupling. The presence of these well-resolved satellites suggested that proof for the existence of $[V(CO)_5]^{3-}$ and $[HV(CO)_5]^{2-}$ in solution might be obtained from spectra of the corresponding 99% ^{13}C enriched species. By employing Calderazzo's procedure (56), 99% ^{13}CO-enriched $[Na(diglyme)_2][V(^{13}CO)_6]$ was easily obtained from VCl_3. As before, this material was converted to a roughly equimolar mixture of the

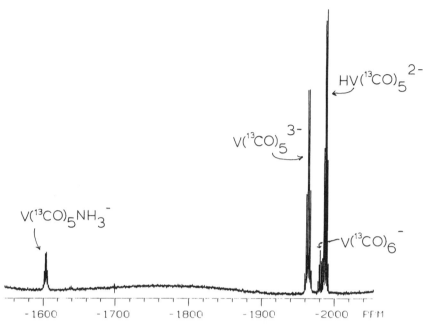

FIG. 7. The 77.87-MHz ^{51}V NMR spectrum showing 99% ^{13}C-enriched carbonylvanadates in liquid ammonia at $-50°C$. Identity of signals: Na[V(CO)$_5$NH$_3$], δ -1605 ppm, $J_{V-C} = 122.4$ Hz; Na$_3$[V(CO)$_5$], δ $- 1965$ ppm, $J_{V-C} = 139$ Hz; Na$_2$[(HV(CO)$_5$], δ $- 1988$ ppm, $J_{V-C} = 124.5$, $J_{V-H} = 27.6$ Hz. Impurity at δ $- 1981$ ppm, $J_{V-C} = 116$ Hz, is V(CO)$_6^-$.

trianion and its monoprotonated derivative. ^{51}V, ^{13}C, and ^{1}H-NMR spectra of Na$_3$[V(^{13}CO)$_5$] and Na$_2$[HV(^{13}CO)$_5$] in liquid ammonia at $- 50°C$ provided truly spectacular results. The ^{51}V and ^{13}C-NMR spectra of the trianion showed unequivocally that it was a nonhydridic mononuclear complex, in which the vanadium is coordinated to five equivalent carbon atoms, and the spectra were entirely consistent with the presence of [V(CO)$_5$]$^{3-}$ units. The highly reduced nature of this anion was indicated by the large 67.8-ppm downfield shift of its ^{13}C resonance relative to that of [V(CO)$_6$]$^-$ (Fig. 8). Confirmation of the hydridic and mononuclear nature of [HV(^{13}CO)$_5$]$^{2-}$ was provided by its ^{1}H-NMR spectrum in NH$_3$ at $- 50°C$: δ $- 4.78$ ppm, a partially overlapped sextet of octets, $J_{V-H} = 27.7$ Hz; $J_{C-H} = 10.8$ Hz (Fig. 9). From the ^{51}V, ^{13}C, and ^{1}H-NMR spectra of the dianion it was evident that the vanadium and hydride were coupled to five equivalent carbons at $- 50°C$ in liquid NH$_3$ (57). In this respect, the vanadium hydride resembles the isoelectronic chromium monoanion, [HCr(CO)$_5$]$^-$, which has been reported to be fluxional at $25°C$ in acetonitrile (58).

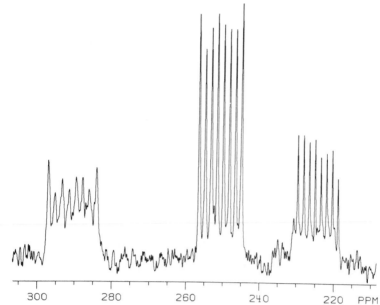

FIG. 8.   The 75.44-MHz ^{13}C NMR spectrum showing 99% ^{13}C-enriched carbonylvanadates in liquid ammonia at $-50°$C. Identity of signals: δ 290.1 ppm, $J_{V-C} = 139$ Hz, Na$_3$-[V(CO)$_5$]; δ 250.3 ppm, $J_{V-C} = 124$ Hz, $J_{C-H} = 11$ Hz (not resolved in this spectrum), Na$_2$[HV(CO)$_5$]; δ 224.2 ppm, $J_{V-C} = 116$ Hz, Na[V(CO)$_6$]. An unsymmetrical octet for V(CO)$_6^-$ is observed due to the presence of overlapping V(CO)$_5$(NH$_3$)$^-$, δ 225.3 ppm, unresolved octet, $J_{V-C} \approx 121$ Hz.

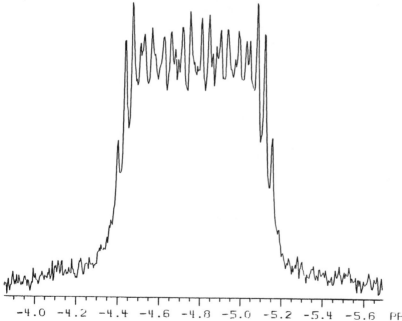

FIG. 9.   The 300-MHz ^{1}H NMR spectrum of 99% ^{13}C-enriched [HV(CO)$_5$]$^{2-}$ in the hydride region in liquid ammonia at $-50°$C. δ -4.78 ppm, $J_{V-H} = 27.7$ Hz, $J_{C-H} = 10.8$ Hz.

D. Reactions of $[V(CO)_5]^{3-}$ with Main-Group Electrophiles and Ph_3PAuCl

Representative reactions of $Na_3[V(CO)_5]$ with several main-group electrophiles are shown in the scheme shown here:

$$[V(CO)_5PPh_3]^- + \tfrac{1}{2}Ph_2 \xleftarrow{\;Ph_4P^+\;} [V(CO)_5]^{3-} \xrightarrow{\;R_3SnCl\;} [R_3SnV(CO)_5]^{2-} \quad (R = Me, Ph)$$

$$[V(CO)_5NH_3]^- + H_2 \xleftarrow[\;CH_3I\;]{2NH_4^+} \qquad \xrightarrow[\;2R_3SnCl\;]{} [(R_3Sn)_2V(CO)_5]^- \quad (Re = C_6H_{11}, Ph)$$

$$[CH_3V(CO)_5]^{2-} \qquad \Big\downarrow {\scriptstyle ROH} \qquad \xrightarrow[\;2Ph_3PbCl\;]{} [(Ph_3Pb)_2V(CO)_5]^- \xrightarrow{\;Na/Hg\;}$$

$$[HV(CO)_5]^{2-} \qquad\qquad\qquad [Ph_3PbV(CO)_5]^{2-} + Ph_4Pb + Pb$$

All reactions are believed to proceed stepwise and in many cases a monosubstituted dianion of the general formula $[EV(CO)_5]^{2-}$ can be isolated. These in turn react with a second equivalent of an electrophile to provide $[(E)(E')V(CO)_5]^-$ or $[E_2V(CO)_5]^-$, which remains uncharacterized for E = H. A single-crystal X-ray study confirmed the presence of seven-coordinate vanadium in $[(Ph_3Sn)_2V(CO)_5]^-$ (Fig. 10). The isoelectronic titanium species, $[(Ph_3Sn)_2Ti(CO)_5]^{2-}$, has also been reported (59). Although no concerted attempts to attach a third Ph_3Sn^+ unit to $[(Ph_3Sn)_2V(CO)_5]^-$ have been made, an eight-coordinate vanadium complex was easily obtained with the much less sterically demanding triphenylphosphinegold ligand. Thus, the reaction of Ph_3PAuCl with $Na_3[V(CO)_5]$ in liquid ammonia provided 10–20% yields of the air-stable and remarkably inert $(Ph_3PAu)_3V(CO)_5$. Up to 60% yields of the same substance were readily obtained from the reaction of insoluble $Cs_3[V(CO)_5]$ with three equivalents of Ph_3PAuCl in tetrahydrofuran, which, in contrast to liquid ammonia, is inert toward Ph_3PAuCl (60). It is noteworthy that, unlike monotriphenylphosphinegold derivatives of vanadium carbonyl, such as $Ph_3PAuV(CO)_6$ or $Ph_3PAuV(CO)_5PPh_3$ (61), the trigold species showed no tendency to react with very polar molecules such as liquid ammonia at $-33°C$, hexamethylphosphoric triamide, nitromethane, soluble halides, or excess triphenylphosphine (all of the latter at ambient temperature). Although we have no evidence for the formation of $[Ph_3PAuV(CO)_5]^{2-}$ or $[(Ph_3PAu)_2V(CO)_5]^-$ during these reactions (even when fewer than three equivalents of Ph_3PAuCl were used), they would appear to be reasonable intermediates and may be isolable species. Figure 11 depicts the core structure of $(Ph_3PAu)_3V(CO)_5$, which contains a very nearly tetrahedral cluster unit, containing formally three Au—Au bonds and six Au—V bonds of about the same length (60). It is undoubtedly the presence of the tetrahedral cluster that is responsible for the surprisingly robust nature of this material (62).

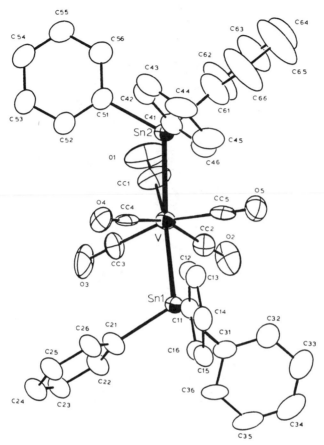

Fig. 10. Molecular structure of $[(Ph_3Sn)_2V(CO)_5]^-$. Selected interatomic distances (Å): V–Sn(1), 2.757(3); V–Sn(2), 2.785(3); V–C (mean), 1.94(3); C–O (mean), 1.16(3). Angle (degrees): Sn(1)–V–Sn(2): 137.9(1).

Of the products shown in the above scheme, only the methyl derivative, $[CH_3V(CO)_5]^{2-}$, has not been isolated as a pure substance. In view of the existence of the well-known acylchromates, $[(RCO)Cr(CO)_5]^-$, and their importance in organochromium chemistry, attempts to prepare corresponding acylvanadates, $[(RCO)V(CO)_5]^{2-}$, will be of obvious interest and possible importance.

E. Chemical Properties of $[V(CO)_5(NH_3)]^-$

Perhaps the most useful reactions of $[V(CO)_5]^{3-}$ examined to date are those with Brønsted acids to ultimately provide the ammine substitution

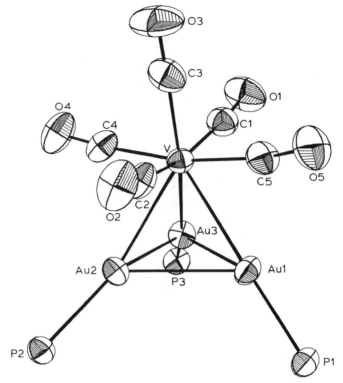

FIG. 11. Molecular structure of $(Ph_3PAu)_3V(CO)_5$; phenyl groups have been omitted for clarity. Selected mean interactomic distances (Å): V–Au, 2.73(2); Au–Au, 2.82(3); V–C, 1.95(4); C–O, 1.16(2).

product $[V(CO)_5NH_3]^-$. As mentioned previously, $[HV(CO)_5]^{2-}$ forms initially during the reaction of NH_4^+ with $Na_3[V(CO)_5]$. Addition of a second equivalent of ammonium chloride caused the orange–yellow color of the monohydride to change to an unstable red–orange intermediate. Within seconds at $-78°C$, this unstable species evolved gas and generated the striking bright red–violet ammine complex, $[V(CO)_5NH_3]^-$, which was first prepared by the photolysis of $[V(CO)_6]^-$ in liquid ammonia (63). Although it is tempting to formulate the red–orange transient as $[H_2V(CO)_5]^-$, which would be isoelectronic with the known and quite labile molecular hydrogen complex, $[(\eta^2\text{-}H_2)Cr(CO)_5]$ (64), no spectroscopic evidence is yet available to justify this formulation.

Amminepentacarbonylvanadate(1−) reacted with a variety of acceptor ligands, L, to provide excellent yields of complexes of the general formula $[V(CO)_5L]^-$, many of which could not be prepared by other routes. For

example, the first isocyanide-substituted carbonylvanadates(1 −) were prepared by this route and are not available by the usual photolytic substitution
route. Some of these syntheses are shown in the scheme below.

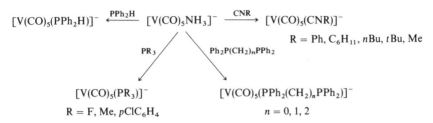

Treatment of $[V(CO)_5NH_3]^-$ with excess CN^- (in liquid ammonia) or acetonitrile provided $[V(CO)_5(CN)]^{2-}$ and $[V(CO)_5(NCCH_3)]^-$, respectively,
which have also been prepared by the photolytic route. Photogenerated and
extremely labile $[V(CO)_5THF)]^-$ has been employed as a useful precursor
to a variety of $[V(CO)_5L]^-$ (55,65) complexes.

F. Derivatives of $[Nb(CO)_5]^{3-}$ and $[Ta(CO)_5]^{3-}$

Hydridopentacarbonylniobate and -tantalate dianions have also been
prepared and isolated in good yields (60–80%) as tetraethylammonium salts
of satisfactory purity by the same method used for the vanadium analogue.
These thermally unstable disodium salts, $Na_2[HM(CO)_5]$, were characterized
by their NMR spectra in liquid ammonia at $-50°C$. Although the hydride
resonance for $Na_2[HTa(CO)_5]$ was easily seen and consisted of a fairly sharp
singlet ($W_{1/2} = 30$ Hz) at -2.2 ppm, initial attempts to detect the corresponding signal for $Na_2[HNb(CO)_5]$ were unsuccessful due to ^{93}Nb ($I = \frac{9}{2}$) and ^1H
coupling. We were able to observe the ^{93}Nb resonance for this substance,
however, at -2122 ppm (broad doublet; $J_{Nb-H} = 80$ Hz). Its $^{93}Nb\{^1H\}$ NMR
spectrum showed the expected singlet. Once the frequency of the ^{93}Nb
resonance was determined, the ^{93}Nb-decoupled ^1H NMR of $[HNb(CO)_5]^{2-}$
was examined and showed a beautiful sharp hydride signal at -1.89 ppm (57).
This last experiment provided an indication of how useful a multinuclear
approach can be in the characterization of carbonylmetallates by NMR
spectroscopy. Table IV (66–68) summarizes ^1H NMR data for known
compounds of the general formula $[HM(CO)_5]^z$.

Recently, protonation reactions of $Na_3[Ta(CO)_5]$ in liquid ammonia have
been studied by ^{13}C NMR (69). Large line widths were observed for $Ta(CO)_6^-$
($W_{1/2} = 830$ Hz) and $HTa(CO)_5^-$ ($W_{1/2} = 126$ Hz) due to the sizable electric
quadrupole moment of ^{181}Ta ($I = \frac{7}{2}$; nearly 100% abundant) (70). For this
reason, 99% ^{13}C enriched samples of the carbonyltantalates were synthesized

TABLE IV
^{1}H NMR Chemical Shifts for $[HM(CO)_5]^z$

Compound	δ, ppm	Reference
$[HV(CO)_5]^{2-}$	$-4.8, J_{H-V} = 27.6$ Hz	55,57
$[HNb(CO)_5]^{2-}$	$-1.9, J_{H-Nb} \approx 80$ Hz	55
$[HTa(CO)_5]^{2-}$	-2.2	55
$[HCr(CO)_5]^-$	-6.9	58
$[HMo(CO)_5]^-$	-4.0	58
$[HW(CO)_5]^-$	$-4.2, J_{H-W} = 53.4$ Hz	58
$HMn(CO)_5$	-7.5	66
$HTc(CO)_5$	—	67
$HRe(CO)_5$	-5.7	67,68

by the method of Dewey *et al.* (*53*) and were used to facilitate observation of these signals. For the less symmetrical anions, $[Ta(CO)_5]^{3-}$ ($W_{1/2} = 10$ Hz) and $[Ta(CO)_5(NH_3)]^-$ (δ 219 ppm, *cis*-CO, and 225 ppm, *trans*-CO; $W_{1/2} = 4$ Hz), however, quite sharp signals were observed. Some of the data from this study and related systems are shown in Table V (*71,72*). Undoubtedly the most surprising entry is the extremely downfield resonance for Na$_3$-$[Ta(CO)_5]$, which is even slightly downfield of the ^{13}C resonance for Na$_3$-$[V(CO)_5]$. This is an unusual and unexpected result because carbonyl ^{13}C resonance positions usually become less positive (i.e., shift upward) as the

TABLE V
^{13}C NMR Chemical Shifts for Selected Anionic and Neutral Metal Carbonyls
of Some Early Transition Elements

Group 5	$\delta(^{13}C)$, ppm	Reference	Group 6	$\delta(^{13}C)$, ppm	Reference
$[V(CO)_6]^-$	224	57	$Cr(CO)_6$	212	71
$[HV(CO)_5]^{2-}$	250	57	$[HCr(CO)_5]^-$	228, 232^c	58
$[V(CO)_5]^{3-}$	290	57	$[Cr(CO)_5]^{2-}$	—	—
$[Nb(CO)_6]^-$	217^a	69	$Mo(CO)_6$	204	71
$[Ta(CO)_6]^-$	211	69	$W(CO)_6$	192	71
$[HTa(CO)_5]^{2-}$	232^b	69	$[HW(CO)_5]^-$	205, 210^c	58
$[Ta(CO)_5]^{3-}$	293	69	$[W(CO)_5]^{2-}$	257^d	72

^{a} CD$_3$CN at 20°C, $J_{^{93}Nb-^{13}C} = 236$ Hz.

^{b} Large line width ($W_{1/2} = 126$ Hz) precluded possible resolution of *cis*- and *trans*-CO.

^{c} Two signals due to *cis*- and *trans*-CO.

^{d} Cation is Na(2.2.1)$^+$. Chemical shift has been found to be a very sensitive function of cation and solvent. For example, Na$_2[W(CO)_5]$ in THF has a ^{13}C δ value of 247 ppm (*72*).

atomic number of the metal increases in a triad for a given type of binary or substituted metal carbonyl (73). For example, the expected trends are observed for $[HV(CO)_5]^{2-}$ and $[HTa(CO)_5]^{2-}$ or $[V(CO)_6]^-$, $[Nb(CO)_6]^-$, and $[Ta(CO)_6]^-$. (In the case of low-valent Group 4 carbonyls, different trends have been reported recently) (74). Although the ^{13}C chemical shift for $[Ta(CO)_5]^{3-}$ seems unusual, it is perhaps significant that the observed ^{13}C resonance positions for the third-row pentacarbonyls, $[M(CO)_5]^Z$, are nearly a linear function of the charge on the complex, i.e., $Os(CO)_5$, δ 182.6 ppm ($CDCl_3$ at $-40°C$) (75); $Na[Re(CO)_5]$, δ 217.7 ppm (DMSO-d_6) (76); $[Na(2.2.1)]_2[W(CO)_5]$, δ 257.3 ppm (THF) (72); and $Na_3[Ta(CO)_5]$, δ 293.0 ppm (NH_3 at $-50°C$) (69).

The orange–red and red crystalline salts, $[Et_4N]_2[HNb(CO)_5]$ and $[Et_4N]_2[HTa(CO)_5]$, like their yellow vanadium analogue, are extremely reactive materials and effectively function as hydride transfer reagents. In this respect they are similar to the Group VI analogues, $[HM(CO)_5]^-$ (M = Cr, Mo, and W), which M. Darensbourg has studied extensively (77). However, these substances appear to be far more reactive. For example, whereas the Group 6 monoanions are stable in acetonitrile, the corresponding Group 5 dianions rapidly deprotonated the solvent according to Eq. (20).

$$[HM(CO)_5]^{2-} + 2CH_3CN \rightarrow [M(CO)_5(NCCH_3)]^- + H_2 + CH_2CN^- \qquad (20)$$

In solvents such as DMSO or HMPA, the anions apparently reacted with the counterion, Et_4N^+, to generate labile substituted monoanions, as shown in Eq. (21).

$$[HM(CO)_5]^{2-} + Et_4N^+ + DMSO \rightarrow [M(CO)_5DMSO]^- + H_2 + Et_3N + CH_2{=}CH_2$$

$$(21)$$

In the presence of a variety of metal carbonyls, these potent metal hydrides formed metal formyls, i.e., Eq. (22).

$$Fe(CO)_5 + [HM(CO)_5]^{2-} \xrightarrow{\text{THF}} [Fe(CO)_4CHO]^- + [M(CO)_5THF]^- \qquad (22)$$

The latter deep-purple THF adducts of niobium and tantalum have been prepared independently by the photolysis of the corresponding $[M(CO)_6]^-$ in tetrahydrofuran at $-70°C$ (54). They are even more reactive than the corresponding ammine complexes (vide infra) with respect to ligand substitution reactions.

We had anticipated that these hydrides would be less reactive and perhaps more useful in chemical syntheses than the parent trianions. In one case this expectation was realized. Although slurries of $Cs_3[Ta(CO)_5]$ in THF did not react with Ph_3PAuCl to provide any of the desired $(Ph_3PAu)_3Ta(CO)_5$

[essentially the same procedure used to prepare the previously discussed $(Ph_3PAu)_3V(CO)_5$], up to 60% yields of this gold–tantalum cluster have been obtained from $[Et_4N]_2[HTa(CO)_5]$ according to Eq. (23).

$$[HTa(CO)_5]^{2-} + 3Ph_3PAuCl \rightarrow (Ph_3PAu)_3Ta(CO)_5 + HCl + 2Cl^- \qquad (23)$$

Spectroscopically this trigold–tantalum species proved to be virtually identical to $(Ph_3PAu)_3V(CO)_5$. As in the case of $[HV(CO)_5]^{2-}$, further protonation of $[HNb(CO)_5]^{2-}$ and $[HTa(CO)_5]^{2-}$ in liquid ammonia led to the formation of dihydrogen and the ammine complexes, $[M(CO)_5(NH_3)]^-$ (vide infra) (54).

Treatment of liquid ammonia solutions of $Na_3[M(CO)_5]$ dropwise with one equivalent of Ph_3SnCl in THF provided, after metathesis, 70–80% yields of orange to orange–red crystalline $[Et_4N]_2[Ph_3SnM(CO)_5]$. These oxygen-sensitive materials have infrared spectra in the $v(CO)$ region that are consistent with the presence of a substituted dianion of C_{4v} symmetry (Fig. 12). Also, nearly quantitative yields of the previously unknown $Na[M(CO)_5NH_3]$ were obtained from the reaction of $Na_3[M(CO)_5]$ with ammonium chloride in liquid ammonia. Although the sodium salts are thermally unstable, 40–45% yields of red–violet (M = Nb) to deep-violet (M = Ta) crystalline tetraphenylarsonium salts, $[Ph_4As][M(CO)_5NH_3]$, were obtained by metathesis.

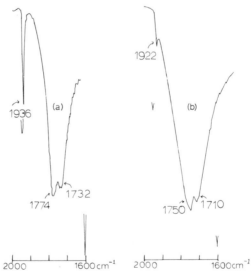

FIG. 12. Mineral oil mull infrared spectra of (a) $[Et_4N]_2[Ph_3SnTa(CO)_5]$ and (b) $[Et_4N]_2$-$[HTa(CO)_5]$ in the $v(CO)$ region.

These ammine complexes, like their vanadium analogue, proved to be very labile in solution and readily reacted at $-20-0°C$ with a variety of π-acceptor ligands such as PR_3, $P(OR)_3$, and RNC to provide 50–80% isolated yields of the corresponding $[M(CO)_5L]^-$. For example, the first isocyanide derivatives of tantalum and niobium carbonyls were obtained by treating ammonia solutions of $Na[M(CO)_5NH_3]$ with t-BuNC followed by cation exchange and crystallization. Orange salts of the composition $[Et_4N][M(CO)_5CNtBu]$ were thereby obtained in 50–60% yields. These reactions are summarized in Eqs. (24) and (25).

$$[Ph_3SnM(CO)_5]^{2-} \xleftarrow{\text{Ph}_3\text{SnCl}} [M(CO)_5]^{3-} \xrightarrow{2NH_4^+} [M(CO)_5NH_3]^- + H_2 \quad (24)$$

$$(M = Nb, Ta)$$

$$[M(CO)_5NH_3]^- + L \xrightarrow{-20-0°C} [M(CO)_5L]^- + NH_3 \quad (25)$$

$$[L = PR_3, P(OR)_3, RNC]$$

New anions of the general formula $[Ta(CO)_5L]^Z$ prepared by this method included $[Ta(CO)_5(AsPh_3)]^-$ and $[Ta(CO)_5(CN)]^{2-}$ (54). The cyanide derivative is of particular interest because its synthesis suggests that other more interesting dianionic species, such as $[Ta(CO)_5(C\equiv CR)]^{2-}$ and related $[Ta(CO)_5(\eta^1\text{-hydrocarbyl})]^{2-}$ species, might be accessible materials. Clearly, much chemistry of $[Nb(CO)_5]^{3-}$ and $[Ta(CO)_5]^{3-}$ remains to be explored. Of particular interest will be a systematic examination of the reactions of these materials with transition metal and organic electrophiles.

IV

TRICARBONYLMETALLATES(3−) OF COBALT, RHODIUM, AND IRIDIUM

A. Synthesis and Characterization of Na₃[Co(CO)₃]

Approximately 25 years after Mond and co-workers had prepared the first carbonyl of cobalt, $Co_2(CO)_8$ (78), Hieber obtained $HCo(CO)_4$ by the acidification of salts containing $[Co(CO)_4]^-$ (79), one of the first established carbonylmetallates (80). In 1941, Behrens, a student of Hieber, initiated his important and extensive investigations on the reduction of metal carbonyls and their derivatives by alkali and alkaline earth metals in liquid ammonia (81). At this time he established that various salts of $[Co(CO)_4]^-$, including $Na[Co(CO)_4]$ and $K[Co(CO)_4]$, could be obtained from the neutralization of $HCo(CO)_4$ or the reduction of $Cd[Co(CO)_4]_2$ by sodium or potassium in liquid ammonia (4). On the basis of Behrens' pioneering studies, we were initially quite pessimistic concerning the possibility that $[Co(CO)_4]^-$ could be reduced further in liquid ammonia, because it seemed likely that Behrens would have reported on such a reaction long ago. However, after we observed that $[Mn(CO)_5]^-$ and $[V(CO)_6]^-$ readily reacted with Na or K in liquid ammonia to provide the carbonyl trianions, $[Mn(CO)_4]^{3-}$ and $[V(CO)_5]^{3-}$, respectively, it was felt that an investigation of the interaction of $[Co(CO)_4]^-$ with alkali metals in liquid ammonia was in order.

Several years ago preliminary results in this study were reported and provided good evidence for the reduction of $[Co(CO)_4]^-$ by sodium in liquid ammonia (82). Although the nature of the reduction product remained obscure, it was formulated to contain $[Co(CO)_3]^{3-}$ on the basis of infrared spectra and derivative studies. More recently, the first isolation of a satisfactorily pure sample of $Na_3[Co(CO)_3]$ was accomplished. Formally, the trisodium salt contains $[Co(CO)_3]^{3-}$, an exceedingly rare example of a three-coordinate, 18-electron metal ion. To our knowledge the only previously well-characterized example of such a compound is $[Ni(C_2H_4)_3]^{2-}$, which is formally present in $[Li(TMEDA)]_2[Ni(C_2H_4)_3]$ (83).

Of the homoleptic carbonylmetallates(1−) we have attempted to reduce, $[Co(CO)_4]^-$ appears to be the most difficult. Although the sodium salts of $[M(CO)_6]^-$ (M = V, Nb, and Ta) were quickly reduced in liquid ammonia by sodium metal to provide the corresponding trianions, $[M(CO)_5]^{3-}$ (vide supra), it seems unlikely that we have ever effected complete reduction of $Na[Co(CO)_4]$ to $Na_3[Co(CO)_3]$. Even after 2 days of refluxing (at $-33°C$) anhydrous ammonia solutions of $Na[Co(CO)_4]$ with excess Na, considerable amounts of the tetracarbonylcobaltate(1−) remained. Low yields of a heterogeneous-appearing brown to olive–brown insoluble solid were isolated; this solid has been shown to contain $Na_3[Co(CO)_3]$ (vide infra). As in the case of $[Re(CO)_5]^-$, we found that solutions of potassium in liquid ammonia were far more effective at reducing $[Co(CO)_4]^-$. However, unlike $[Re(CO)_5]^-$, $[Rh(CO)_4]^-$, or $[Ir(CO)_4]^-$ (vide infra), there was no evidence that $[Co(CO)_4]^-$ was reduced by sodium or potassium metal in hexamethylphosphoric triamide. We observed that excess sodium naphthalenide slowly (over a period of 40–50 hr at room temperature) converted $Na[Co(CO)_4]$ in THF to an impure and insoluble brown powder that contained $Na_3[Co(CO)_3]$, but this synthesis appeared to be of little or no utility.

Reduction of $[Co(CO)_4]^-$ by sodium or potassium in liquid ammonia proceeded according to the Eq. (26) (shown for the potassium reaction).

$$KCo(CO)_4 + 3K \rightarrow K_3[Co(CO)_3] + \tfrac{1}{2}[K_2C_2O_2] \tag{26}$$

The presence of $K_2C_2O_2$ in the reaction product was established by comparison of the mineral mull IR spectrum of bonafide $K_2C_2O_2$, prepared from the reaction of free CO and potassium metal in liquid ammonia (84), and that of the impure $K_3[Co(CO)_3]$. Both contained a weak but sharp absorption at about 2140 cm^{-1}, which is characteristic of $K_2C_2O_2$. Although efforts to separate the $K_2C_2O_2$ from $K_3[Co(CO)_3]$ have been unsuccessful to date, we did succeed in purifying a small sample of $Na_3[Co(CO)_3]$ by a rather surprising procedure. This method was inspired by Corbett's studies on Zintl ions (85). He found that $[Na(2.2.2)]_3Sb_7$ could be prepared by the solubilization of NaSb alloy with cryptand 2.2.2 in neat ethylenediamine. In an attempt to prepare $[Na(2.2.2)]_3[Co(CO)_3]$, an impure sample of Na_3-$[Co(CO)_3]$, containing both $Na_2C_2O_2$ and some $NaNH_2$, was treated with cryptand 2.2.2 in ethylenediamine at room temperature. Rather quickly, a deep red–brown color formed in the ethylenediamine and we thought that we had solubilized the $Na_3[Co(CO)_3]$! Surprisingly, however, instead the cryptand apparently dissolved all of the impurities! Filtration of the aforementioned solution provided a insoluble homogeneous yellow to olive solid which proved to be analytically pure $Na_3[Co(CO)_3]$!! Unfortunately, this method seems unlikely to be of any importance in the routine purification of $Na_3[Co(CO)_3]$ in view of the high price of cryptand 2.2.2. But we were quite amused at this rather unusual use of a cryptand in cleaning up a material that had defied previous, more conventional, attempts at its purification. The nature of the materials dissolved in the ethylenediamine solution remain largely unknown. No $Na_3[Co(CO)_3]$ and very little $Na[Co(CO)_4]$ were present. One attempt to prepare $[K(2.2.2)]_3[Co(CO)_3]$ by reduction of performed $[K(2.2.2)][Co(CO)_4]$ with three equivalents of $[K(2.2.2)]^+e^-$ (86) in liquid ammonia was also carried out. Very interestingly, no reduction of $[Co(CO)_4]^-$ was observed to occur under these conditions. If this experiment and our other observations are of any consequence, it may well indicate that the formation and/or existence of $[Co(CO)_3]^{3-}$ depends on strong cation–anion interactions in the solid state. If this is true, then attempts to solubilize salts containing $[Co(CO)_3]^{3-}$ may be doomed to failure. Indeed, it has been observed that treatment of $Na_3[Co(CO)_3]$ with the very polar solvent HMPA resulted in decomposition and the formation of $Na[Co(CO)_4]$ as the only observable carbonylmetallate in solution. Surprisingly, however, solutions of $Na_3[Rh(CO)_3]$ and $Na_3[Ir(CO)_3]$ in HMPA have been obtained (vide infra).

Progress of the reduction of $K[Co(CO)_4]$ by potassium metal in liquid ammonia to form $K_3[Co(CO)_3]$ has been monitored by taking mineral oil mull infrared spectra of the isolated product after addition two and three equivalents of potassium metal. The resulting spectra in the $\nu(CO)$ region demonstrated that the reduction was still substantially incomplete after consumption of two equivalents of K. However, the third equivalent of potassium metal caused all but traces of the $[Co(CO)_4]^-$ to disappear and left mainly $K_3[Co(CO)_3]$. Figure 13 shows mineral oil mull infrared spectra of $K_3[Co(CO)_3]$ and $Na_3[Co(CO)_3]$ in the $\nu(CO)$ region. The most intense

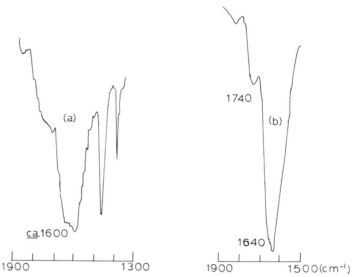

FIG. 13.   Mineral oil mull infrared spectra of (a) $Na_3[Co(CO)_3]$ and (b) $K_3[Co(CO)_3]$ in the $\nu(CO)$ region.

bands, centered at about 1600 cm^{-1} for $Na_3[Co(CO)_3]$ or 1640 cm^{-1} for $K_3[Co(CO)_3]$, fall in the same region as do those of corresponding bands observed for mull IR spectra of other carbonylmetallate(3 −) salts, including $K_3[V(CO)_5]$ (1812 m, 1600 vs, br cm^{-1}) and $Na_3[Mn(CO)_4]$ (1790 w, 1600 vs, br cm^{-1}).

B. *Derivatives of [Co(CO)$_3$]$^{3-}$*

Treatment of a suspension of $A_3[Co(CO)_3]$ (A = Na or K) in THF or liquid ammonia with two to three equivalents of Ph_3ECl (E = Ge, Sn, Pb), followed by cation exchange, provided 50–70% isolated yields of colorless (E = Ge, Sn) or yellow (E = Pb) fairly air- and moisture-stable tetraethylammonium salts containing anions of the composition $[(Ph_3E)_2Co(CO)_3]^-$, as shown by Eq. (27). Formation of these derivatives in

$$2Ph_3ECl + [Co(CO)_3]^{3-} \rightarrow [trans\text{-}(Ph_3E)_2Co(CO)_3]^- + 2Cl^- \qquad (27)$$

good yields gave important additional evidence for the existence of $Na_3[Co(CO)_3]$ and $K_3[Co(CO)_3]$. Although it seems likely that the reaction proceeds by the intermediate formation of tetrahedral dianions of the type $[(Ph_3E)Co(CO)_3]^{2-}$, and we have IR spectral evidence to suggest that such materials do exist, attempts to isolate satisfactorily pure samples of these dianions have not been successful to date. These monoanions are analogous to the previously known cations $trans$-$[(Ph_3Y)_2Co(CO)_3]^+$ (Y = P, As, or Sb) (87). Because IR solution spectra of $[Et_4N][(Ph_3E)_2Co(CO)_3]$ showed only one band in the $v(CO)$ region, it is very likely that these anions also contain trans-disubstituted trigonal–bipyramidal units. Reactions of $[trans-(Ph_3E)_2Co(CO)_3]^-$ with electrophiles have not been examined. It seems possible that neutral octahedral, formally Co(III), complexes of the type $(Ph_3E)_2Co(CO)_3(E)$ may be accessible, where E is a main-group or transition metal electrophilic unit.

Although the protonation reactions of the carbonyl trianions of the Group 5 and Group 7 elements have been examined in some detail, little is presently known about corresponding reactions of the Group 9 carbonyl trianions. Treatment of a slurry of $K_3[Co(CO)_3]$ in liquid ammonia at $-78°C$ with two equivalents of ethanol provided an orange–yellow solution. The ^{1}H-NMR spectrum of this solution at $-50°C$ in the hydride region showed a very broad band ($W_{1/2} = 720$ Hz) centered at $\delta - 7.8$ ppm. Provisionally, this band has been assigned to $[H_2Co(CO)_3]^-$, but the possibility that it is some other cobalt hydride cannot be ruled out presently. Clearly, more work needs to be done to properly characterize this material. It is of interest that low-temperature matrix isolation studies have provided evidence for the existence of $H_3Co(CO)_3$, which may be considered to be the conjugate acid of $[H_2Co(CO)_3]^-$ (88).

Addition of three equivalents of NH_4Cl to a liquid ammonia slurry of $K_3[Co(CO)_3]$, followed by treatment with PPh_3 or $P(C_6H_{11})_3$, and cation exchange resulted in the isolation of 40 and 45% isolated yields of satisfactorily pure $[Et_4N][Co(CO)_3PPh_3]$ and $[Ph_4As][Co(CO)_3P(C_6H_{11})_3]$, respectively. No attempts to improve these yields were made because the identical materials are more readily obtained by the method of Hieber, which involves reduction of the respective dimers, $[Co(CO)_3(PR_3)]_2$, by Na–Hg, followed by cation exchange (89). However, isocyanide-substituted carbonyl-cobaltate ions are not available by Hieber's route because the precursors, $Co_2(CO)_8(CNR)_2$, are unknown. Attempts to prepare these materials by the substitution of $Co_2(CO)_8$ with isocyanides resulted in the disproportionation of the carbonyl in accordance with Eq. (28) (90).

$$Co_2(CO)_8 + 5RNC \xrightarrow{-4CO} [Co(CNR)_5][Co(CO)_4] \qquad (28)$$

For this reason, it was felt that the synthesis of anions of the type $[Co(CO)_3(CNR)]^-$ would be a significant extension of our general procedure. In fact, treatment of $K_3[Co(CO)_3]$ with three equivalents of NH_4Cl, followed by the addition of RNC (R = tBu, C_6H_{11}, and 2,6-dimethylphenyl), provided complexes of the type $[Co(CO)_3CNR]^-$. All were characterized by IR solution spectra. In addition, $[Co(CO)_3(CNtBu)]^-$ was isolated as a satisfactorily pure tetraethylammonium salt and further characterized by its ^{1}H-NMR spectrum. The other anions were not isolated as pure substances, but have IR spectra very similar to that of $[Co(CO)_3(CNtBu)]^-$; i.e., $[Et_4N][Co(CO)_3(CN-2,6-Me_2C_6H_3)]$, IR(THF), ν(CN, CO): 2040 w, 1900 s, 1860 s, and 1586 w cm^{-1}; $[Et_4N][Co(CO)_3(CNC_6H_{11})]$, IR(THF), ν(CN, CO): 2040 w, 1915 s, and 1850 s cm^{-1}. A plausible sequence of reactions to account for the formation of these monoanions, $[Co(CO)_3L]^-$, is depicted in the scheme below. Although none of the intermediates, $[HCo(CO)_3]^{2-}$, $[H_2Co(CO)_3]^-$, or $[Co(CO)_3NH_3]^-$, has been unambiguously identified, such a sequence is attractive in terms of what is known about the instability of $H_3Co(CO)_3$, (88), the high lability of the known amminecarbonylmetallates-$(1-)$, and related studies on the protonation reactions of other carbonyl trianions, $Na_3[M(CO)_5]$ (M = V, Nb, and Ta) and $Na_3[M(CO)_4]$ (M = Mn and Re) (vide supra).

$$K_3[Co(CO)_3] + \xrightarrow[-KCl]{+NH_4Cl} K_2[HCo(CO)_3] \xrightarrow[-KCl]{+NH_4Cl} K[H_2Co(CO)_3]$$

$$K[H_2Co(CO)_3] \xrightarrow[-KCl]{+NH_4Cl} [H_3Co(CO)_3] \xrightarrow[-H_2, -NH_4^+]{+2NH_3} [Co(CO)_3NH_3]^-$$

$$[Co(CO)_3NH_3]^- \xrightarrow[-NH_3]{+L} [Co(CO)_3L]^-$$

C. Syntheses, Characterization, and Derivative Chemistry of $Na_3[Rh(CO)_3]$ and $Na_3[Ir(CO)_3]$

Of the known carbonylmetallates$(3-)$, those of rhodium and iridium have received by far the least attention. However, unlike $Na_3[Co(CO)_3]$, solutions of $Na_3[Rh(CO)_3]$ and $Na_3[Ir(CO)_3]$ have been prepared by reduction of the corresponding $Na[M(CO)_4]$ in HMPA in accord with Eq. (29).

$$Na[M(CO)_4] + 3Na \xrightarrow[25°C]{HMPA} Na_3[M(CO)_3] + \cdots \qquad (29)$$

The procedure is essentially identical to that reported recently for Na_3-$[Mn(CO)_4]$ (91), and because good routes to the tetracarbonylmetallates$(1-)$ of rhodium and iridium have been developed recently (92), there has been renewed interest in investigating the properties of these unusual materials.

In a typical preparation, $Na[Ir(CO)_4]$ was stirred with 3.2 equivalents of sodium sand in HMPA for 12 hr at room temperature under a dynamic vacuum (~ 0.5 mm Hg). A deep yellow–brown solution of $Na_3[Ir(CO)_3]$ was thereby obtained. Infrared spectral analysis of the solution showed a broad intense absorption in the $v(CO)$ region at about 1670 cm^{-1}. Addition of excess liquid ammonia to the solution followed by reflux at about $-33°C$ provided a yellow–orange precipitate in a deep yellow–red solution. Low-temperature ($-78°C$) filtration of the slurry, followed by extensive washing with liquid ammonia and drying *in vacuo*, gave a 90% yield of a yellow and thermally stable but highly air-sensitive powdery $Na_3[Ir(CO)_3]$, for which satisfactory elemental analyses (C, H, Na) were obtained (69). A mineral oil mull spectrum in the $v(CO)$ region (1800 w, 1620 vs, broad cm^{-1}) closely resembled that of $Na_3[Co(CO)_3]$. No corresponding attempts to isolate pure samples of $Na_3[Rh(CO)_3]$ have been carried out to date, but in view of the practically identical mineral oil mull and HMPA solution spectra of $Na_3[Rh(CO)_3]$ and $Na_3[Ir(CO)_3]$ (Fig. 14), respectively, there is little doubt that these materials are entirely analogous to one another.

^{1}H-NMR spectral studies on the protonation reactions of $Na_3[Ir(CO)_3]$ in HMPA were carried out and conclusively demonstrated that the freshly

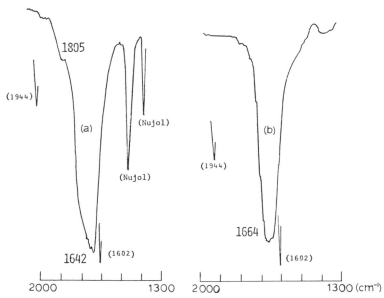

FIG. 14. Mineral oil mull infrared spectrum of (a) $Na_3[Ir(CO)_3]$ and (b) HMPA solution infrared spectrum of $Na_3[Rh(CO)_3]$ in the $v(CO)$ region.

prepared tricarbonyliridate was largely nonhydridic in nature. Initially only a weak singlet in the hydride region (δ -15.2 ppm) was observed. However, introduction of an equivalent of ethanol caused this signal, attributed to $[HIr(CO)_3]^{2-}$, to grow dramatically. Addition of a second equivalent of ethanol caused the signal at -15.2 ppm to disappear and a new intense singlet at δ -12.3 ppm to grow in. Attempts to isolate the latter species, which is believed to be $[trans\text{-}H_2Ir(CO)_3]^-$, as a pure substance have not been successful to date (69). Infrared mineral oil mull spectra of an impure and thermally unstable Ph_4As^+ salt showed an intense band at 1880 cm^{-1}, which was at lower energy than that observed for bona fide $[Ph_4As][Ir(CO)_4]$. Mineral oil mull infrared spectra of the latter salt in the $v(CO)$ region showed bands at 1908 sh and 1893 vs, br cm^{-1}. Efforts to unambiguously determine the composition of these and related rhodium species by ^1H and ^{13}C NMR spectra of the 99% enriched ^{13}CO-labeled compounds should be carried out in the near future. The rhodium study should be especially informative due to the presence of ^{103}Rh, which is 100% abundant and has a nuclear spin of $1/2$ (93).

Bimetallic products have been obtained from the reactions of Na_3-$[M(CO)_3]$ with two equivalents of Ph_3ECl (E = Ge, Sn). These reactions proceeded according to Eq. (30). Much reduction of Ph_3ECl to $(Ph_3E)_2$ accompanied these reactions, which resulted in poor to moderate (unoptimized) yields of products.

$$Na_3[M(CO)_3] + 2Ph_3ECl \xrightarrow{\text{HMPA}} Na[trans\text{-}(Ph_3E)_2M(CO)_3] + 2NaCl \qquad (30)$$
$$M = Rh, Ir; E = Ge, Sn$$

After cation exchange and recrystallization, low (10–20%) yields of pure rhodium complexes were obtained, whereas somewhat better (35–40%) yields of the corresponding iridium complexes have been isolated as colorless and fairly air-stable crystalline tetraethylammonium salts. These quite thermally stable materials (dec > 130°C without melting) all showed one strong absorption in the $v(CO)$ region in THF solution at about 1930 cm^{-1} and on this basis have been proposed to have the same D_{3h} geometry established for $[(Ph_3P)_2Co(CO)_3]^+$ (87) and proposed earlier for $[(Ph_3Sn)_2Co(CO)_3]^-$ (vide supra). Recently, a formally analogous neutral trigonal–bipyramidal nickel(II) complex, $trans\text{-}(Cl_3Si)_2Ni(CO)_3$, a derivative of the presently unknown $[Ni(CO)_3]^{2-}$, was structurally characterized (94). Clearly, much remains to be done in exploring the basic chemical reactivity patterns of the tricarbonylmetallates(3−) of cobalt, rhodium, and iridium. In particular, the reactions of these materials with organic electrophiles promise to be quite interesting in view of a recent report on corresponding reactions with $Na_3[Re(CO)_4]$ (19).

V

TETRACARBONYLMETALLATES(4−) OF CHROMIUM, MOLYBDENUM, AND TUNGSTEN AND RELATED MATERIALS

A. *Syntheses and Characterizations; The Reductive Labilization Strategy for the Preparation of Highly Reduced Organometallics*

Because $Fe(CO)_5$ and $[Mn(CO)_5]^-$ had been shown to undergo facile reductions by alkali metals to provide salts containing $[Fe(CO)_4]^{2-}$ and $[Mn(CO)_4]^{3-}$, it was of interest to determine whether the isoelectronic $[Cr(CO)_5]^{2-}$ might be similarly reduced to yield salts containing $[Cr(CO)_4]^{4-}$. In fact, $Na_2[Cr(CO)_5]$ (4) appeared to be totally inert toward further reduction by sodium metal in liquid ammonia or hexamethylphosphoramide. On this basis, it was clear that if $[Cr(CO)_4]^{4-}$ was to be prepared, another synthetic route had to be developed. In view of the prior existence of the isoelectronic $Cr(NO)_4$ (95), it was felt that such a substance should be accessible. Because reductions of transition metal complexes containing both good and poor or nonacceptor ligands had been shown previously to generally cause preferential loss of the poorest acceptor group (i.e., the ligand expected on electronic grounds to be most repelled or labilized by a more electron-rich or lower valent metal center), it was felt that this "reductive labilization" method or strategy could be a route to the desired material. Accordingly, the reduction of $Cr(CO)_4(TMEDA)$ by sodium in liquid ammonia was examined. The reaction represented by Eq. (31) required about 6–8 hr at −33°C. Nearly quantitative yields of a homogeneous yellow and pyrophoric powder were obtained, which gave satisfactory analyses (C, H, Cr, Na) for the composition $Na_4CrC_4O_4$ with no further purification. In this process,

$$Cr(CO)_4(TMEDA) + 4Na \xrightarrow[-33°C]{NH_3 (\ell)} Na_4[Cr(CO)_4]\downarrow + TMEDA \qquad (31)$$

N,N,N',N'-tetramethylethylenediamine was effectively replaced by four electrons to produce the first example of a tetrasodium carbonylmetallate (16). This reaction undoubtedly proceeded stepwise, as suggested by Eqs. (32) and (33); however, no conclusive evidence for the existence of $[Cr(CO)_4(\eta^1\text{-}TMEDA)]^{2-}$ has yet been obtained. Satisfactorily pure and brightly colored yellow–orange to orange precipitates of the

$$\left(\begin{array}{c}N\\N\end{array}\right)Cr(CO)_4 + 2Na \longrightarrow 2Na^+[N\underset{\smile}{\quad}N-Cr(CO)_4]^{2-}2Na^+ \qquad (32)$$

$$[N\underset{\smile}{\quad}N-Cr(CO)_4]^{2-}2Na^+ + 2Na \longrightarrow Na_4[Cr(CO)_4] + TMEDA \qquad (33)$$

tetracarbonylmetallates(4−) of molybdenum and tungsten have been obtained by essentially the same procedure. In important variations on this theme, Cooper (96) and M. Darensbourg and co-workers (97) used mixed amine phosphine carbonyls to synthesize $[W(CO)_4(PR_3)]^{2-}$, where R = iPr, OMe, C_6H_5, and CH_3 and L = NH_3 and piperidine [Eq. (34)].

$$W(CO)_4(PR_3)(L) + 2NaC_{10}H_8 \xrightarrow[-78°C]{THF} Na_2[W(CO)_4(PR_3)] + L + 2C_{10}H_8 \quad (34)$$

One very important and obvious limitation of this reductive labilization method is that the ligand to be expelled should not be susceptible to reduction either in a coordinated or free state. For this reason, complexes of the type $M(CO)_4L_2$, containing the preformed tetracarbonylmetal unit and bearing ligands that are not attacked by sodium in liquid ammonia, were initially examined. Our best results (yields >90% of $Na_4[M(CO)_4]$) were obtained by the reduction of $M(CO)_4(TMED)$ in liquid ammonia. These reductions were remarkably clean compared to those of other systems we examined, in which acetylenediolate dianion, $C_2O_2^{2-}$, and small amounts of $NaNH_2$ invariably contaminated the final product, unless they could be removed by filtration. It was observed that the reduction of $M(CO)_4$(norbornadiene) gave very poor yields (2–3% for M = W) of $Na_4M(CO)_4$. It is known that norbornadiene reacts with Na–NH_3, but the complexed ligand may also have suffered reduction. Reductions of $M(CO)_4$(ethylenediamine) generally gave substantially lower yields of $Na_4[M(CO)_4]$ than did $M(CO)_4(TMED)$. Although free ethylenediamine is inert to Na–NH_3 at −33°C, the coordinated en group might have been attacked (e.g., deprotonated) under these conditions. In contrast to $Na_2[Cr(CO)_5]$, the disodium molybdenum and tungsten pentacarbonylates appeared to react with Na/NH_3, but the dark brown–black product(s) from these reactions apparently did not contain any Na_4-$[M(CO)_4]$; the natures of these reduction products remain unknown.

Infrared spectra of $Na_4[M(CO)_4]$ in the carbonyl stretching frequency region showed intense broad bands centered at about 1460–1480 cm^{-1} (Table VI). Figure 15 shows the silicone fluid mull infrared spectrum of $Na_4[Cr(CO)_4]$, which is quite similar to those of the molybdenum and tungsten analogues. The $v(CO)$ positions for these tetrasodium compounds are almost 200 cm^{-1} lower than those reported previously for $Na_3[M(CO)_4]$ (M = Mn and Re) and $Cs_3[V(CO)_5]$ and thus are formally consistent with a metal center bearing substantially more negative charge than that present on the metals in the carbonyl trianions. On this basis these insoluble tetrasodium compounds were formulated as $[M(CONa)_4]_x$, in which extensive and strong sodium ion–carbonyl oxygen interactions were proposed to be responsible for the very low $v(CO)$ values (16). Because these materials have not been obtained

TABLE VI

INFRARED DATA FOR THE TETRACARBONYLMETALLATES(4−) OF CHROMIUM, MOLYBDENUM, AND TUNGSTEN AND SELECTED DERIVATIVES

Compound	Medium	$v(CO)$, cm^{-1}	Reference
$Na_4[Cr(CO)_4]$	Silicone fluid[a]	1657 w, 1462 vs, br	16
$Na_4[Mo(CO)_4]$	Silicone fluid[a]	1680 w, 1471 vs, br	16
$Na_4[W(CO)_4]$	Silicone fluid[a]	1679 w, 1529 sh, 1478 vs, br	16
$[Et_4N]_2[H_2Cr_2(CO)_8]$	CH_3CN	1958 w, 1877 s, 1825 m, 1789 m-s	101
$[Et_4N]_2[H_2Mo_2(CO)_8]$	CH_3CN	1977 w, 1893 s, 1833 w, 1799m	101
$[Et_4N]_2[H_2W_2(CO)_8]$	CH_3CN	1976 w, 1884 s, 1829 m, 1800 m	101
$[Et_4N]_2[(Ph_3Sn)_2Cr(CO)_4]$	CH_3CN	1918 m, 1822 vs, 1782 s	99
$[Et_4N]_2[(Ph_3Sn)_2Mo(CO)_4]$	CH_3CN	1948 m, 1844 vs, 1791 s	99
$[Et_4N]_2[(Ph_3Sn)_2W(CO)_4]$	CH_3CN	1947 m, 1838 vs, 1790 s	99
$[Et_4N][(Ph_3Sn)_3Cr(CO)_4]$	CH_3CN	1956 m, 1901 s, 1860 vs	99
$[Et_4N][(Ph_3Sn)_3Mo(CO)_4]$	CH_3CN	1999 m, 1897 vs, 1832 w	99
$[Et_4N][(Ph_3Sn)_3W(CO)_4]$	CH_3CN	1997 m, 1894 vs, 1828 w	99
$[nPr_4N]_4[HMo(CO)_3]_4$[b]	CH_3CN	1869 vs, 1769 s	107

[a] Mull spectra.

[b] Known tungsten analogue has not been prepared from $Na_4[W(CO)_4]$ (see Ref. 107).

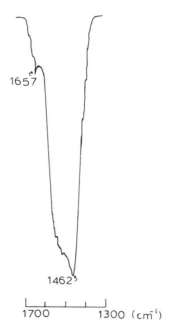

FIG. 15.   Silicone fluid mull infrared spectrum of $Na_4[Cr(CO)_4]$ in the $v(CO)$ region.

as single crystals suitable for X-ray analysis or as solutions in any unreactive solvent, solid-state MAS ^{13}C-NMR spectra may be useful in providing more information on the structural units present in these compounds. The intriguing possibility that reductively coupled carbonyl units are present in these materials has not been ruled out (16,98). As is true with $Na_2[C_2O_2]$, these tetrasodium compounds are active reducing agents and are vigorously (sometimes explosively) oxidized by perfluorinated hydrocarbons, such as perfluorokerosine. The solid tetrasodium compounds also have been reported to explode on contact with metal spatulas and must therefore be handled with utmost caution and respect.

B. Derivatives of $Na_4[M(CO)_4]$

Additional evidence for the existence of the tetrasodium compounds was obtained by their reactions with chlorotriphenylstannane (99). These reactions proceeded according to Eqs. (35) and (36) and provided moderate to high yields of heterometallic products containing six- and seven-coordinate Group 6 metals.

$$2Ph_3SnCl + Na_4[M(CO)_4] \rightarrow Na_2[(Ph_3Sn)_2M(CO)_4] + 2NaCl \qquad (35)$$

$$3Ph_3SnCl + Na_4[M(CO)_4] \rightarrow Na[(Ph_3Sn)_3M(CO)_4] + 3NaCl \qquad (36)$$

On the basis of spectroscopic data and elemental analyses, the bis-tin adducts were characterized as containing the dianionic units, $[cis\text{-}(Ph_3Sn)_2M\text{-}(CO)_4]^{2-}$ (Table VI), whereas $[(Ph_3Sn)_3Cr(CO)_4]^-$ was established to contain seven-coordinate chromium by a single-crystal X-ray study (100). The latter compound was the first seven-coordinate chromium complex to contain only monodentate ligands (Fig. 16).

By analogy with the Ph_3SnCl reactions, it was hoped that reactions of $Na_4M(CO)_4$ with acids would lead to the formation of the new species, $[H_2M(CO)_4]^{2-}$ and $[H_3M(CO)_4]^-$. Four equivalents or an excess of NH_4Cl reacted with slurries of $Na_4M(CO)_4$ in liquid ammonia to provide 50–80% yields of the diamine complexes, $M(CO)_4(NH_3)_2$ (101). Although the intermediates in this reaction have not yet been identified, a plausible sequence of reactions to account for the formation of these neutral diamine complexes is depicted in Scheme 2. Certainly, in view of the very strong reducing nature of $Na_4M(CO)_4$, it seems very possible that electron transfer reactions may also be involved. A significant difficulty in attempting to isolate intermediates from these protonation reactions was that even when fewer than four equivalents of NH_4Cl were added, the final product contained much $M(CO)_4(NH_3)_2$ in

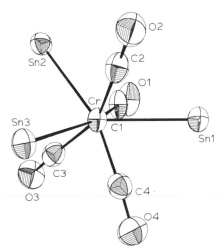

FIG. 16.   Molecular structure of the core atoms of $[(Ph_3Sn)_3Cr(CO)_4]^-$; phenyl groups have been omitted for clarity. Selected mean interatomic distances (Å): Cr–Sn, 2.71(2); Cr–C, 1.86(2); C–O, 1.15(1).

addition to unreacted $Na_4M(CO)_4$. Evidently, the insoluble tetrasodium compound was converted to soluble intermediates (e.g., $[H(NH_3)M(CO)_4]^-$) that reacted much more rapidly with NH_4Cl than with $Na_4M(CO)_4$. Attempts to trap potential intermediates such as $[HM(CO)_4]^-$ with phosphines and other nucleophiles were unsuccessful. However, it is noteworthy that anions of the general formula $[HW(CO)_4PR_3]^-$ have been obtained by protonation of the corresponding $[W(CO)_4(PR_3)]^{2-}$ (97). Also, although $[H_2W(CO)_4]^{2-}$ has not been isolated from the reaction of $Na_4[W(CO)_4]$ with Brønsted acids, it has been obtained as a dipotassium salt from the reaction of $W(CO)_4(TMEDA)$ with excess $K[sec\text{-}Bu_3BH]$ in refluxing THF (101).

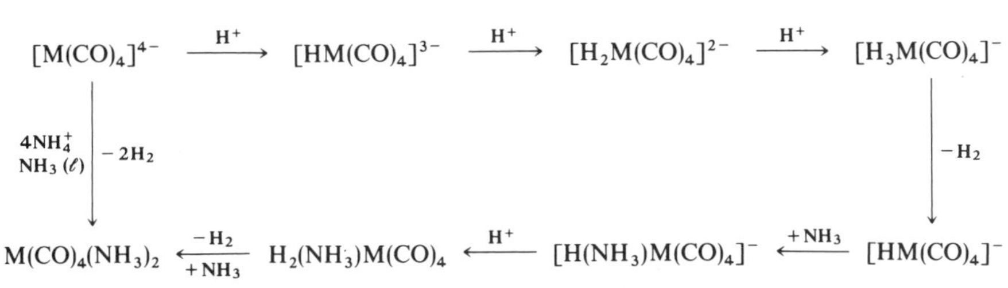

$$[M(CO)_4]^{4-} \xrightarrow{H^+} [HM(CO)_4]^{3-} \xrightarrow{H^+} [H_2M(CO)_4]^{2-} \xrightarrow{H^+} [H_3M(CO)_4]^-$$

$$4NH_4^+ \quad NH_3(\ell) \quad \Big| \quad -2H_2 \qquad\qquad\qquad\qquad\qquad\qquad\qquad -H_2 \quad \Big|$$

$$M(CO)_4(NH_3)_2 \xrightleftharpoons[+NH_3]{-H_2} H_2(NH_3)M(CO)_4 \xleftarrow{H^+} [H(NH_3)M(CO)_4]^- \xleftarrow{+NH_3} [HM(CO)_4]^-$$

SCHEME 2.   Proposed pathway for the conversion of $Na_4[M(CO)_4]$ to $M(CO)_4(NH_3)_2$.

Treatment of liquid ammonia slurries of $Na_4[M(CO)_4]$ with about 2.5 equivalents of NH_4Cl or excess acetonitrile gave 20–50% yields of $[H_2M_2(CO)_8]^{2-}$, which have been isolated as Et_4N^+ or $(Ph_3P)_2N^+$ salts. The chromium and molybdenum dianions were new species and have been characterized by their elemental compositions and infrared (Table VI) and ^1H NMR spectra. The ditungsten complex was previously obtained by the reactions of $W(CO)_6$ and $[Et_4N][BH_4]$ in THF (102,103) or of [PPN]-$[W(CO)_5I]$ with $[PPN][BH_4]$ in THF, the latter of which provided up to 65% isolated yields of the PPN salt $[PPN = (Ph_3P)_2N^+]$ (104). The first well-defined reactions of $[H_2W_2(CO)_8]^{2-}$ with nucleophiles were identified. Compounds formed in these reactions included $[Et_4N]_2[W_2(CO)_8(PMe_3)_2]$, $[Et_4N]_2[W_2(CO)_8(PMe_2Ph_2)_2]$, $[Et_4N]_2[W_2(CO)_8(P(OMe)_3)_2]$, and $K_2[H_2W(CO)_4]$. The former were the first reported bis-phosphine-substituted derivatives of $[M_2(CO)_{10}]^{2-}$ dianions, whereas $[H_2W(CO)_4]^{2-}$ is formally a diprotonated derivative of $[W(CO)_4]^{4-}$. High yields (70%) of $[H_2W(CO)_4]^{2-}$ were also obtained by the reaction of $W(CO)_4(TMED)$ with excess $K[sec\text{-}Bu_3BH]$ in THF. Treatment of $[W_2(CO)_8L_2]^{2-}$ with water provided the hydride anions, $[HW_2(CO)_8L_2]^-$, which were isolated as Et_4N^+ salts and were characterized by elemental analyses and infrared and ^1H-NMR spectra. M. Darensbourg, and co-workers reported on alternative routes, not involving the dimetal dianions, $[M_2(CO)_8L_2]^{2-}$, to similar substituted bridging hydrides of Mo and W (97,105,106).

One of the most interesting developments of this work on Group 6 carbonyl anions was the synthesis and characterization of the first carbonylhydrido clusters of molybdenum and tungsten. Treatment of $[H_2W(CO)_4]^{2-}$ with water or of (diethylenetriamine)$W(CO)_3$ with excess HBR_3^- provided about 30–40% yields of the 56-electron tetrahedral cluster anion, $[HW(CO)_3]_4^{4-}$. The analogous molybdenum complex could be obtained in up to 60% yields by reaction of one equivalent of HBR_3^- with $(TMED)Mo(CO)_4$ or (diethylenetriamine)$Mo(CO)_3$. The single-crystal X-ray structure of $[n\text{-}Pr_4N]_4[HMo(CO)_3]_4$ was obtained (107) (Fig. 17) and found to be essentially identical to that of the previously established formally unsaturated and isoelectronic $[HRe(CO)_3]_4$ (108). Hydrogen atoms were not observed directly but are believed to symmetrically bridge each tetrahedral face. In this way the carbonyl groups are forced to assume otherwise unfavorable eclipsed orientations with respect to the edges of the tetrahedron. In view of the unsaturated nature of the $[HM(CO)_3]_4^{4-}$ molecules, one would expect them to be quite reactive. The only well-characterized reaction we have examined was that with carbon monoxide, in which clean cluster degradation occurred to form $[H_2M_2(CO)_8]^{2-}$ and finally $[M_2(CO)_{10}]^{2-}$. However, these clusters appeared to be surprisingly robust in the presence of a variety of bases, including refluxing acetonitrile. Also indicative of the relative inertness

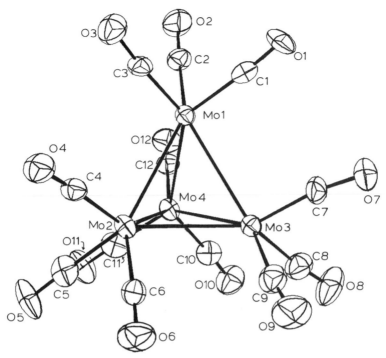

Fig. 17. Molecular structure of the tetrahedral cluster, $[HMo(CO)_3]_4^{4-}$, which was obtained in low yield from the protonation of $Na_4[Mo(CO)_4]$. Selected mean interatomic distances (Å): Mo—Mo, 3.11(2); Mo—C, 1.91(3); C—O, 1.19(2).

of the molybdenum tetramer was its formation under a variety of conditions, as shown in Scheme 3. Indeed, it is now known that the "intensely purple" carbonylmolybdates obtained by Hayter (109) and Behrens (110) in the 1960s contained the Mo tetramer. After completing research in this area, we were informed by Professor R. Bau that he and his associates also obtained similar and convincing evidence in this regard (111).

C. Reduction of W(CO)₃(PMTA) (PMTA = 1,1,4,7,7-Pentamethyldiethylenetriamine): A Route to [W(CO)₃]⁶⁻?

Because reduction of $M(CO)_4(TMEDA)$ by $Na-NH_3$ effectively resulted in the substitution of TMEDA by four electrons to produce $Na_4[M(CO)_4]$ in high (>90%) yields, it was of interest to determine whether the readily available $M(CO)_3(PMTA)$ (M = Cr, Mo, and W) (112) would undergo

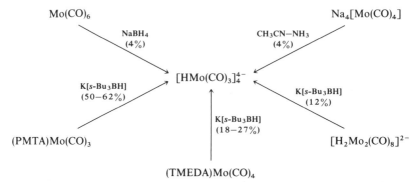

SCHEME 3. A summary of routes to $[HMo(CO)_3]_4^{4-}$. Yields of the tetramer are in parentheses; PMTA = 1,1,4,7,7-pentamethyldiethylenetriamine.

analogous reduction to give what would be an incredible species, $Na_6[M(CO)_3]$! Potassium and cesium metal reductions of $W(CO)_3(PMTA)$ in liquid ammonia resulted in the precipitation of heterogeneous and rather unattractive greenish–tan solids that contained about the same alkali metal-to-tungsten ratio (~ 3.5 to 1) and had quite similar mineral oil mull infrared spectra in the $\nu(CO)$ region (e.g., for cesium product, 1785 m and 1633 vs br cm^{-1}). Much alkali metal amide also formed during these reductions. The infrared spectra indicated that the products contained less reduced carbonylmetallate units than those present in $Na_4[M(CO)_4]$. Indeed, the bands were in the region characteristic of carbonyltrianions and at lower energies than those of carbonyl dianions, e.g., $[W(CO)_5]^{2-}$ (1775 and 1731 cm^{-1} in HMPA) (30) or $[H_2W(CO)_4]^{2-}$ (1851, 1837, 1729, and 1704 cm^{-1} in HMPA) (101). On this basis we believe that incompletely reduced materials, perhaps containing the unreported $[H_3W(CO)_3]^{3-}$, were formed in these reductions. Infrared spectra of the corresponding molybdenum and chromium products looked very similar to those of the tungsten compound but gave abysmal yields of derivatives (113,114). Although the nature of the carbonyltungstate(s) from these reductions remains unknown, these compounds undoubtedly contain $W(CO)_3$ units because only tungsten tricarbonyl derivatives, i.e., $W(CO)_3(NH_3)_3$ or $[HW(CO)_3(SnPh_3)_3]^{2-}$, were obtained (in 10–30% yields), following reaction of the carbonyltungstate precursor with NH_4Cl or Ph_3SnCl, respectively. Due to these less than satisfactory results, however, we have at least temporarily abandoned plans to examine whether similarly noteworthy (or outrageous?) species can be synthesized, such as $Na_5[V(CO)_4]$ or $Na_{12}Cr$ (115).

VI

CONCLUDING REMARKS

Elemental analyses, infrared and NMR spectra, and derivative studies have provided excellent support for the existence of a series of previously unknown highly reduced carbonylmetallates, which contain transition metals in their lowest known oxidation states. As we have gained experience on how best to handle these electron-rich substances, it has become quite clear that they have a bright future as precursors to new classes of organometallic compounds. Most poorly explored are reactions of these substances with organic and transition metal electrophiles, but many aspects of their general reaction chemistry and structural properties, especially in the solid state, remain unknown. What are the prospects for the synthesis of other "superreduced" carbonylmetallates? Undoubtedly the most reasonable possibility is Na_3-$[Tc(CO)_4]$, since the manganese and rhenium analogues are quite robust materials in solution and the solid state. In view of the current interest in technetium chemistry (116), attempts to synthesize this material and examine its basic chemistry are clearly warranted. Other and perhaps more problematic candidates include $[Sc(CO)_6]^{3-}$ (which would truly be a "scandalous" compound) and $[Ti(CO)_5]^{4-}$. In view of the existence of $[Ti(CO)_6]^{2-}$, $[V(CO)_5]^{3-}$, and $[Cr(CO)_4]^{4-}$, the latter may not be totally unreasonable possibilities, however.

Although the number of valence electrons present on an atom places definite restrictions on the maximum formal oxidation state possible for a given transition element in chemical combination, in condensed phases, at least, there seem to be no *a priori* restrictions on minimum formal oxidation states. In future studies we hope to arrive at some definitive conclusions on how much negative charge can be added to a metal center before reduction and/or loss of coordinated ligands occur. Answers to these questions will ultimately define the boundaries of "superreduced" transition metal chemistry and also provide insight on the relative susceptibility of coordinated ligands to reduction, an area that has attracted substantial interest ($98,117–119$).

ACKNOWLEDGMENTS

I wish to express gratitude and appreciation to all of my co-workers who are listed as joint authors in the references, to the Petroleum Research Fund, administered by the American Chemical Society, and to the National Science Foundation for continuing generous support of this research on highly reduced organometallics. Special thanks are due to Kelly V. Beamer for expert assistance with the manuscript.

REFERENCES

1. W. Hieber, W. Beck, and G. Braun, *Angew. Chem.* **72,** 795 (1960), and references cited therein.
2. H. Hock and H. Stuhlmann, *Chem. Ber.* **62,** 431 (1929).
3. F. Feigl and P. Krumholz, *Monatsh. Chem.* **59,** 314 (1932).
4. H. Behrens, *Adv. Organomet. Chem.* **18,** 1 (1980), and references cited therein.
5. F. Hein and H. Pobloth, *Z. Anorg. Allg. Chem.* **248,** 84 (1941).
6. E. O. Fischer and R. Böttcher, *Z. Naturforsch. B: Anorg. Chem., Org. Chem., Biochem., Biophys., Biol.* **10B,** 600 (1955).
7. T. S. Piper and G. Wilkinson, *J. Inorg. Nucl. Chem.* **3,** 104 (1956).
8. F. Calderazzo, R. Ercoli, and G. Natta, *in* "Organic Syntheses via Metal Carbonyls" (I. Wender and P. Pino, eds.), Vol. 1, pp. 1-270, and references cited therein. Wiley (Interscience), New York, 1968.
9. R. B. King, *Adv. Organomet. Chem.* **2,** 157 (1964), and references cited therein.
10. M. I. Bruce and F. G. A. Stone, *Angew. Chem., Int. Ed. Engl.* **7,** 747 (1966), and references cited therein.
11. A. A. Blanchard, *Chem. Rev.* **21,** 3 (1937); J. S. Anderson, *Q. Rev. Chem. Soc.* **1,** 331 (1947); E. W. Abel, *ibid.* **17,** 133 (1963); E. W. Abel and F. G. A. Stone, *ibid.* **24,** 498 (1970); D. F. Shriver, *Acc. Chem. Res.* **3,** 231 (1970); R. B. King, *ibid.*, p. 417. The above, as well as refs. 1, 4, 8-10, represent excellent reviews on early progress in metal carbonyl anion chemistry.
12. J. E. Ellis, *J. Organomet. Chem.* **86,** 1 (1975).
13. J. P. Collman, *Acc. Chem. Res.* **8,** 342 (1975).
14. J. E. Ellis and R. A. Faltynek, *J. Chem. Soc., Chem. Commun.*, 966 (1975).
15. F. A. Cotton and G. Wilkinson, "Advanced Inorganic Chemistry," 5th ed., p. 1041. Wiley, New York, 1988.
16. J. E. Ellis, C. P. Parnell, and G. P. Hagen, *J. Am. Chem. Soc.* **100,** 3605 (1978).
17. R. G. Finke and T. N. Sorrell, *Org. Synth., Collect. Vol.* **6,** 807 (1988).
18. J. Lacoste and F. Schue, *J. Organomet. Chem.* **231,** 279 (1982).
19. G. F. Warnock, L. C. Moodie, and J. E. Ellis, *J. Am. Chem. Soc.* **111,** 2131 (1989).
20. J. A. Gladysz and W. Tam, *J. Org. Chem.* **43,** 2279 (1978).
21. G. F. Warnock, Ph. D. Thesis, University of Minnesota, Minneapolis, 1985.
22. N. M. Boag and H. D. Kaesz, *in* "Comprehensive Organometallic Chemistry" (G. Wilkinson, F. G. A. Stone, and E. W. Abel, eds.), Vol. 4, pp. 162–167, and references cited therein. Pergamon, Oxford, 1982.
23. W. Büchner, *Helv. Chim. Acta* **46,** 2111 (1963).
24. M. Szwarc, ed., "Ions and Ion Pairs in Organic Reactions," Vols. 1 and 2. Wiley (Interscience) New York, 1972 and 1974.
25. M. Y. Darensbourg, *Prog. Inorg. Chem.* **33,** 221 (1985).
26. R. G. Teller, R. G. Finke, J. P. Collman, H. B. Chin, and R. Bau, *J. Am. Chem. Soc.* **99,** 1104 (1977).
27. J. E. Ellis and R. A. Faltynek, unpublished research.
28. H. B. Chin and R. Bau, *J. Am. Chem. Soc.* **98,** 2434 (1976).
29. W. F. Edgell, M. T. Yang, R. Bayer, and J. Koizumi, *J. Am. Chem. Soc.* **87,** 3080 (1965).
30. J. E. Ellis and G. P. Hagen, *J. Am. Chem. Soc.* **96,** 7825 (1974).
31. P. P. Edwards, S. H. Guy, D. M. Holton, and W. J. McFarlane, *J. Chem. Soc., Chem. Commun.*, 1185 (1981).
32. W. Hieber and F. Leutert, *Chem. Ber.* **64,** 2832 (1931).
33. G. Ciani, G. D'Alfonso, M. Freni, P. Romiti, and A. Sironi, *J. Organomet. Chem.* **152,** 85 (1978).

34. P. E. Riley and R. E. Davis, *Inorg. Chem.* **19,** 159 (1980).
35. R. F. Jordan and J. R. Norton, *J. Am. Chem. Soc.* **104,** 1255 (1982).
36. W. D. Jones, J. M. Huggins, and R. G. Bergman, *J. Am. Chem. Soc.* **103,** 4415 (1981).
37. W. J. Carter, J. W. Kelland, S. J. Okrasinski, K. E. Warner, and J. R. Norton, *Inorg. Chem.* **21,** 3955 (1982).
38. E. Lindner, E. Schauss, W. Killer, and R. Fawzi, *Angew. Chem., Int. Ed. Engl.* **23,** 711 (1984).
39. E. Lindner, R-M. Jansen, and H. A. Mayer, *Angew. Chem., Int. Ed. Engl.* **25,** 1008 (1986).
40. C. P. Casey and C. A. Bunnell, *J. Chem. Soc., Chem. Commun.,* 733 (1974).
41. C. M. Lukehart, G. P. Torrence, and J. V. Zeile, *J. Am. Chem. Soc.* **97,** 6903 (1975).
42. C. P. Casey and C. A. Bunnell, *J. Am. Chem. Soc.* **98,** 436 (1976).
43. C. P. Casey and D. M. Scheck, *J. Am. Chem. Soc.* **102,** 2723, 2728 (1980).
44. C. M. Lukehart, *Acc. Chem. Res.* **14,** 109 (1981).
45. J. E. Ellis and R. A. Faltynek, *J. Am. Chem. Soc.* **99,** 1801 (1977).
46. A. L. Casalnuovo, L. H. Pignolet, J. W. A. van der Welden, J. J. Bour, and J. J. Steggarda, *J. Am. Chem. Soc.* **105,** 5957 (1983).
47. P. D. Boyle, B. J. Johnson, A. Buehler, and L. H. Pignolet, *Inorg. Chem.* **25,** 7 (1986).
48. J. E. Ellis and M. C. Palazzotto, *J. Am. Chem. Soc.* **98,** 8264 (1976).
49. J. E. Ellis, K. L. Fjare, and T. G. Hayes, *J. Am Chem. Soc.* **103,** 6100 (1981).
50. J. E. Ellis, *ACS Symp. Ser.* **357,** 34 (1987).
51. J. E. Ellis and K. M. Chi, unpublished research; also see K. M. Chi, S. R. Frerichs, S. B. Philson, and J. E. Ellis, *J. Am. Chem. Soc.* **110,** 303 (1988).
52. P. A. Bianconi, R. N. Vrtis, C. P. Rao, I. D. Williams, M. P. Engeler, and S. J. Lippard, *Organometallics* **6,** 1968 (1987).
53. C. G. Dewey, J. E. Ellis, K. L. Fjare, K. M. Pfahl, and G. F. Warnock, *Organometallics* **2,** 388 (1983).
54. G. F. Warnock, K. L. Fjare, and J. E. Ellis, *J. Am. Chem. Soc.* **105,** 672 (1983).
55. G. F. Warnock and J. E. Ellis, *J. Am. Chem. Soc.* **106,** 5016 (1984).
56. F. Calderazzo and G. Pampaloni, *J. Organomet. Chem.* **250,** C33 (1983).
57. G. F. Warnock, S. B. Philson, and J. E. Ellis, *J. Chem. Soc., Chem. Commun.,* 893, (1984).
58. M. Y. Darensbourg and S. Slater, *J. Am. Chem. Soc.* **103,** 5914 (1981).
59. J. E. Ellis, T. G. Hayes, and R. E. Stevens, *J. Organomet. Chem.* **216,** 191 (1981); K. M. Chi, S. R. Frerichs, and J. E. Ellis, *J. Chem. Soc., Chem. Commun.,* 1013 (1988).
60. J. E. Ellis, *J. Am. Chem. Soc.* **103,** 5016 (1981).
61. A. Davison and J. E. Ellis, *J. Organomet. Chem.* **36,** 113 (1972).
62. D. G. Evans and D. M. P. Mingos, *J. Organomet. Chem.* **232,** 171 (1982).
63. D. Rehder, *J. Organomet. Chem.* **37,** 303 (1972).
64. R. K. Opmacis, M. Poliakoff, and J. J. Turner, *J. Am. Chem. Soc.* **108,** 3645 (1986).
65. D. Rehder and K. Ihmels, *Inorg. Chim. Acta* **76,** L313 (1983).
66. F. A. Cotton, J. L. Down, and G. Wilkinson, *J. Chem. Soc.,* 833 (1959).
67. J. C. Hileman, D. K. Huggins, and H. D. Kaesz, *Inorg. Chem.* **1,** 933 (1962).
68. A. Davison, J. A. McCleverty, and G. Wilkinson, *J. Chem. Soc.,* 1133 (1963).
69. G. F. Warnock and J. E. Ellis, unpublished research.
70. D. Rehder, *in* "Multinuclear NMR" (J. Mason, ed.), p. 497. Plenum, New York, 1987.
71. B. E. Mann, *J. Chem. Soc., Dalton Trans.,* 2012 (1973).
72. J. M. Maher, R. P. Beatty, and N. J. Cooper, *Organometallics* **4,** 1354 (1985).
73. B. E. Mann and B. E. Taylor, "¹³C NMR Data for Organometallic Compounds." Academic Press, London, 1981.
74. S. R. Frerichs and J. E. Ellis, *J. Organomet. Chem.* **359,** C41 (1989).
75. P. Bushman, G. N. Van Buuren, M. Shiralian, and R. K. Pomeroy, *Organometallics* **2,** 693 (1983).

76. L. C. Moodie and J. E. Ellis, unpublished research.
77. P. L. Gaus, S. C. Kao, M. Y. Darensbourg, and L. W. Arndt, *J. Am. Chem. Soc.* **106**, 4752 (1984); P. L. Gaus, S. C. Kao, K. Youngdahl, and M. Y. Darensbourg, *ibid.* **107**, 2428 (1985), and references cited therein.
78. L. Mond, H. Hirtz, and M. D. Cowap, *J. Chem. Soc.*, 798 (1910).
79. W. Hieber and H. Schulten, *Z. Anorg. Allg. Chem.* **232**, 29 (1937).
80. W. Hieber and H. Schulten, *Z. Anorg. Allg. Chem.* **232**, 17 (1937).
81. H. Behrens, *Angew. Chem.* **61**, 444 (1949).
82. J. E. Ellis, P. T. Barger, and M. L. Winzenburg, *J. Chem. Soc., Chem. Commun.*, 686 (1977).
83. K. Jonas, *Angew. Chem., Int. Ed. Engl.* **14**, 752 (1975).
84. E. Weiss and W. Buchner, *Helv. Chim. Acta* **46**, 1121 (1963).
85. D. A. Adolphson, J. D. Corbett, and D. J. Marryman, *J. Am. Chem. Soc.* **98**, 7234 (1976).
86. J. L. Dye, *Sci. Am.* **257**(3), 66 (1987).
87. W. Hieber and W. Freyer, *Chem. Ber.* **93**, 462 (1960).
88. R. L. Sweany, *J. Am. Chem. Soc.* **104**, 3740 (1982).
89. W. Hieber and E. Lindner, *Z. Naturforsch. B: Anorg. Chem., Org. Chem., Biochem., Biophys., Biol.* **16B**, 127 (1961).
90. A. Sacco, *Gazz. Chim. Ital.* **83**, 632 (1953); W. Hieber and J. Sedlmeier, *Chem. Ber.* **87**, 787 (1954).
91. J. E. Ellis and G. F. Warnock, *Organomet. Synth.* **4**, 100 (1988).
92. L. Garlaschelli, R. D. Pergola, and S. Martinengo, *Inorg. Synth.*, (in press).
93. R. J. Goodfellow, *in* "Multinuclear NMR" (J. Mason, ed.), pp. 521–561. Plenum, New York, 1987.
94. S. K. Janikowski, L. J. Radonovich, T. J. Groshens, and K. J. Klabunde, *Organometallics* **4**, 396 (1985).
95. S. K. Satija and B. L. Swanson, *Inorg. Synth.* **16**, 2 (1976).
96. J. M. Maher, R. P. Beatty, and N. J. Cooper, *Organometallics* **1**, 215 (1982).
97. S. G. Slater, R. Lusk, B. F. Schumann, and M. Y. Darensbourg, *Organometallics* **1**, 1662 (1982).
98. R. N. Vrits, C. P. Rao, S. G. Bott, and S. J. Lippard, *J. Am. Chem. Soc.* **110**, 7564 (1988).
99. J. T. Lin, G. P. Hagen, and J. E. Ellis, *Organometallics* **2**, 1145 (1983).
100. J. T. Lin, G. P. Hagen, and J. E. Ellis, *Organometallics* **3**, 1288 (1984).
101. J. T. Lin, G. P. Hagen, and J. E. Ellis, *J. Am. Chem. Soc.* **105**, 2296 (1983).
102. M. R. Churchill, S. W. Chang, M. L. Berch, and A. Davison, *J. Chem. Soc., Chem. Commun.*, 691 (1973).
103. M. R. Churchill and S. W. Chang, *Inorg. Chem.* **25**, 145 (1979).
104. C. Y. Wei, M. W. Marks, R. Bau, S. W. Kirtley, D. Bisson, M. E. Henderson, and T. F. Koetzle, *Inorg. Chem.* **21**, 2556 (1982).
105. M. Y. Darensbourg, J. L. Atwood, W. E. Hunter, and R. R. Burch, *J. Am. Chem. Soc.* **102**, 3290 (1980).
106. M. Y. Darensbourg, R. El Mehdawi, T. J. Delord, R. F. Fronczek, and S. F. Watkins, *J. Am. Chem. Soc.* **106**, 2583 (1984).
107. J. T. Lin and J. E. Ellis, *J. Am. Chem. Soc.* **105**, 6252 (1983).
108. D. K. Higgins, W, Fellman, J. M. Smith, and H. D. Kaesz, *J. Am. Chem. Soc.* **86**, 4841 (1964).
109. R. G. Hayter, *J. Am. Chem. Soc.* **88**, 4376 (1966).
110. H. Behrens and W. Haag, *Chem. Ber.* **94**, 320 (1961); H. Behrens and J. Yogl, *ibid.* **96**, 2220 (1963).
111. S. W. Kirtley, D. L. Tipton, and R. Bau, unpublished research (1979–1981); personal communication from Professor R. Bau described in footnote 19b of ref. 107.
112. J. E. Ellis and G. L. Rochfort, *Organometallics* **1**, 682 (1982).

113. G. L. Rochfort and J. E. Ellis, *J. Organomet. Chem.* **250,** 265 (1983).
114. G. L. Rochfort and J. E. Ellis, *J. Organomet. Chem.* **250,** 277 (1983).
115. M. Y. Darensbourg and K. Youngdahl, private communication (1984); B. Bursten, private communication (1986).
116. A. Davison, J. F. Kronauge, A. G. Jones, R. M. Pearlstein, and J. R. Thornback, *Inorg. Chem.* **27,** 3245 (1988), and references cited therein.
117. R. Hoffman, C. N. Wilker, S. J. Lippard, J. L. Templeton, and D. C. Brower, *J. Am. Chem. Soc.* **105,** 146 (1983).
118. P. A. Bianconi, I. D. Williams, M. P. Engeler, and S. J. Lippard, *J. Am. Chem. Soc.* **108,** 311 (1986).
119. W. J. Evans, *Polyhedron* **6,** 803 (1987).

The Interplay of Alkylidyne and Carbaborane Ligands in the Synthesis of Electronically Unsaturated Mixed-Metal Complexes

F. GORDON A. STONE

Department of Inorganic Chemistry
The University
Bristol BS8 1TS, England

I

INTRODUCTION

The chemistry of CR fragments ligating metal centers has been a topic of considerable interest since the discovery of the first alkylidyne–metal complexes in E.O. Fischer's Laboratory in 1973 (*1*). These ligands have been implicated in Fischer–Tropsch reactions (*2*) and in alkyne metathesis (*3*). Moreover, the isolation of stable compounds containing carbon–metal triple bonds completed the matrix of bond types represented here:

$$C—C \qquad C{=}C \qquad C{\equiv}C$$

$$C—M \qquad C{=}M \qquad C{\equiv}M$$

$$M—M \qquad M{=}M \qquad M{\equiv}M$$

The connectivity between organic chemistry on the one hand, and organometal and polynuclear metal chemistry on the other, has thus become obvious.

In 1978, Ashworth and the present author recognized that molecules containing either unsaturated carbon–metal or unsaturated metal–metal bonds should "complex" with other metal–ligand systems, as do alkenes or alkynes, provided the metal centers were electron rich yet had a vacant coordination site (*4*). This concept is illustrated in Scheme 1, with platinum (d^{10}) as the paradigm metal center.

Relevant to the results described in this chapter is the synthesis of the tungsten–platinum species **1** (*5*). In this reaction an ethylene molecule

$$[W({\equiv}CC_6H_4Me\text{-}4)(CO)_2(\eta\text{-}C_5H_5)] + [Pt(C_2H_4)(PMe_2Ph)_2] \xrightarrow[25°C]{pentane}$$

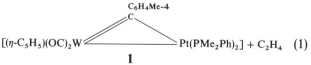

$$[(\eta\text{-}C_5H_5)(OC)_2W{=}\!\!\!=\!\!\!=\!Pt(PMe_2Ph)_2] + C_2H_4 \quad (1)$$

1

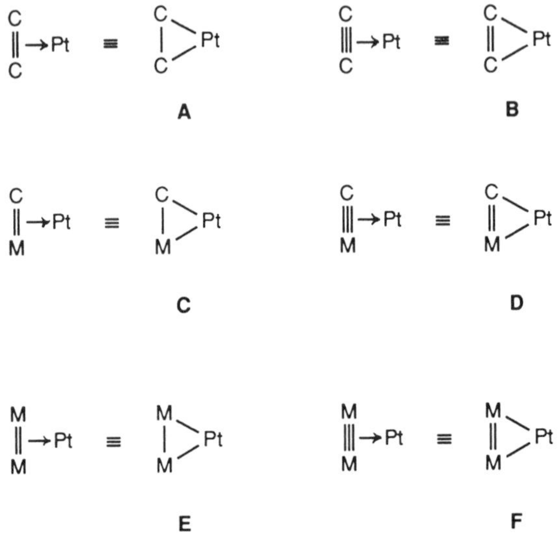

C
‖→Pt   ≡   C
C | >Pt
               C

A

C
‖‖→Pt   ≡   C
C   ‖ >Pt
               C

B

C
‖→Pt   ≡   C
M | >Pt
               M

C

C
‖‖→Pt   ≡   C
M   ‖ >Pt
               M

D

M
‖→Pt   ≡   M
M | >Pt
               M

E

M
‖‖→Pt   ≡   M
M   ‖ >Pt
               M

F

SCHEME 1. Alkene–and alkyne–metal complexes may be formulated with metallacyclo-propane (**A**) or -propene (**B**) structures, respectively. Similarly, complexation of a metal atom with an alkylidene or alkylidyne group gives rise to a dimetallacyclopropane (**C**) or -propene (**D**) ring system.

coordinated to the platinum in the precursor is displaced by a $RC\equiv W$-$(CO)_2(\eta-C_5H_5)$ group functioning as a ligand. The resulting product is best formulated with a dimetallacyclopropene structure. Morever, the methodology leading to **1** has wide applicability in other syntheses. This is because there is an isolobal mapping between $M(CO)_2(\eta-C_5H_5)$ (M = W, Mo, or Cr) fragments and CR (R = alkyl or aryl) groups (6), so that the molecules $[M(\equiv CR)(CO)_2(\eta-C_5H_5)]$ display a coordination chemistry akin to alkynes (Scheme 2) (7).

By employing the principles indicated in Scheme 2, we have reported in ~100 primary journal articles the synthesis of a plethora of di- and tri-nuclear metal complexes containing bonds between dissimilar transition elements, and between these elements and Cu, Ag, or Au (8). Moreover, polynuclear metal compounds are readily obtained by treating molecules with structures of type **A** or **B** with further quantities of the precursors $[M(\equiv CR)(CO)_2(\eta-C_5H_5)]$ or $M'L'_n$. Added versatility is provided by replacement of the $\eta-C_5H_5$ ligands with the groups $\eta-C_5Me_5$, H_2Bpz_2, or $HBpz_3$ (pz = pyrazol-1-yl), or by employing as starting materials the di-

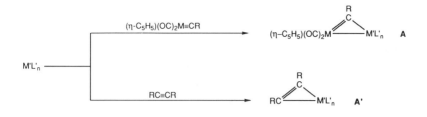

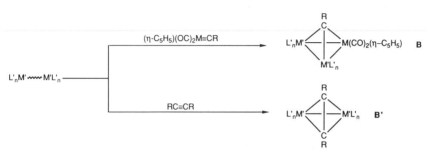

SCHEME 2. M'L'$_n$ represents a metal ligand fragment having a vacant coordination site; this is a species generally formed by loss of a labile ligand (e.g., CO, C$_2$H$_4$, THF) from a complex. L'$_n$M'∿M'L'$_n$ symbolizes a dimetal species containing a M—M, M=M, or M≡M bond and π acceptor ligands L'$_n$.

metal compounds [L'$_n$M'M(≡CR)(CO)$_4$] [M = Cr, Mo, or W, M'L'$_n$ = Co(CO)$_4$, Re(CO)$_5$, or Mo(CO)$_3$(η-C$_5$H$_5$)].

A merger between alkylidyne–metal chemistry and carbametallaborane chemistry became evident in our laboratory from an isolobal mapping of the species [W(≡CR)(CO)$_2$(η^5-C$_5$R'$_5$)] with [W(≡CR)(CO)$_2$(η^5-C$_2$B$_9$H$_9$R'$_2$)]$^-$ (R' = H or Me). The relationship between the ligands η^5-C$_5$H$_5^-$ and η^5-C$_2$B$_9$H$_{11}^{-2}$ was first experimentally demonstrated by Hawthorne and his co-workers (9) with the synthesis of [Fe(η^5-C$_2$B$_9$H$_{11}$)$_2$]$^{-2}$, an analogue of [Fe(η-C$_5$H$_5$)$_2$]. It follows that because the Fischer (10) compounds [W(≡CR)(CO)$_2$(η-C$_5$H$_5$)] (R = alkyl or aryl) exist, it was likely that salts of the anionic complexes [W(≡CR)(CO)$_2$(η^5-C$_2$B$_9$H$_9$R'$_2$)]$^-$ would also be stable. Moreover, salts of these anions would be desirable synthons combining the functionalities of both CR and η^5-C$_2$B$_9$H$_9$R'$_2$ groups, and thereby opening up a new vista in organometallic chemistry.

II

SYNTHESIS OF SALTS CONTAINING THE ANIONS
$[W(\equiv CR)(CO)_2(\eta^5\text{-}C_2B_9H_9R'_2)]^-$ (R = ALKYL OR ARYL; R' = H OR Me),
$[Mo(\equiv CC_6H_4Me\text{-}4)(CO)\{P(OMe)_3\}(\eta^5\text{-}C_2B_9H_9Me_2)]^-$, AND
$[W(\equiv CR)(CO)_2(\eta^6\text{-}C_2B_{10}H_{10}Me_2)]^-$ (R = ARYL)

Addition of tetrahydrofuran (THF) solutions of $Na_2[7,8\text{-}C_2B_9H_9Me_2]$ to the compounds $[W(\equiv CR)Br(CO)_4]$ in the same solvent at $\sim -40°C$, followed by treatment of the mixtures with one or other of the salts XCl [X = NEt_4, $N(PPh_3)_2$, $P(CH_2Ph)Ph_3$, or PPh_4], affords the stable complexes $[X][W(\equiv CR)(CO)_2(\eta^5\text{-}1,2\text{-}C_2B_9H_9Me_2)]$ (2) (11,12a). The salt $[N(PPh_3)_2]$-$[W(\equiv CC_6H_4Me\text{-}4)(CO)_2(\eta^5\text{-}1,2\text{-}C_2B_9H_{11})]$ (3)[1] was similarly obtained from $[W(\equiv CC_6H_4Me\text{-}4)Br(CO)_4]$ and $Na_2[7,8\text{-}C_2B_9H_{11}]$, in the presence of

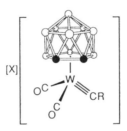

●CMe   ○BH

	X	R
(2a)	NEt_4	Me
(2b)	NEt_4	Ph
(2c)	NEt_4	$C_6H_4Me\text{-}4$
(2d)	$N(PPh_3)_2$	$C_6H_4Me\text{-}4$
(2e)	$P(CH_2Ph)Ph_3$	$C_6H_4Me\text{-}4$
(2f)	PPh_4	$C_6H_4Me\text{-}4$
(2g)	NEt_4	$C_6H_4Me\text{-}2$
(2h)	NEt_4	$C_6H_3Me_2\text{-}2,6$
(2i)	PPh_4	$C_6H_3Me_2\text{-}2,6$

$[N(PPh_3)_2]Cl$ (11). The molybdenum compound $[NEt_4][Mo(\equiv CC_6H_4\text{-}Me\text{-}4)(CO)\{P(OMe)_3\}(\eta^5\text{-}C_2B_9H_9Me_2)]$ (4) has also been prepared. The reaction between $Na_2[7,8\text{-}C_2B_9H_9Me_2]$ and $[Mo(\equiv CC_6H_4Me\text{-}4)Cl(CO)\{P(OMe)_3\}_3]$, followed by addition of $[NEt_4]Cl$, gives the salt 4 (13).

---

[1] Not shown; formulas 7, 17, 46, 52, and 54 are also not shown.

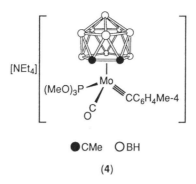

●CMe ○BH

(4)

An X-ray diffraction study of **2f** confirmed the pentahapto ligation of the tungsten atom by the nido-icosahedral $C_2B_9H_9Me_2$ fragment (*12b*). The $W{\equiv}C$ separation [1.826(7) Å] compares well with that in the neutral species $[W({\equiv}CC_6H_4Me\text{-}4)(CO)_2(\eta\text{-}C_5H_5)]$ [1.82(2) Å] (*10*). ^{13}C-NMR spectroscopy is an important diagnostic tool in this area, and in this context the alkylidyne-carbon nuclei in the complexes **2** resonate in the range δ 297–310 ppm, whereas the signal for the $CC_6H_4Me\text{-}4$ nucleus in the NMR spectrum of **3** occurs at δ 293.9 ppm. These chemical shifts are very similar to those observed for alkylidyne-carbon resonances in the neutral compounds $[W({\equiv}CR)(CO)_2(\eta\text{-}C_5R_5')]$; the presence of the carbaborane ligand evidently has little influence on the observed shifts.

Alkylidyne–tungsten complexes have been prepared in which the tungsten atom is part of a 13-vertex docosahedral $C_2B_{10}W$ framework, rather than the 12-vertex icosahedral C_2B_9M (M = W or Mo) group found in **2–4**. Addition of THF solutions of the salt $Na_2[C_2B_{10}H_{10}Me_2]$ to THF solutions of the compounds $[W({\equiv}CR)Cl(CO)_2L_2]$ (R = $C_6H_4Me\text{-}4$ or $C_6H_3Me_2\text{-}2,6$; L = pyridine or 4-methylpyridine) in the presence of $[NEt_4]Cl$ affords the complexes $[NEt_4][W({\equiv}CR)(CO)_2(\eta^6\text{-}C_2B_{10}H_{10}Me_2)]$ (**5**) (*14*). An X-ray diffraction study on **5b** reveals that the tungsten atom is ligated by two CO groups, the alkylidyne moiety [$W{\equiv}C$, 1.84(1) Å], and the carbaborane cage. The open face of the latter is η^6-coordinated to the tungsten but is decidedly nonplanar. The requirement of a six-atom $\overline{BCBBBC}$ ring above a pentagonal belt of five borons results in two faces of the cage distorting from a triangular to an essentially square arrangement. Variable-temperature NMR studies showed that in solution the anions of the salts **5** undergo a dynamic exchange process. This property is interpreted by the process shown in Scheme 3. The cage CMe fragments change their relationship with the tungsten atom via a diamond–square–diamond polytopal rearrangement of the cage. This is accompanied by a concomitant rotation of the $W({\equiv}CR)(CO)_2$ (R = $C_6H_4Me\text{-}4$ or $C_6H_3Me_2\text{-}2,6$) moiety. The latter process enables the two CMe units to

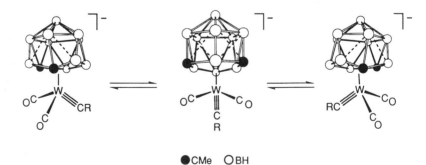

●CMe ○BH

	R
(5a)	C_6H_4Me-4
(5b)	$C_6H_3Me_2$-2,6

alternate between transoid and nontransoid sites with respect to the alkylidyne group.

The availability of the salts **2–5** has allowed their reactions to be studied with a variety of halo-metal compounds, cationic species, and neutral metal complexes possessing readily displaceable ligands. At the time of writing this review, however, most of the reported syntheses have employed the salts **2** or **3** as precursors. The chemistry of the molybdenum reagent **4** and that of the compounds **5**, and other species containing nonicosahedral carborane fragments, has yet to be developed. For **2–5**, the C≡M (Mo or W) bonds are the initial site of reactivity. However, frequently subsequent reactions occur as a result of the carborane ligands adopting a nonspectator role. Reactions involving neutral complexes afford salts, whereas reactions with halo-metal or cationic species afford neutral products.

●CMe ○BH

SCHEME 3.   Possible pathway for dynamic behavior in solution of the salts (**5**).

III

THE CARBABORANE GROUP AS A SPECTATOR LIGAND

Treatment of $[AuCl(PPh_3)]$ with the appropriate salt **2** in THF, in the presence of $TlPF_6$ to remove chloride as insoluble TlCl, affords the tungsten–gold complexes $[WAu(\mu\text{-}CR)(CO)_2(PPh_3)(\eta^5\text{-}C_2B_9H_9R_2')]$ (**6a**, R = $C_6H_4Me\text{-}4$, R' = Me; **6b**, R = $C_6H_4Me\text{-}4$, R' = H; **6c**, R = R' = Me) (*11*). Similarly, the reaction between **4** and $[AuCl(PPh_3)]$, in the presence of KPF_6, affords the molybdenum–gold compound $[MoAu(\mu\text{-}CC_6H_4Me\text{-}4)(CO)\{P(OMe)_3\}\text{-}(PPh_3)(\eta^5\text{-}C_2B_9H_9Me_2)]$ (**6d**) (*13*). An X-ray diffraction study on **6a** served to

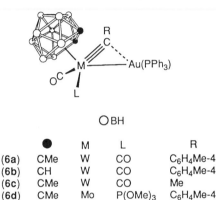

		M	L	R
(**6a**)	CMe	W	CO	$C_6H_4Me\text{-}4$
(**6b**)	CH	W	CO	$C_6H_4Me\text{-}4$
(**6c**)	CMe	W	CO	Me
(**6d**)	CMe	Mo	$P(OMe)_3$	$C_6H_4Me\text{-}4$

place the structures of this group of complexes on a firm basis. The η^5 coordination of the $C_2B_9H_9Me_2$ group to the W atom was established. However, an interesting feature of the structure was the semibridging of the W–Au bond by the $\mu\text{-}CC_6H_4Me\text{-}4$ ligand, the latter lying appreciably closer to the tungsten [$\mu\text{-}C$—W, 1.88(3) Å; $\mu\text{-}C$—Au, 2.19(3) Å]. The $\mu\text{-}C$—W distance lies between the C≡W separations found in (**2f**) and $[W(\equiv CC_6H_4Me\text{-}4)\text{-}(CO)_2(\eta\text{-}C_5H_5)]$ (see previously). Other structures with semibridging alkylidyne groups will be mentioned later. The discovery of this structural feature is important in understanding the chemistry of CR fragments. The asymmetric spanning of metal–metal bonds by CO or alkylidene ligands is well established, but as a result of this work it is now confirmed for alkylidyne groups also.

Treatment of the salts **5** with $[AuCl(PPh_3)]$, in the presence of KPF_6, affords the complexes $[WAu(\mu\text{-}CR)(CO)_2(PPh_3)(\eta^6\text{-}C_2B_{10}H_{10}Me_2)]$ (**7a**, R = $C_6H_4Me\text{-}4$; **7b**, R = $C_6H_3Me_2\text{-}2,6$) (*14*). The species **7** are formed as mixtures of two isomers, with each isomer undergoing fluxional behavior of the kind discussed above for the anions **5**. The isomerism may be the result of different orientations of the $C_2B_{10}H_{10}Me_2$ cage, so that in one isomer a cage-carbon atom is transoid to the μ-CR group, and in the other a cage-boron atom is transoid to this group (14).

The reaction between **2a** and [AuCl(tht)] (tht, tetrahydrothiophene) in CH_2Cl_2 yields the trimetal compound $[N(PPh_3)_2][W_2Au(\mu$-CC_6H_4Me-4)$_2$-$(CO)_4(\eta^5$-$C_2B_9H_9Me_2)_2]$ (**8**) (11). The anion of **8** is isolobally mapped with the cation $[W_2Au(\mu$-CC_6H_4Me-4)$_2(CO)_4(\eta$-$C_5H_5)_2]^+$ (15), and compound **6a** is similarly related to $[WAu(\mu$-CC_6H_4Me-4)$(CO)_2(PPh_3)(\eta$-$C_5H_5)]$-$[PF_6]$ (16).

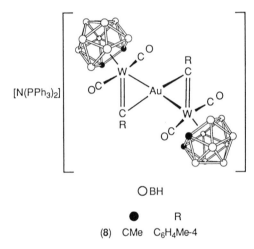

○ BH

● R

(8)   CMe C_6H_4Me-4

Treatment of **2b**, **2e**, or **2i** with [Pt(cod)$_2$] (cod, cycloocta-1,5-diene) in THF gives, respectively, the complexes $[X][WPt(\mu$-$CR)(CO)_2(cod)(\eta^5$-$C_2B_9H_9Me_2)]$ [**9a**, R = Ph, X = NEt_4; **9b**, R = C_6H_4Me-4, X = $P(CH_2Ph)$-Ph_3; **9c**, R = $C_6H_3Me_2$-2,6, X = PPh_4] (12a,17). The NMR data show that

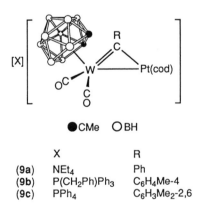

● CMe   ○ BH

	X	R
(9a)	NEt_4	Ph
(9b)	$P(CH_2Ph)Ph_3$	C_6H_4Me-4
(9c)	PPh_4	$C_6H_3Me_2$-2,6

the $C_2B_9H_9Me_2$ fragment in these products is η^5 coordinated to the tungsten. The complexes **9** react with $[AuCl(PPh_3)]$ in the presence of $TIPF_6$ to give the trimetal compounds $[WPtAu(\mu_3\text{-}CR)(CO)_2(PPh_3)(cod)(\eta^5\text{-}C_2B_9H_9Me_2)]$ (**10a**, R = Ph; **10b**, R = C_6H_4Me-4; **10c**, R = $C_6H_3Me_2$-2,6). However, on the basis of NMR evidence it appears that in **10c** the μ-CR group edge-bridges a WPtAu triangle, rather than capping it as in **10a** or **10b** (*17*). This difference in the structures is probably due to the bulky nature of the $C_6H_3Me_2$-2,6 substituent.

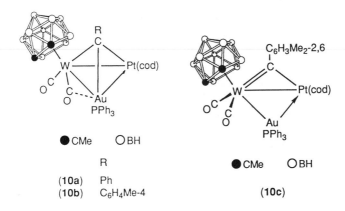

(10a)   Ph
(10b) C_6H_4Me-4

(10c)

Reactions between various Rh(I) complexes and the salts **2d**, **3**, **4**, and **5** have also been investigated. By employing the reagents $[RhCl(PPh_3)_3]$, $[Rh(PPh_3)_2(cod)][PF_6]$, or $[Rh(dppe)(nbd)][PF_6]$ (dppe, $Ph_2PCH_2CH_2$-PPh_2; nbd, norbornadiene) with **2d** and **3**, the dimetal compounds $[WRh$-$(\mu\text{-}CC_6H_4Me$-4$)(CO)_2L_2(\eta^5\text{-}C_2B_9H_9R'_2)]$ (**11a**, L = PPh_3, R' = Me; **11b**, L = PPh_3, R' = H; **11c**, L_2 = dppe, R' = Me) have been prepared (*11*).

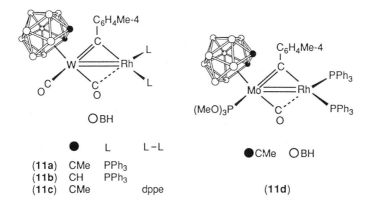

	●	L	L–L
(11a)	CMe	PPh_3	
(11b)	CH	PPh_3	
(11c)	CMe		dppe

(11d)

The structure of **11a** has been established by X-ray diffraction. The compounds **11** are 30-cluster valence-electron (CVE) dimetal species, and are therefore electronically unsaturated. The electron distribution within the W(μ-C)Rh rings may be represented by various canonical forms. However, that shown, which implies that the W≡C bond in the $(\eta^5$-C$_2$B$_9$H$_9$R$_2'$)(OC)$_2$-W≡CC$_6$H$_4$Me-4 moiety formally contributes three electrons to the rhodium center (16-electron valence shell), accords with other results from our laboratory involving electronically unsaturated dimetal compounds with bridging alkylidyne ligands (*18*).

The compound [MoRh(μ-CC$_6$H$_4$Me-4)(μ-CO){P(OMe)$_3$}(PPh$_3$)$_2$(η^5-C$_2$B$_9$H$_9$Me$_2$)] (**11d**), akin to the species **11a–11c**, has been prepared by treating [Rh(cod)(PPh$_3$)$_2$][PF$_6$] in CH$_2$Cl$_2$ with **4** (*13*). The tungsten–rhodium complexes [WRh(μ-CR)(CO)$_2$(PPh$_3$)$_2$(η^6-C$_2$B$_{10}$H$_{10}$Me$_2$)] (**12a**, R = C$_6$H$_4$Me-4; **12b**, R = C$_6$H$_3$Me$_2$-2,6) have been synthesized by treating THF solutions of the salts **5** with [Rh(cod)(PPh$_3$)$_2$][PF$_6$] (*14*). The products **12** show dynamic NMR spectra, like the tungsten–gold compounds **7**, mentioned above. Moreover, the NMR properties of **12a** indicated the presence in solution of an isomeric species containing an exopolyhedral B—H ⟶ Rh bond, a property discussed further below.

CMe   BH   B

(12a)

It was during studies on reactions between the tungsten salts **2d** and **3** and the rhodium compounds that we discovered our first definitive examples of products which revealed that the carbaborane cage could be activated to adopt a nonspectator role. Although not unknown, such behavior is rare for the related η-C$_5$R$_5'$ (R' = H or Me) ligands. Treatment of [Rh(PPh$_3$)$_2$(nbd)]-[PF$_6$] with **2d** and **3** afforded, respectively, the complexes [WRh(μ-CC$_6$H$_4$-Me-4)(CO)$_2$(PPh$_3$)$_2$\{η^5-C$_2$B$_9$(C$_7$H$_9$)H$_8$R$_2'$\}] (**13a**, R' = Me; **13b**, R' = H). In these species a novel nortricyclene (tricyclo[2.2.1.0^{2,6}] heptane) fragment is attached to the cage, and this was confirmed by an X-ray diffraction study

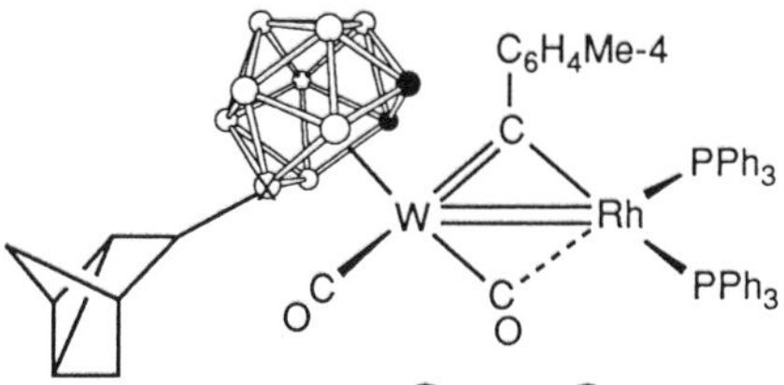

OBH ⊗B

●

(13a) CMe
(13b) CH

on **13a** (*11*). The formation of the C_7H_9 substituent represents an unprecedented hydroboration of the nbd ligand in the precursor. The formation of
the compounds **13** probably proceeds via intermediates having exopolyhedral
B—H→Rh bonds of the kind described in Section IV. Interestingly,
examination of NMR data revealed that an isomer of **13b** existed in which
the C_7H_9 fragment is attached to a boron atom adjacent to a CH group in
the face of the cage. This type of isomerism is also discussed further in the
following sections.

IV

EXOPOLYHEDRAL B—H→METAL BONDING IN DI- AND TRIMETAL COMPOUNDS

Treatment of the salts **2a** or **2d** with $[Ru(CO)(NCMe)_2(\eta-C_5H_5)][BF_4]$
in CH_2Cl_2 affords the dimetal complexes $[WRu(\mu-CR)(CO)_3(\eta-C_5H_5)(\eta^5-C_2B_9H_9Me_2)]$ (**14a**, R = Me; **14b**, R = C_6H_4Me-4) (*19*). Examination of the

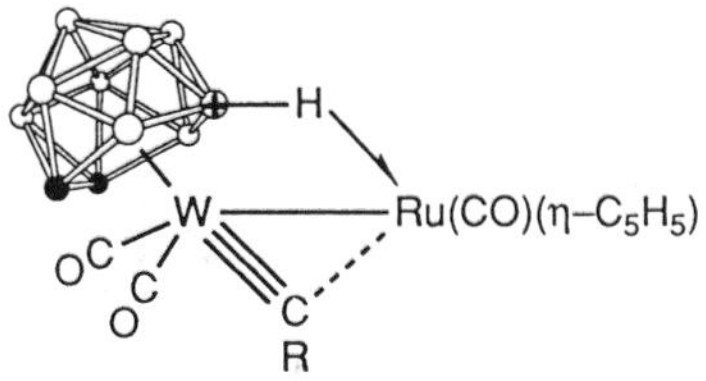

● CMe   OBH   ⊕ B

R
(14a)   Me
(14b) C_6H_4Me-4

^{1}H-NMR spectra of these products revealed a high-field resonance at $\delta \sim$ -11.5 ppm, appearing as a quartet [J(BH) ~ 70 Hz]. These signals are diagnostic for the presence of a three-center two-electron B—H $\rightarrow$ Ru bond, a feature substantiated from the observation in the ^{11}B-$\{^1$H$\}$ NMR spectra of a singlet peak at $\delta \sim 20$ ppm. The remaining ^{11}B nuclei of the cage display broad peaks at ~ -8 to -17 ppm. An X-ray diffraction study on **14b** confirmed the presence of the B—H $\rightarrow$ Ru bridged system, and also showed that the μ-CC$_6$H$_4$Me-4 ligand semibridged the metal–metal bond [μ-C–W, 1.890(6) Å; μ-C–Ru, 2.220(7) Å], as in **6a**.

In similar chemistry, salts **2d** and **3** react with the complexes [M(CO)$_2$-(NCMe)$_2$(η^5-C$_9$H$_7$)][BF$_4$] (M = Mo or W; C$_9$H$_7$, indenyl) to give the dimetal compounds [MW(μ-CC$_6$H$_4$Me-4)(CO)$_3$(η^5-C$_9$H$_7$)(η^5-C$_2$B$_9$H$_9$R$_2'$)] **15a**, M = Mo, R$'$ = Me; **15b**, M = W, R$'$ = Me; **15c**, M = Mo, R$'$ = H). X-Ray diffraction studies on **15a** and **15b** showed that they were isostructural,

	M	L	●
(15a)	Mo	CO	CMe
(15b)	W	CO	CMe
(15c)	Mo	CO	CH
(15d)	W	PMe$_3$	CMe

with the metal–metal bond bridged by the p-tolylmethylidyne group and by a three-center two-electron B—H $\rightarrow$ M bond (20). In toluene at 80°C, compounds **15a** and **15b** undergo a polytopal rearrangement of the cage system, thereby affording, respectively, the species **16a** and **16b** in which the carbon atoms in the face of the cage are no longer adjacent; this result was confirmed by X-ray crystallography. Generally, such polytopal rearrangements require temperatures of ~ 200°C or higher, although Hawthorne et al. (21) observed that the mononuclear metal complex [IrH(PPh$_3$)$_2$(η^5-C$_2$B$_9$H$_{10}$Ph)] undergoes a similar isomerization in refluxing toluene.

Treatment of [Mo(NCMe)(CO)$_2$(η^7-C$_7$H$_7$)][BF$_4$] with **2c** in THF affords the dimetal compound [MoW(μ-CC$_6$H$_4$Me-4)(CO)$_2$(η^7-C$_7$H$_7$)(η^5-C$_2$B$_9$H$_9$-Me$_2$)] (**17**). The latter species, like complex **15a**, contains a B—H $\rightarrow$ Mo three-center two-electron bond (22).

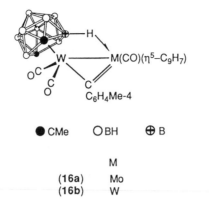

● CMe ○ BH ⊕ B

	M
(16a)	Mo
(16b)	W

Compounds **15** and **16** are electronically unsaturated 32-CVE species. This unsaturation is reflected in the dimensions of the $\overline{W(\mu\text{-CC}_6H_4Me\text{-}4)M}$ rings and in the relatively deshielded resonances (δ 364–382 ppm) of the μ-C nuclei in the $^{13}C-\{^1H\}$ NMR spectra (*18*). The compounds **15a** and **15b** undergo a number of interesting reactions. Thus the diazoalkanes N_2CR_2 (R = C_6H_4Me-4 or Ph) react in diethyl ether at room temperature to yield the complexes $[MW(\mu\text{-CC}_6H_4Me\text{-}4)(CO)_2(N_2CR_2)(\eta^5\text{-}C_9H_7)(\eta^5\text{-}C_2B_9H_9Me_2)]$ [R = C_6H_4Me-4, M = Mo (**18a**) or W (**18b**); R = Ph, M = Mo (**18c**) or W (**18d**)] (*23*). Complex **15b** reacts with aqueous H_2O_2 and $N_3C_6H_4Me$-4, respectively, to give the compounds $[W_2(\mu\text{-CC}_6H_4Me\text{-}4)(CO)_2(L)(\eta^5\text{-}C_9H_7)\text{-}(\eta^5\text{-}C_2B_9H_9Me_2)]$ (**18e**, L = O; **18f**, L = NC_6H_4Me-4). The structures of **18b** and **18e** have been established by X-ray diffraction (*23*). The W—N

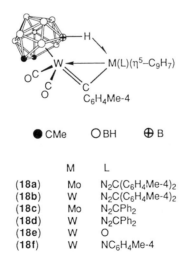

● CMe ○ BH ⊕ B

	M	L
(18a)	Mo	$N_2C(C_6H_4Me\text{-}4)_2$
(18b)	W	$N_2C(C_6H_4Me\text{-}4)_2$
(18c)	Mo	N_2CPh_2
(18d)	W	N_2CPh_2
(18e)	W	O
(18f)	W	$NC_6H_4Me\text{-}4$

separation [1.768(5) Å] in **18b** and the W—O distance [1.699(6) Å] in **18c**
suggest that lone-pair electrons on the N and O atoms, respectively, are
engaged in $p\pi - d\pi$ donor bonding with the W atom. Thus the molecules **18** are
probably best regarded as electronically saturated 34-CVE dimetal species,
with the ligands L contributing four electrons. In agreement with this
supposition (*18*), in the $^{13}C-\{^1H\}$ NMR spectra of the compounds **18**, the
resonances for the μ-C nuclei occur in the range δ 303–321 ppm, whereas in
the spectra of the 32-CVE complexes **15a**–**15c**, the corresponding signals are
appreciably more deshielded, occurring in the range δ 367.6–381.9 ppm (*20*).

Treatment of **18b** with PMe$_3$ displaces the diazoalkane ligand and affords
the complex $[W_2(\mu\text{-}CC_6H_4Me\text{-}4)(CO)_2(PMe_3)(\eta^5\text{-}C_9H_7)(\eta^5\text{-}C_2B_9H_9Me_2)]$
(**15d**). In the $^{13}C-\{^1H\}$ NMR spectrum of this product, the μ-C nucleus res-
onates at δ 353.1 ppm. Because the PMe$_3$ ligand can only contribute two
electrons to the dimetal system, compound **15d** is a 32-CVE species, and, in
agreement, the μ-C signal is more deshielded than that observed (δ 316.4 ppm)
in the $^{13}C-\{^1H\}$ NMR spectrum of **18b**.

In both series of compounds, **15** and **18**, exopolyhedral B—H $\rightarrow$ M
(M = Mo or W) bonds are present. However, in certain circumstances this
interaction may be lifted. Thus treatment of **15a** or **15b** with PMe$_3$ affords,
respectively, the complexes $[MW(\mu\text{-}CC_6H_4Me\text{-}4)(CO)_3(PMe_3)(\eta^5\text{-}C_9H_7)\text{-}(\eta^5\text{-}C_2B_9H_9Me_2)]$ (**19a**, M = Mo; **19b**, M = W). In contrast, the reaction

● CMe   ○ BH

	M
(19a)	Mo
(19b)	W

between PMe$_3$ and **15c** yields $[MoW(\mu\text{-}CC_6H_4Me\text{-}4)(CO)_2(PMe_3)(\eta^5\text{-}C_9H_7)(\eta^5\text{-}C_2B_9H_{11})]$ (**20**), a process that preserves the B—H $\rightarrow$ Mo linkage
present in the precursor. The difference in reactivity pattern shown by **15a**
and **15c** toward PMe$_3$ may reflect the greater steric demands of the
C$_2$B$_9$H$_9$Me$_2$ cage versus C$_2$B$_9$H$_{11}$, so that with **15a**, addition of PMe$_3$
occurs at Mo, and with **15c** it occurs at W. Both **19a** and **20** are 32-valence-
electron species.

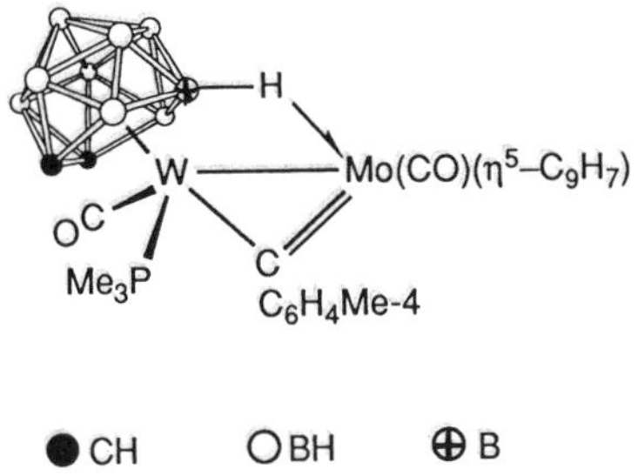

● CH   ○ BH   ⊕ B

(20)

The delicate balance between structures having B—H → Metal bridge bonds and those that do not is well illustrated by studies with tungsten–iridium (24) and tungsten–cobalt systems (25). The reaction between the compounds [Ir(PPh$_3$)$_2$(cod)][PF$_6$] and **2d** in THF affords the complex [WIr(μ-CC$_6$H$_4$Me-4)(CO)$_2$(PPh$_3$)$_2$(η^5-C$_2$B$_9$H$_9$Me$_2$)] (**21a**). The PPh$_3$ groups in the latter can be displaced with PEt$_3$, P(OPh)$_3$, or P(OMe)$_3$ to give the compounds [WIr(μ-CC$_6$H$_4$Me-4)(CO)$_2$L$_2$(η^5-C$_2$B$_9$H$_9$Me$_2$)] [**21b**, L = PEt$_3$; **21c**, L = P(OPh)$_3$; **21d**, L = P(OMe)$_3$]. An X-ray diffraction study on **21b** revealed that in the crystal the molecule adopts the structure **I** shown. The B—H → Ir bond involves the boron atom in the face of the C$_2$B$_9$ cage, which is not adjacent to a carbon atom. However, spectroscopic studies (IR and NMR) on **21b** indicate that in solution three isomers (**I–III**) exist, the proportions in equilibrium depending on the solvent used and the temperature. The interconversion between isomers **I** and **III** could readily occur via the intermediacy of **II**. In the latter, the η^5-C$_2$B$_9$H$_9$Me$_2$ group could rotate to allow a BH group either in the α or β position to the carbon atoms to form

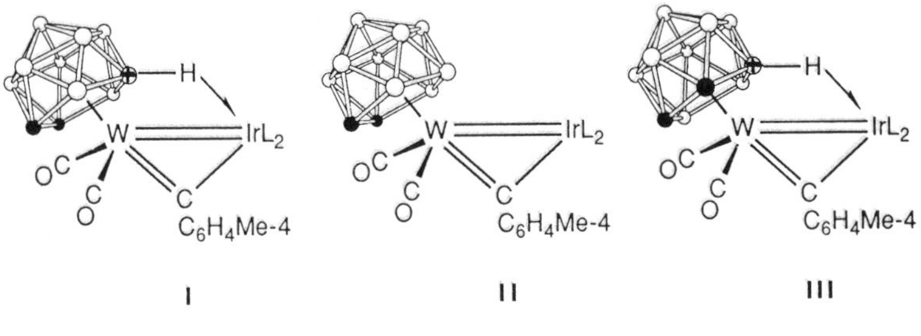

I   II   III

● CMe   ○ BH   ⊕ B

	L
(21a)	PPh$_3$
(21b)	PEt$_3$
(21c)	P(OPh)$_3$
(21d)	P(OMe)$_3$

a B—H → Ir bond. Solutions of **21a** also show spectroscopic properties, indicating an equilibrium between isomers of types **I** and **II**. In contrast, **21c** and **21d** appear in solution to exist as one isomer having the structure **I**.

It is interesting to note in passing that whereas compound **21a** in solution displays spectroscopic properties indicating an equilibrium between isomers of types **I** and **II**, the rhodium analogue **11a** appears to exist solely in one isomeric form, with the carbaborane cage adopting a spectator role. However, as mentioned earlier, the tungsten–rhodium compound **12a**, containing the docosahedral $C_2B_{10}W$ group, appears to exist in solution as a mixture of two isomers, one form (**I**) having a B—H → Rh bond present and the other (**II**) without such a linkage. It should be noted that where a B—H → M (Rh or Ir) bond is present, it corresponds to incipient oxidative addition at the metal center, and that such agostic interactions are more likely to occur at iridium than at rhodium. This is supported by results described in Section V, where complexes are described that very probably result from a completed oxidative-addition process: B—H → Ir → B—Ir—H.

It should also be noted that there is a delicate balance between structures of types **I** and **III**, well illustrated by the existence of the compounds **14** and $[WRu(\mu\text{-}CC_6H_3Me_2\text{-}2,6)(CO)_3(\eta\text{-}C_5H_5)(\eta^5\text{-}C_2B_9H_9Me_2)]$ (**22**). The latter is prepared by treating **2i** with $[Ru(CO)(NCMe)_2(\eta\text{-}C_5H_5)][BF_4]$ (*17*). It exists

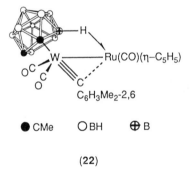

(22)

primarily as the isomer, in which it is a BH group in the face of the cage adjacent to a CMe fragment that partakes in the B—H → Ru exopolyhedral bonding. There is, however, some slender evidence for the presence in solution of a species with a more symmetrical structure akin to **14b**.

An interesting example of two B—H → metal bridge-bonds formed by the same carbaborane cage[2] is found in the anions of the salts $[NEt_4][WCo_2\text{-}$

---

[2] Hawthorne and his co-workers (*26*) have earlier reported the synthesis of the compounds $[Rh_2(PR_3)_2(\eta^5\text{-}C_2B_9H_{11})_2]$ (PR_3 = PPh₃, PEt₃, or PMe₂Ph). In these molecules there are two B—H→Rh bridge bonds. However, these linkages involve both $C_2B_9H_{11}$ cages, each partaking in one B—H→Rh unit.

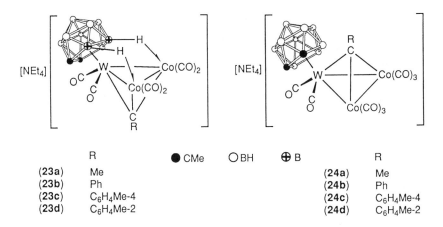

	R			R
	● CMe	○ BH	⊕ B	
(23a)	Me		**(24a)**	Me
(23b)	Ph		**(24b)**	Ph
(23c)	C₆H₄Me-4		**(24c)**	C₆H₄Me-4
(23d)	C₆H₄Me-2		**(24d)**	C₆H₄Me-2

$(\mu_3\text{-CR})(CO)_6(\eta^5\text{-}C_2B_9H_9Me_2)]$ (**23a**, R = Me; **23b**, R = Ph; **23c**, R = C₆H₄-Me-4; **23d**, R = C₆H₄Me-2) (25). These complexes are obtained by treating the appropriate reagents **2** with [Co₂(CO)₈]. They form via the intermediacy of the compounds $[NEt_4][WCo_2(\mu_3\text{-CR})(CO)_8(\eta^5\text{-}C_2B_9H_9Me_2)]$ (**24a**, R = Me; **24b**, R = Ph; **24c**, R = C₆H₄Me-4; **24d**, R = C₆H₄Me-2), in which the carbaborane group is a spectator ligand. In the presence of CO the hexacarbonyl trimetal compounds **23** revert to the octacarbonyl trimetal complexes **24**, a process that is readily reversed *in vacuo*.

Another example of a trimetal complex with two exopolyhedral B—H → metal bonds has been discovered. The reaction between **2c** and [Mo(CO)₃(NCMe)₃] in THF affords $[NEt_4][Mo_2W(\mu_3\text{-CC}_6H_4Me\text{-}4)(\mu\text{-CO})(CO)_8(\eta^5\text{-}C_2B_9H_9Me_2)]$ (**25a**). The latter with PMe₃ affords compound **25b**, the structure of which has been established by X-ray diffraction

	L
● CMe	○ BH	⊕ B
(25a)	CO
(25b)	PMe₃

(*22*). The complexes **25** are thus structurally akin to **23**, except for the presence of the μ-CO group in the former, which is necessary so as to maintain an 18-electron shell for the molybdenum atoms.

As will be seen in later sections, the complexes in which the heteronuclear metal–metal bonds are bridged by agostic B—H $\rightarrow$ metal interactions are intermediates in a variety of reactions involving the carbaborane cages and the μ-alkylidyne ligands.

V

FORMATION OF EXOPOLYHEDRAL σ BONDS BETWEEN CAGE BORON ATOMS AND TRANSITION ELEMENTS

During the synthesis of compound **14b** (*19*) it was observed that if solutions of this species were chromatographed on basic alumina, the complex was deprotonated, yielding $[N(PPh_3)_2][WRu(\mu\text{-}CC_6H_4Me\text{-}4)(\mu\text{-}\sigma,\eta^5\text{-}C_2B_9H_8\text{-}Me_2)(CO)_3(\eta\text{-}C_5H_5)]$ (**26**) in the presence of $N(PPh_3)_2^+$. A more rational synthesis of **26**, which contains a two-center B—Ru σ bond, involves treating **14b** with LiBu^n and $[N(PPh_3)_2]Cl$.

[N(PPh₃)₂]

● CMe   ○ BH   ⊕ B

(26)

Reactions between several of the tungsten reagents **2** and $[Fe_2(CO)_9]$ or $[Fe_3(CO)_{12}]$ have been investigated (*17,27*). The course of these reactions is critically dependent on the nature of the substituent R on the alkylidyne-carbon atom. Thus **2h**, containing the sterically demanding $C_6H_3Me_2\text{-}2,6$ group, affords one product: the dimetal species $[NEt_4][WFe(\mu\text{-}CC_6H_3Me_2\text{-}2,6)(CO)_5(\eta^5\text{-}C_2B_9H_9Me_2)]$ (**27a**). In the latter the carbaborane adopts a spectator role. Compound **27a** belongs to the family of dimetal compounds with bridging alkylidyne groups in which the C≡W fragment donates formally four electrons to a metal center (*18*). In agreement, in the $^{13}C\text{-}\{^1H\}$ NMR spectrum of **27a**, the μ-C resonance is highly deshielded (δ 388.4 ppm).

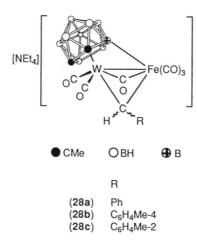

● CMe ○ BH

R

(27a) $C_6H_3Me_2$-2,6
(27b) C_6H_4Me-2

The reagents **2b** and **2c** react with iron carbonyls to afford chromato-graphically separable mixtures of the compounds $[NEt_4][WFe(\mu\text{-}CHR)(\mu\text{-}\sigma,\eta^5\text{-}C_2B_9H_8Me_2)(\mu\text{-}CO)(CO)_5]$ (**28a**, R = Ph; **28b**, R = C_6H_4Me-4) and $[NEt_4][WFe_2(\mu_3\text{-}CR)(\mu_3\text{-}\sigma:\sigma',\eta^5\text{-}C_2B_9H_7Me_2)(CO)_8]$ (**29a**, R = Ph; **29b**,

● CMe ○ BH ⊕ B

R

(28a) Ph
(28b) C_6H_4Me-4
(28c) C_6H_4Me-2

R = C_6H_4Me-4). Compound **2a** and $[Fe_2(CO)_9]$ give $[NEt_4][WFe_2(\mu_3\text{-}CMe)(\mu_3\text{-}\sigma:\sigma',\eta^5\text{-}C_2B_9H_7Me_2)(CO)_8]$ (**29c**) as the only isolable product. It should be noted that an X-ray diffraction study on **28b** established that it is a boron atom adjacent to a CMe group that forms the σ bond to the iron.

Reactions employing **2g** and iron carbonyls shed light on the pathways followed in these syntheses. The reagent **2g** contains the CC_6H_4Me-2 group, which, although sterically more demanding than the Me, Ph, or C_6H_4Me-4 substituents present in **2a**–**2c**, is less bulky than the $CC_6H_3Me_2$-2,6 moiety present in **2h**. Treatment of **2g** with $[Fe_3(CO)_{12}]$ yields a mixture of the three

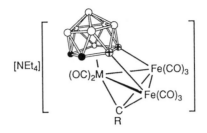

● CMe   ○ BH   ⊕ B

	M	R
(29a)	W	Ph
(29b)	W	C$_6$H$_4$Me-4
(29c)	W	Me
(29d)	W	C$_6$H$_4$Me-2
(29e)	Mo	C$_6$H$_4$Me-4

compounds **27b**, **28c**, and **29d**. Moreover, the important observation (*17*) was made that in the presence of CO, complex **27b** was converted into the μ-alkylidene compound **28c**, and that the former with iron carbonyls afforded **29d**. It is thus apparent that these reactions have as a common intermediate an Fe(CO)$_3$ adduct of type **27**. If the R group is not too bulky, these species with CO can transform into products of type **28**, or with iron carbonyl in excess, into the trimetal compounds **29**. Both processes very likely involve the intermediacy of other complexes containing B—H → Fe bonds, but these were not isolated. Combination of B—H → Fe and μ-CR (R = Ph, C$_6$H$_4$-Me-4, or C$_6$H$_4$Me-2) fragments would give the complexes **28**. However, loss of hydrogen is required for the formation of the trimetal compounds **29**. This process could readily occur via an oxidative-addition mechanism: B—H → Fe → B—Fe—H. This would be followed by a reductive release of molecular hydrogen via a combination of two H ligands on adjacent iron centers.

The tungsten–iron compounds **27–29** may become suitable precursors to other mixed-metal compounds. Thus treatment of **29c** with [AuCl(PPh$_3$)] affords the tetranuclear metal cluster [WFe$_2$Au(μ$_3$-CMe)(μ$_3$-σ:σ′,η^5-C$_2$B$_9$-H$_7$Me$_2$)(CO)$_8$(PPh$_3$)] (**30**). There is NMR evidence that in **30** the μ$_3$-CMe group caps the WFe$_2$ triangle in an asymmetric manner (*27*).

Treatment of **4** with [Fe$_2$(CO)$_9$] affords [NEt$_4$][MoFe$_2$(μ$_3$-CC$_6$H$_4$Me-4)(μ$_3$-σ:σ′,η^5-C$_2$B$_9$H$_7$Me$_2$)(CO)$_8$] (**29e**), together with a mononuclear molybdenum complex [NEt$_4$][Mo{σ,η^5-CH(C$_6$H$_4$Me-4)C$_2$B$_9$H$_8$Me$_2$}(CO)$_3$] (**31**), discussed later (*13*). The structure of **29e**, established by X-ray diffraction, is similar to that of **29a**, which has also been determined by this technique (*27*).

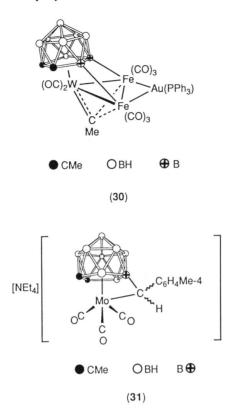

● CMe ○ BH ⊕ B

(30)

(31)

Reactions between the salts 5 and $[Fe_2(CO)_9]$ have also been investigated (14). These studies have revealed the different influences of the C_2B_9 and C_2B_{10} cages on the nature of the products obtained. Treatment of THF solutions of the reagent 5a with $[Fe_2(CO)_9]$ affords the trimetal complex $[NEt_4][WFe_2(\mu_3-CC_6H_4Me-4)(\mu-\sigma:\sigma',\eta^6-C_2B_{10}H_8Me_2)(CO)_8]$ (32). In this product the carbaborane group is η^6 coordinated to the tungsten atom, but the cage has slipped over the face of the metal triangle so as to also form two σ B—Fe bonds. The structure thus resembles that of the species 29. In contrast with the reactions between the salts 2 and iron carbonyls, discussed above, no dimetal products similar to the compounds 27 or 28 were isolated in the reaction with 5a. However, the reaction between $[Fe_2(CO)_9]$ and 5b, the latter containing the sterically demanding $CC_6H_3Me_2$-2,6 group, yields a dimetal complex as the only product (14). This compound $[NEt_4][WFe(\mu-CC_6H_3Me_2$-2,6$)(CO)_4(\eta^6-C_2B_{10}H_{10}Me_2)]$ (33) has a novel structure. The molecule possesses a relatively rare $Fe(CO)_2$ group, and there is an exo-polyhedral B—H → Fe bond involving the unique boron in the $\overline{CBCBBB}$

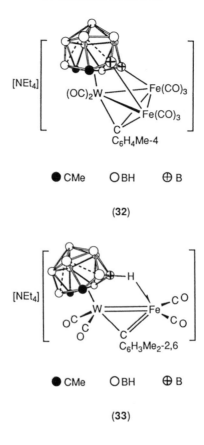

(32)

(33)

face of the cage having connectives only to other boron nuclei. As in **27a**, the xylyl ring in **33** is approximately orthogonal to the $\overline{W(\mu\text{-}C)Fe}$ ring. Like **27a**, complex **33** is an electronically unsaturated 32-CVE species, but in the latter W=Fe bond [2.512(2) Å] is even shorter than in **27a** [2.600(1) Å], a feature perhaps due to the presence of the B—H → Fe bridge. The bonding within the $\overline{W(\mu\text{-}C)Fe}$ ring in **33** is best viewed in terms of the C≡W group acting as a four-electron donor (*18*). This not only accounts for the short tungsten–iron separation, but also for the relative lengthening of the μ-C—W [2.02(1) Å] and shortening of the μ-C—Fe [1.82(1) Å] distances (*14*). Moreover, in the $^{13}C-\{^1H\}$ NMR spectrum of **33**, the μ-C nucleus is relatively deshielded (δ 369.2 ppm), in accord with the electronic unsaturation (*18,28*).

The intermediacy of B—H → metal linkages in the formation of direct boron–metal σ bonds is well illustrated by studies in carbaboranetungsten–platinum chemistry. The reaction between the salts [PtH(Me$_2$CO)(PEt$_3$)$_2$]-

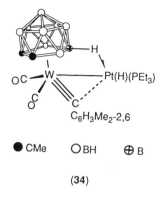

● CMe   ○ BH   ⊕ B

(34)

[BF$_4$] and **2h** in acetone at $\sim -30°C$ affords the dimetal compound [WPtH(μ-CC$_6$H$_3$Me$_2$-2,6)(CO)$_2$(PEt$_3$)(η^5-C$_2$B$_9$H$_9$Me$_2$)] (**34**) *(29)*. This product readily releases hydrogen, giving [WPt(μ-CC$_6$H$_3$Me$_2$-2,6)(μ-σ,η^5-C$_2$B$_9$-H$_8$Me$_2$)(CO)$_2$(PEt$_3$)] (**35**), formed as a mixture of two isomers. The structure of the dominant isomer **35a** ($\sim 90\%$) was established by X-ray diffraction. Evidence from NMR spectroscopy suggests that the minor isomer **35b** has a

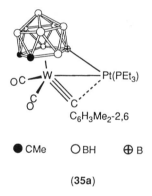

● CMe   ○ BH   ⊕ B

(35a)

very similar structure, but with the Pt atom σ bonded to a boron atom in the face of the cage adjacent to a CMe group. Interestingly, the formation of **35** from **34** can be partially reversed by treating the former with hydrogen gas. These processes are, as far as we are aware, the first example of reversible oxidative addition and reductive elimination of a B–H group at a dimetal center. Moreover, studies using [PtD(Me$_2$CO)(PEt$_3$)$_2$][BF$_4$] or D$_2$ in these reactions revealed scrambling of BH and BD groups in the cage system. This would be expected if the B—H $\longrightarrow$ Pt interaction in **34** migrated between the BH sites in the face of the cage, which are, respectively, adjacent and nonadjacent to the CMe groups. Other mechanisms need to be invoked to explain the D and H exchange observed at the other sites in the cage *(29)*.

The formation of compound **35** from **34** is similar to processes observed in tungsten–iridium chemistry (*24*). Treatment of the complexes [IrL$_2$(cod)]-[PF$_6$][L$_2$ = dppe (Ph$_2$PCH$_2$CH$_2$PPh$_2$) or bipy (2,2′-bipyridine) with **2d** yields the compounds [WIrH(μ-CC$_6$H$_4$Me-4)(μ-σ,η^5-C$_2$B$_9$H$_8$Me$_2$)(CO)$_3$L$_2$] (**36a**, L$_2$ = dppe; **36b**, L$_2$ = bipy) with terminal Ir—H and σ B—Ir bonds. It is very likely that the compounds **36** form via intermediates containing

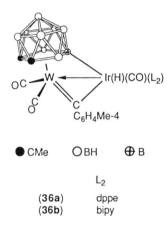

● CMe   ○ BH   ⊕ B

	L$_2$
(36a)	dppe
(36b)	bipy

B—H ⟶ Ir linkages bridging W—Ir bonds. In support of this idea it was observed that PMe$_3$ displaces the PPh$_3$ groups in **21a** to give [WIrH(μ-CC$_6$H$_4$Me-4)(μ-σ,η^5-C$_2$B$_9$H$_8$Me$_2$)(CO)$_2$(PMe$_3$)$_3$] (**37**).

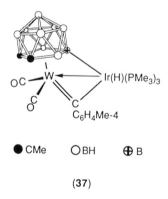

● CMe   ○ BH   ⊕ B

(**37**)

Other examples of compounds with exopolyhedral boron–metal σ bonds mentioned in Section VI.

VI

REACTIONS LEADING TO TRANSFER OF ALKYLIDYNE GROUPS
TO THE CARBABORANE CAGE

The first reaction of this type to be discovered involved that between
EtC≡CEt and **15a**, which yielded the complex [MoW{μ-σ,η^5-CH(C_6H_4Me-
4)·$C_2B_9H_8Me_2$}(CO)$_3$(η-EtC$_2$Et)(η^5-C$_9$H$_7$)] (**38**), the structure of which was

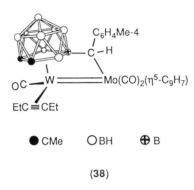

(**38**)

established by X-ray diffraction (30). The metal–metal connectivity is very
short [2.604(1) Å] and very probably corresponds to a Mo=W bond. In the
^{13}C-{^1H} NMR spectrum, the chemical shifts of the ligated carbon nuclei of
the EtC$_2$Et ligand (δ 172.0 and 178.2 ppm) are in the region associated with
an alkyne donating three or four electrons rather than two, to a metal center
(28). It is debatable whether in **38** the C$_2$B$_9$H$_8$Me$_2$ group is η^5 or η^3
coordinated to the W atom. In the former bonding mode it contributes four
electrons and with the EtC$_2$Et ligand also donating four electrons an
18-electron shell is attained. If, however, the cage is η^3 coordinated it
contributes three electrons, and with the EtC$_2$Et similarly contributing three
electrons, there would be a 16-electron configuration at the W center.

Evidently compound **38** is formed by insertion of the μ-CC$_6$H$_4$Me-4 group
in **15a** into the B—H → Mo bond, a process promoted by substitution of a
CO ligand in the precursor by EtC≡CEt. Several compounds with structures
similar to that of **38** have been discovered. Treatment of (**14b**) with PMe$_3$
gives a complex formulated as [WRu{μ-σ,η^5-CH(C$_6$H$_4$Me-4)·C$_2$B$_9$H$_8$Me$_2$}-
(CO)$_3$(PMe$_3$)(η-C$_5$H$_5$)] (**39**) (19). We referred earlier to the reaction between
21a and PMe$_3$ that gave **37**. Another product of this reaction is the complex
[WIrH{μ-σ,η^5-CH(C$_6$H$_4$Me-4)·C$_2$B$_9$H$_7$Me$_2$}(CO)$_2$(PMe$_3$)$_4$] (**40**).

The complete transfer of an alkylidyne group to the carbaborane cage
was first observed with tungsten–platinum species. In acetone, the salts **2d**

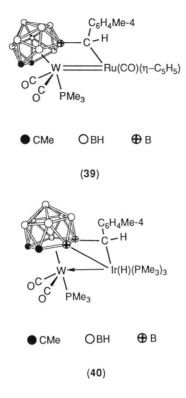

● CMe   ○ BH   ⊕ B

(39)

● CMe   ○ BH   ⊕ B

(40)

and [PtH(Me₂CO)(PEt₃)₂][BF₄] yield the dimetal compound [WPt(CO)₂ (PEt₃)₂{η⁶-C₂B₉H₈(CH₂C₆H₄Me-4)Me₂}] (**41a**), the structure of which has been established by X-ray diffraction (*31*). The Pt=W bond [2.602(1) Å] is semibridged by the two CO ligands, and a novel η⁶-C₂B₉H₈- (CH₂C₆H₄Me-4)Me₂ group ligates the W atom. Six atoms (BCBBBC) in the

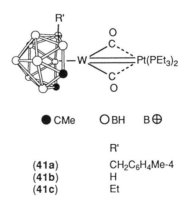

● CMe   ○ BH   B ⊕

	R'
(**41a**)	CH₂C₆H₄Me-4
(**41b**)	H
(**41c**)	Et

face of the ligand are within bonding distance of the metal, but the C···C separation (1.88 Å) within the ring is nonbonding. Treatment of **2a** with [PtH-(Me$_2$CO)(PEt$_3$)$_2$][BF$_4$] produces an inseparable mixture of the two complexes [WPt(CO)$_2$(PEt$_3$)$_2${η^6-C$_2$B$_9$H$_8$(R')Me$_2$}] (**41b**, R' = H; **41c**, R' = Et). The molecules **41** are novel in having hyper-closo structures for the carbaborane–metal cage (*32–34*). They appear to be the first examples of species of this class based on a 12-vertex cage system and necessitate modification of the classical skeletal electron pair theory.

The pathways by which the compounds **41** are formed very probably involve μ-alkylidene dimetal species as intermediates. A suggested route for the synthesis of the mixture of **41b** and **41c** is shown in Scheme 4. The agostic C—H → W interaction proposed in intermediate **A** would allow via hydride transfer a route to the μ-vinyl species **C**. From the latter an ethylene ligand could be formed as in **D**, and then released to afford **41b**. Alternatively, migration of the alkylidene ligand from a bridging metal–metal site in **A** to the terminal metal site in **B** could provide a pathway for subsequent hydroboration of the W=C(H)Me group and its transfer to the cage in **41c**. The hyper-closo structures would result in a final step from a need to

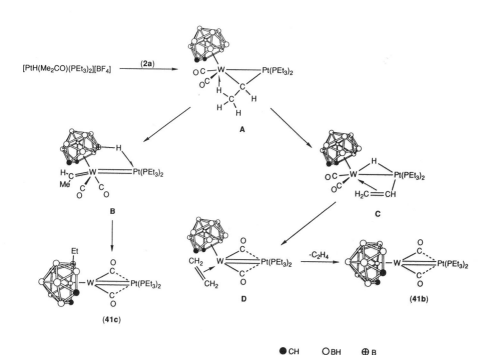

● CH   ○ BH   ⊕ B

SCHEME 4.   Possible pathway to the complexes **41b** and **41c**.

contribute six electrons to the W centers to complete an 18-electron shell. Complex **41a** could form via a route similar to that shown for **41c**. However, in this case it is likely that intermediate **A** would have a structure in which the μ-C(H)C$_6$H$_4$Me-4 group adopts a σ,η^3-bonding mode, with two carbon atoms of the arene moiety ligating the tungsten atom. Such a stucture has ample precedent (*31*). As described below, insertion of alkylidene groups into cage B—H bonds is also observed in the protonation of the salts **2**.

It is interesting to compare the nature of the product **41a** obtained from the reaction between [PtH(Me$_2$CO)(PEt$_3$)$_2$][BF$_4$] and **2d** with the complex recently isolated from the reaction between [PtCl(Me)(PMe$_2$Ph)$_2$] and **2c** in the presence of TlBF$_4$. The latter process affords (*35*) the compound [WPt(μ-CC$_6$H$_4$Me-4)(μ-σ,η^5-C$_2$B$_9$H$_8$Me$_2$)(CO)$_2$(PMe$_2$Ph)$_2$] (**42**), containing an exopolyhedral B—Pt bond but with the carbaborane cage η^5 co-ordinated to the tungsten center rather than η^6 coordinated as in **41a**. In **42**,

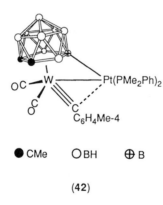

(42)

the *p*-tolylmethylidyne ligand semibridges the W—Pt bond, as in the compounds **6a** and **14b**. The pathway for the formation of **42** probably involves a transient tungsten–platinum species in which the metal–metal bond is bridged by a B—H $\longrightarrow$ Pt moiety. With a Pt–Me group also present, elimination of CH$_4$ at the platinum center could occur, with concomitant formation of the B—Pt σ bond. This process might well be more facile than an alternative pathway involving initial migration of the Me group to the CC$_6$H$_4$Me-4 ligand, so as to form a C(Me)C$_6$H$_4$Me-4 alkylidene fragment for subsequent transfer to the cage via insertion into an activated B—H $\longrightarrow$ Pt bond. As mentioned above, **41a** may well form via the intermediacy of a species containing μ-C(H)C$_6$H$_4$Me-4 and B—H $\longrightarrow$ Pt groups, with the alkylidene ligand arising from addition of the Pt—H bond in the platinum precursor to the C$\equiv$W bond in **2d**.

The formally unsaturated 30-CVE complexes **41** readily react with donor molecules. Thus treatment of **41a** with PMe₃ or CO affords, respectively, the compounds [WPt(μ-H){μ-σ,η^5-C₂B₉H₇(CH₂C₆H₄Me-4)Me₂}(CO)₂(L)-(PEt₃)₂] (**43a**, L = PMe₃; **43b**, L = CO) (*31*). Interestingly, solutions of **41a** slowly give **43b** as a decomposition product. The compounds **43** have closo-cage structures, and thus their formation parallels the reaction between the 10-vertex carbametallaborane [*hyper-closo*-6,6-(PEt₃)₂-6,2,3-RuC₂B₇H₉] and PEt₃, which gives [*closo*-6,6,6-(PEt₃)₃-6,2,3-RuC₂B₇H₉] (*32*).

Treatment of a mixture of **41b** and **41c** with CO produced a complex mixture of the three compounds [WPt(μ-H)μ-σ,η^5-C₂B₉H₇(R′)Me₂(CO)₃-(PEt₃)₂] (**43c**, R′ = H; **43d**, R′ = Et) and [WPt(CO)₃(PEt₃)₂(η^5-C₂B₉H₉-Me₂)] (**44**), all identified by X-ray crystallography (*31*). Compound **43d** is

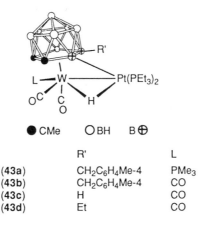

● CMe   ○ BH   B ⊕

	R′	L
(43a)	CH₂C₆H₄Me-4	PMe₃
(43b)	CH₂C₆H₄Me-4	CO
(43c)	H	CO
(43d)	Et	CO

structurally analogous to **43b** but with an Et substituent on the cage instead of a CH₂C₆H₄Me-4 group. Thus **43d** is the expected product from treating **41c** with CO, and correspondingly **43c** is the anticipated product from **41b** and CO. Compound **44** is an isomer of **43c**. Interestingly, spectroscopic studies reveal that **43c** and **44** exist in equilibrium in solution. This requires breaking a B—H → Pt bond and forming a B—Pt, bond and in this context the reader is reminded that similar reversible behavior is observed between **34** and **35** in the presence of hydrogen. In the case of the conversion **43c** ⇌ **44**, rotation of the C₂B₉H₉Me₂ ligand is required, because boron atoms occupying different sites in the face of the cage are involved. These observations suggest that formation of the compounds **43** from **41** involve intermediates with exopolyhedral B—H → Pt bonds.

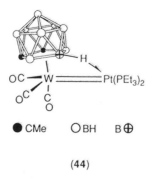

● CMe   ○ BH   B ⊕

(44)

A stepwise transfer of an alkylidyne group from a metal center to a ligating carbaborane cage has been observed in the reaction between $[Fe_2(CO)_9]$ and the salt **4** (*13*). As mentioned earlier, the products of this reaction are the trimetal species **29e** and the mononuclear molybdenum compound **31**. It is plausible to envisage that the latter complex forms via a dimetal molybdenum–iron compound having a structure similar to that of the tungsten–iron compound **28b** but with the B—Fe bond involving the boron atom in the face of the cage β to the CMe groups. Ejection of an $Fe(CO)_3$ group from such a molybdenum–iron intermediate, with concomitant linking of the μ-C(H)C$_6$H$_4$Me-4 ligand with the boron atom in the face of the cage, would yield **31**. Interestingly, protonation of **31** in the presence of CO or PMe$_3$ gives the complexes $[Mo(CO)_3(L)\{\eta^5\text{-}C_2B_9H_8(CH_2C_6H_4Me\text{-}4)\text{-}Me_2\}]$ (**45a**, L = CO; **45b**, L = PMe$_3$), respectively (*13*).

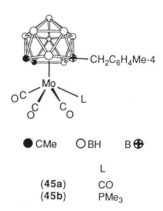

● CMe   ○ BH   B ⊕

	L
(45a)	CO
(45b)	PMe$_3$

Transfer of alkylidyne ligands from one metal center to a carbaborane cage ligating another transition element has been observed in reactions involving the salts $[NEt_4][Rh(CO)(PPh_3)(\eta^5\text{-}C_2B_9H_{11})]$ (**46a**) and $[NEt_4][Rh(CO)_2\text{-}$

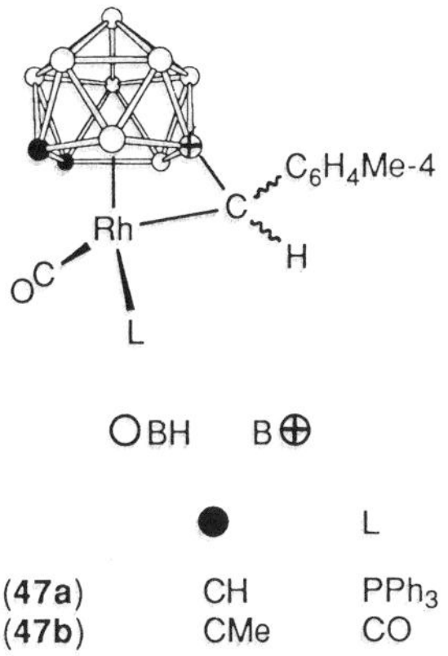

		L
(47a)	CH	PPh$_3$
(47b)	CMe	CO

○ BH   B ⊕   ●

$(\eta^5\text{-}C_2B_9H_9Me_2)]$ **(46b)** *(36)*. The reaction between $[Mn(\equiv CC_6H_4Me\text{-}4)\text{-}(CO)_2(\eta\text{-}C_5H_4Me)][BCl_4]$ and **46a** in the presence of CO affords $[Rh(CO)(PPh_3)\{\sigma,\eta^5\text{-}CH(C_6H_4Me\text{-}4)\cdot C_2B_9H_{10}\}]$ **(47a)**. A similar reaction employing **46b** yields an analogous product **47b**. Treatment of **47a** with $K[BH(CHMeEt)_3]$ in the presence of $[NEt_4]Cl$ gives the salt $[NEt_4]\text{-}[Rh(CO)(PPh_3)\{\eta^5\text{-}C_2B_9H_{10}(CH_2C_6H_4Me\text{-}4)\}]$ **(48)**.

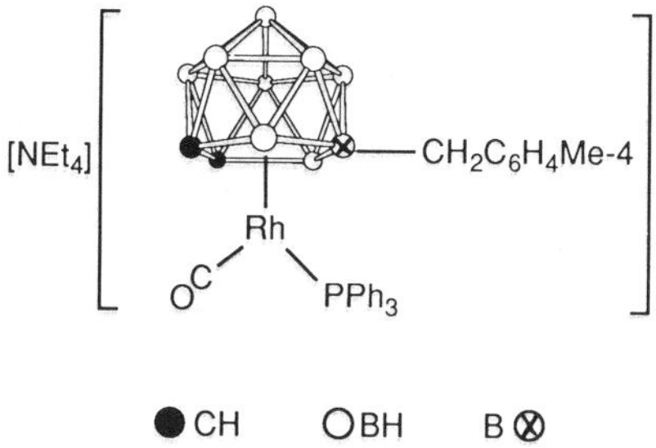

● CH   ○ BH   B ⊗

(48)

Reactions of **46a** or **46b** with the rhenium species $[Re(\equiv CC_6H_4Me\text{-}4)\text{-}(CO)_2(\eta\text{-}C_5H_4R')][BF_4]$ ($R'{=}H$ or Me) afford the dimetal compounds $[ReRh\{\mu\text{-}\sigma,\eta^5\text{-}C_2B_9H_7(CH_2C_6H_4Me\text{-}4)R_2\}(CO)_3(L)(\eta\text{-}C_5H_5)]$ **(49a**, R = H, L = PPh$_3$; **49b**, R = Me, L = CO) and $[ReRh\{\mu\text{-}\sigma,\eta^5\text{-}C_2B_9H_7(CH_2C_6\text{-}H_4Me\text{-}4)Me_2\}(CO)_4(\eta\text{-}C_5H_4Me)]$ **(49c)**. The structure of the latter species has been established by X-ray diffraction *(36)*. As in the reactions involving the manganese complex $[Mn(\equiv CC_6H_4Me\text{-}4)(CO)_2(\eta\text{-}C_5H_4Me)][BCl_4]$, the p-tolylmethylidyne group attached to rhenium in the precursor becomes

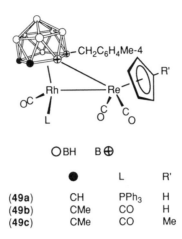

	●	L	R'
(49a)	CH	PPh₃	H
(49b)	CMe	CO	H
(49c)	CMe	CO	Me

hydroborated and linked to the cage in the products (**49**). Clearly the reactions with the manganese and rhenium salts take place, at least in the final steps, by different pathways in view of the different structures of compounds **47** and **49**.

VII

PROTONATION OF THE ANIONIC COMPLEXES
$[W(\equiv CR)(CO)_2(\eta^5\text{-}C_2B_9H_9R'_2)]$ (R = ALKYL OR ARYL, R' = H OR Me)

Protonation (HBF₄·Et₂O) of the neutral alkylidyne–tungsten compound $[W(\equiv CC_6H_4Me\text{-}4)(CO)_2(\eta\text{-}C_5H_5)]$ affords the salt $[W_2(\mu\text{-}H)\{\mu\text{-}C_2(C_6H_4\text{-}Me\text{-}4)_2\}(CO)_4(\eta\text{-}C_5H_5)_2][BF_4]$ (**50**) (*37*). The reaction is believed to proceed through the intermediacy of a cationic alkylidene–tungsten complex $[W\{=C(H)C_6H_4Me\text{-}4\}(CO)_2(\eta\text{-}C_5H_5)]^+$. The latter is electronically unsaturated (16-electron W) and could capture a $[W(\equiv CC_6H_4Me\text{-}4)(CO)_2(\eta\text{-}C_5H_5)]$

$$\left[\begin{array}{c} 4\text{-MeC}_6\text{H}_4 \qquad\qquad \text{C}_6\text{H}_4\text{Me-4} \\ \text{C}\!=\!\!=\!\text{C} \\ (\eta\text{-C}_5\text{H}_5)(\text{OC})_2\text{W} \qquad \text{W(CO)}_2(\eta\text{-C}_5\text{H}_5) \\ \text{H} \end{array} \right] [BF_4]$$

(**50**)

molecule to produce a μ-C(R)=C(H)R (R = C_6H_4Me-4) ditungsten species. This in turn could afford **50** following a hydrogen migration process (37). This result made it important to study the protonation of the salts **2**, in order to establish whether the reaction follows a pathway similar to that which occurs for [W(≡CC_6H_4Me-4)(CO)$_2$(η-C_5H_5)], or whether the η^5-$C_2B_9H_9Me_2$ cage system might play a noninnocent role.

Protonation of the salts **2** can yield a variety of different products depending on reaction temperatures, the presence of other species in solution, the stoichiometry, and, most importantly, the nature of the acid used (38). The first studies were carried out using HBF_4, the conjugate base of which BF_4^- cannot coordinate to the tungsten center. The presence in solution of donor ligands (CO, PPh_3, or PhC≡CPh) helps in the isolation of stable products, because these molecules occupy vacant sites on the tungsten atom.

At $-78°$C in the presence of CO, PPh_3, or PhC≡CPh, equivalent amounts of **2c** and $HBF_4 \cdot Et_2O$ afford, respectively, the complexes [W(CO)$_2$L$_2$-{η^5-$C_2B_9H_8(CH_2C_6H_4$Me-4)Me_2}] (**51a**, L = CO; **51b**, L = PPh_3; **51c**, L = PhC≡CPh). It is likely that an initial step in the reaction involves formation

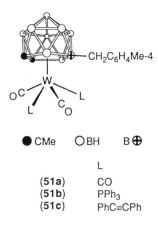

	L
(**51a**)	CO
(**51b**)	PPh_3
(**51c**)	PhC≡CPh

of a transient alkylidene–tungsten complex [W{=C(H)C_6H_4Me-4}(CO)$_2$-(η^n-$C_2B_9H_9Me_2$)] in which the $C_2B_9H_9Me_2$ group may ligate the tungsten atom in a pentahapto ($n = 5$) or hexahapto ($n = 6$) fashion, i.e., function as a 4- or 6-electron donor, respectively, so that the tungsten atom has a 16- or 18-electron configuration. The alkylidene fragment could then insert into the B—H bond in the face of the cage, a process that may be assisted by addition of the ligands L at the tungsten center. If an intermediate is present in which the cage adopts a hyper-closo η^6 bonding mode, it must collapse, because in the products **51**, η^5 coordination of the carbaborane ligand occurs. The structure of **51b** has been established by X-ray diffraction (38).

Interestingly, above $\sim -20°C$, compound **51c** releases a CO molecule to give $[W(CO)(PhC{\equiv}CPh)_2\{\eta^5\text{-}C_2B_9H_8(CH_2C_6H_4Me\text{-}4)Me_2\}]$ (**52**). In the latter the alkyne ligands probably formally contribute three electrons each to the tungsten valence shell (*28*).

The interesting ditungsten product $[W_2(\mu\text{-}CMe)(CO)_3\{\eta^5\text{-}C_2B_9H_8(CH_2\text{-}C_6H_4Me\text{-}4)Me_2\}(\eta\text{-}C_5H_5)]$ (**53**) is obtained by treating **2c** in CH_2Cl_2 at $-78°C$ with $HBF_4 \cdot Et_2O$ in the presence of $[W({\equiv}CMe)(CO)_2(\eta\text{-}C_5H_5)]$. In

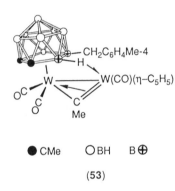

● CMe   ○ BH   B⊕

(53)

this reaction it is reasonable to assume that a $[W(CO)_2\{\eta^n\text{-}C_2B_9H_8(CH_2C_6\text{-}H_4Me\text{-}4)Me_2\}]$ fragment, formed during the protonation process, captures a molecule of $[W({\equiv}CMe)(CO)_2(\eta\text{-}C_5H_5)]$, which then functions formally as a four-electron donor (*18,28*) in **53**, the structure of which has been established by X-ray diffraction (*38*). A somewhat related reaction involves treating the salt **3** with a deficiency of $HBF_4 \cdot Et_2O$ (0.4 equivalents) (*39*). The product is the μ-alkyne ditungsten complex $[N(PPh_3)_2][W_2(\mu\text{-}H)\{\mu\text{-}C_2\text{-}(C_6H_4Me\text{-}4)_2\}(CO)_4(\eta^5\text{-}C_2B_9H_{11})_2]$ (**54**), the anion of which is structurally akin to **50**. Indeed, compound **54** probably forms via a pathway similar to that of **50**.

If HCl is used in the protonation of **2c**, instead of $HBF_4 \cdot Et_2O$, the nature of the product obtained $[NEt_4][WCl(CO)_3\{\eta^5\text{-}C_2B_9H_8(CH_2C_6H_4Me\text{-}4)\text{-}Me_2\}]$ (**55**) is very different. The structure of **55** has been established by X-ray diffraction (*38*) and it will be observed that there has been an unprecedented rearrangement of the carbaborane cage so that the CMe groups are no longer adjacent. Such polytopal rearrangements in boranes and carbaboranes have generally only previously been observed as a result of heating compounds at relatively high temperatures, although, as mentioned above, formation of the species **16** and **41** occur at $\sim 80°C$ and room temperature, respectively.

It is proposed that addition of HCl to **2c** involves a step in which a hyper-

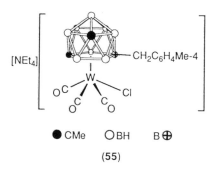

(55)

closo C_2B_9W cage system is present, in which the C—C bond is broken to allow a boron atom in the upper ring to become bonded to the tungsten in an η^6-$C_2B_9H_9Me_2$ bonding mode, as in **41**. When the cage collapses to its η^5 bonding mode, the CMe groups need no longer be adjacent, as is observed. The fact that the acid used has a conjugate base (Cl^-) with strong ligating properties clearly influences the course of reaction. A detailed mechanism for the various protonation reactions is not yet clear, but the relative stability of the tungsten–alkylidene intermediates and whether the metal is associated with a closo or hyper-closo C_2B_9W cage system are very probably critically important factors in determining the nature of the final products.

VIII

CONCLUSION

It will be evident from the many new reactions described above that the discovery of the salts **2–5** has opened up new areas for study. The chemistry emerging serves to build a bridge between different areas of inorganic and organometallic chemistry. The linkage between main-group element chemistry and that of the transition elements is obvious. Also important are the diverse interactions of the carbaborane fragments with the alkylidyne groups, which lead to molecular structures that would have been beyond imagination only a few years ago. It is well established that substitution of η-C_5Me_5 for η-C_5H_5 groups can substantially change reactivity patterns in organometallic chemistry. From the results summarized in this article, it is now abundantly clear that replacement of the cyclopentadienyl groups in the reagents $[M(\equiv CR)(CO)_2(\eta\text{-}C_5R_5')]$ (M = Mo or W, R = alkyl or aryl, R' = H or Me) by the carbaborane ligands present in **2–5** has resulted in the identification of new reactions for C≡M bonds.

Further developments are likely as the chemistry of the compounds described above is explored. Moreover, entirely new dimensions may be added. For example, the synthesis of tungsten–alkylidyne complexes with carbaborane ligands with cage structures smaller than the icosahedral C_2B_9 fragment should result in the isolation of new electronically unsaturated metal cluster and electron-deficient molecules of types as yet unknown.

REFERENCES

1. E. O. Fischer, G. Kreis, C. G. Kreiter, J. Müller, G. Huttner, and H. Lorenz, *Angew. Chem., Int. Ed. Engl.* **12**, 564 (1973); E. O. Fischer, *Adv. Organomet. Chem.* **14**, 1 (1976).
2. R. P. A. Sneedon, *in* "Comprehensive Organometallic Chemistry" (E. W. Abel, F. G. A. Stone, and G. Wilkinson, eds.), Vol. 8, Chapter 50, pp. 40–73. Pergamon, New York, 1982; R. B. Anderson, "The Fischer–Tropsch Synthesis." Academic Press, London, 1984; M. Röper, *in* "Catalysis in C_1 Chemistry" (W. Keim, ed.), pp. 41–88. Reidel, Dordrecht, Netherlands, 1983.
3. R. H. Grubbs, *in* "Comprehensive Organometallic Chemistry" E. W. Abel, F. G. A. Stone, and G. Wilkinson, eds.), Vol. 8, Chapter 54, p. 548. Pergamon, New York, 1982; R. R. Schrock, *J. Organomet. Chem.* **300**, 249 (1986).
4. T. V. Ashworth, J. A. K. Howard, and F. G. A. Stone, *J. Chem. Soc., Chem. Commun.,* 42 (1979).
5. T. V. Ashworth, J. A. K. Howard, and F. G. A. Stone, *J. Chem. Soc., Dalton Trans.,* 1609 (1980). (Manuscript received September 1978.)
6. R. Hoffmann, *Angew. Chem., Int. Ed. Engl.* **21**, 711 (1982).
7. F. G. A. Stone, *Angew. Chem., Int. Ed. Engl.* **23**, 89 (1984); *ACS Symp. Ser.* **211**, 383 (1983); *Pure Appl. Chem.* **58**, 529 (1986).
8. S. J. Davies and F. G. A. Stone, *J. Chem. Soc., Dalton Trans.,* 785 (1989); T. P. Spaniol and F. G. A. Stone, *Polyhedron* **8**, 2271 (1989); A. F. Hill, F. Marken, B. A. Nasir, and F. G. A. Stone, *J. Organomet. Chem.* **363**, 311 (1989), and references cited therein.
9. M. F. Hawthorne, D. C. Young, and P. A. Wegner, *J. Am. Chem. Soc.* **87**, 1818 (1965); K. P. Callahan and M. F. Hawthorne, *Adv. Organomet. Chem.* **14**, 145 (1976).
10. E. O. Fischer, T. L. Lindner, G. Hüttner, P. Friedrich, F. R. Kreissl, and J. O. Besenhard *Chem. Ber.* **110**, 3397 (1977).
11. M. Green, J. A. K. Howard, A. P. James, C. M. Nunn, and F. G. A. Stone, *J. Chem. Soc., Dalton Trans.,* 61 (1987).
12a. F.-E. Baumann, J. A. K. Howard, O. Johnson, and F. G. A. Stone, *J. Chem. Soc., Dalton Trans.,* 2661 (1987).
12b. O. Johnson, J. A. K. Howard, M. Kapan, and G. M. Reisner, *J. Chem. Soc., Dalton Trans.,* p. 2903 (1988).
13. D. D. Devore, C. Emmerich, J. A. K. Howard, and F. G. A. Stone, *J. Chem. Soc., Dalton Trans.,* 797 (1989).
14. S. J. Crennell, D. D. Devore, S. J. B. Henderson, J. A. K. Howard, and F. G. A. Stone, *J. Chem. Soc., Dalton Trans.,* 1363 (1989).
15. G. A. Carriedo, J. A. K. Howard, K. Marsden, F. G. A. Stone, and P. Woodward, *J. Chem. Soc., Dalton Trans.,* 1589 (1984).
16. G. A. Carriedo, J. A. K. Howard, F. G. A. Stone, and M. J. Went, *J. Chem. Soc., Dalton Trans.,* 2545 (1984).
17. F.-E. Baumann, J. A. K. Howard, R. J. Musgrove, P. Sherwood, and F. G. A. Stone, *J. Chem. Soc., Dalton Trans.,* 1879 (1988).
18. S. J. Dossett, A. F. Hill, J. C. Jeffery, F. Marken, P. Sherwood, and F. G. A. Stone, *J. Chem. Soc., Dalton Trans.,* 2453 (1988).

19. M. Green, J. A. K. Howard, A. N. de M. Jelfs, O. Johnson, and F. G. A. Stone, *J. Chem. Soc., Dalton Trans.*, 73 (1987)
20. M. Green, J. A. K. Howard, A. P. James, A. N. de M. Jelfs, C. M. Nunn, and F. G. A. Stone, *J. Chem. Soc., Dalton Trans.*, 81 (1987).
21. J. A. Doi, E. A. Mizusawa, C. B. Knobler, and M. F. Hawthorne, *Inorg. Chem.* **23**, 1482 (1984).
22. S. J. Dossett, I. J. Hart, M. U. Pilotti, and F. G. A. Stone, *J. Chem. Soc., Dalton Trans.* (1990) (in press).
23. J. A. K. Howard, A. P. James, A. N. de M. Jelfs, C. M. Nunn, and F. G. A. Stone, *J. Chem. Soc., Dalton Trans.*, 1221 (1987).
24. J. C. Jeffery, M. A. Ruiz, P. Sherwood, and F. G. A. Stone, *J. Chem. Soc., Dalton Trans.*, 1845 (1989).
25. F-E. Baumann, J. A. K. Howard, R. J. Musgrove, P. Sherwood, and F. G. A. Stone, *J. Chem. Soc., Dalton Trans.*, 1891 (1988).
26. R. T. Baker, R. E. King, C. Knobler, C. A. O'Con, and M. F. Hawthorne, *J. Am. Chem. Soc.* **100**, 8266 (1978); P. E. Behnken, T. B. Marder, R. T. Baker, C. B. Knobler, M. R. Thompson, and M. F. Hawthorne, *ibid.* **107**, 932 (1985).
27. F.-E. Baumann, J. A. K. Howard, O. Johnson, and F. G. A. Stone, *J. Chem. Soc., Dalton Trans.*, 2917 (1987).
28. J. L. Templeton, *Adv. Organomet. Chem.* **29**, 1 (1989).
29. D. D. Devore, J. A. K. Howard, J. C. Jeffery, M. U. Pilotti, and F. G. A. Stone, *J. Chem. Soc., Dalton Trans.*, 303 (1989).
30. M. Green, J. A. K. Howard, A. P. James, A. N. de M. Jelfs, C. M. Nunn, and F. G. A. Stone, *J. Chem. Soc., Chem. Commun.*, 1778 (1985).
31. M. J. Attfield, J. A. K. Howard, A. N. de M. Jelfs, C. M. Nunn, and F. G. A. Stone, *J. Chem. Soc., Dalton Trans.*, 2219 (1987).
32. C. W. Jung, R. T. Baker, C. B. Knobler, and M. F. Hawthorne, *J. Am. Chem. Soc.* **102**, 5782 (1980); C. W. Jung, R. T. Baker, and M. F. Hawthorne, *ibid.* **103**, 810 (1981).
33. R. T. Baker, *Inorg. Chem.* **25**, 109 (1986).
34. R. L. Johnston and D. M. P. Mingos, *Inorg. Chem.* **25**, 3321 (1986).
35. N. Carr, M. C. Gimeno, and F. G. A. Stone, unpublished.
36. M. U. Pilotti, I. Topalŏglu, and F. G. A. Stone, *J. Chem. Soc., Dalton Trans.* (1990) (in press).
37. J. A. K. Howard, J. C. Jeffery, J. C. V. Laurie, I. Moore, F. G. A. Stone, and A. Stringer, *Inorg. Chim. Acta* **100**, 23 (1985).
38. S. A. Brew, N. Carr, D. D. Devore, M. U. Pilotti, and F. G. A. Stone, unpublished results.
39a. S. A. Brew, J. C. Jeffery, M. U. Pilotti, and F. G. A. Stone, *J. Am. Chem. Soc.* (1990) (in press).
39b. A. P. James and F. G. A. Stone, *J. Organomet. Chem.* **310**, 47 (1986).

ADVANCES IN ORGANOMETALLIC CHEMISTRY, VOL. 31

Transition Metal Complexes Incorporating Atoms of the Heavier Main-Group Elements

NEVILLE A. COMPTON,
R. JOHN ERRINGTON, and NICHOLAS C. NORMAN

Department of Chemistry
The University of Newcastle
Newcastle upon Tyne, NE1 7RU, England

I

INTRODUCTION

The interface between transition metal coordination chemistry and main-group chemistry has now become an area of active investigation, with a considerable growth in knowledge having taken place in recent years. In many ways, this field is a logical extension from both its progenitors, but, whereas some of the newly discovered complexes present few surprises, the subject abounds with interesting and unanticipated examples, many of which challenge the tenets of chemical orthodoxy. This is a healthy state of affairs for science in general and inorganic chemistry in particular and attests to the importance of unfettered research in synthesis and the value of serendipitous discovery.

The title subject is now too large to be covered fully in a review of this nature, which takes as its starting point discrete molecular complexes containing transition metal and main-group metal bonds. We shall restrict ourselves, therefore, to those that interest us the most, namely, complexes incorporating organotransition metal fragments (which we define as to include homoleptic carbonyls as well as those with organic ligands) and atoms of the heavier main-group elements in which the latter are bonded only to transition metal centers. Specifically, we will look at complexes containing gallium, indium, thallium, germanium, tin, lead, antimony, bismuth, and tellurium, although reference to compounds involving other elements will be made where appropriate. Our decision to ignore the lighter main-group elements, although somewhat arbitrary, is again dictated by considerations of article length, but also by the fact that many recent reviews have dealt with these elements,

91

notably those of Fehlner (*1*), Scherer (*2*), Stoppioni (*3*), Herrmann (*4*), and Whitmire (*5*). This review is not intended to be fully comprehensive but will concentrate on general aspects of synthesis, structure, bonding, reactivity, and trends.

II

GALLIUM, INDIUM, AND THALLIUM

A.   Complexes Involving Molybdenum and Tungsten

The molecules that have been reported in this category all involve the $M(CO)_3(C_5H_5)$ fragment (M = Cr, Mo, W) and exist as monomeric species of the general formula $[E\{M(CO)_3(C_5H_5)\}_3]$ (E = Ga, In, Tl). The gallium complexes $[Ga\{M(CO)_3(C_5H_5)\}_3]$ (**1**, M = Mo; **2**, M = W) were prepared

2, E = Ga, M = W; 3, E = In, M = Mo

from the corresponding transition metal hydride and Me_3Ga (*6,7*), whereas the analogous indium complexes $[In\{M(CO)_3(C_5H_5)\}_3]$ (**3**, M = Mo; **4**, M = W) (see Fig. 1) were obtained from reactions between $InCl_3$ and three equivalents of the metal carbonylate anion $[M(CO)_3(C_5H_5)]^-$ (*8,8a*). The structures of **2** and **3** have been determined and reveal a trigonal–planar gallium and indium bonded to three $M(CO)_3(C_5H_5)$ fragments such that the molecules possess approximate C_{3h} symmetry. In both cases, however, although the angles and planarity at gallium and indium are close to idealized values, there is considerable variation in the M—E bond lengths (*7–8a*) (see Table I). This may be due simply to crystal packing effects, but another possibility involves a second-order Jahn–Teller distortion. More examples and some theoretical studies on these molecules are clearly warranted.

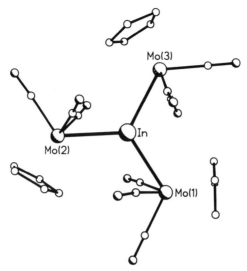

FIG. 1. Structure of $[In\{Mo(CO)_3(C_5H_5)\}_3]$ **(3)**.

TABLE I

TRANSITION METAL BOND LENGTHS TO GALLIUM, INDIUM, AND THALLIUM

Complex	Bond	$r(M—E)$ (Å)	Ref.
$[Ga\{W(CO)_3(C_5H_5)\}_3]$, **2**	Ga—W	2.744(3)	7
		2.758(3)	
		2.716(3)	
$[In\{Mo(CO)_3(C_5H_5)\}_3]$, **3**	In—Mo	2.884(1)	8
		2.887(1)	
		2.862(1)	
$[Tl\{Mo(CO)_3(C_5H_5)\}_3]$, **5**	Tl—Mo	3.001(3)	10
		2.955(3)	
		2.938(2)	
$[Mn_2(CO)_8\{\mu\text{-}GaMn(CO)_5\}_2]$, **9**[a,b]	Ga—Mn(1)	2.445(1)	20a
	Ga—Mn(2)	2.449(1)	
	Ga—Mn(3)	2.460(1)	
$[Re_2(CO)_8\{\mu\text{-}GaRe(CO)_5\}_2]$, **10**[a,c]	Ga—Re(1)	2.599(4)	21
	Ga—Re(2)	2.599(4)	
	Ga—Re(3)	2.569(5)	
$[Mn_2(CO)_8\{\mu\text{-}InMn(CO)_5\}_2]$, **11**[a,d]	In—Mn(1)	2.596(1)	20a
	In—Mn(2)	2.610(1)	
	In—Mn(3)	2.610(1)	

(continued)

TABLE I (*continued*)

Complex	Bond	r(M—E) (Å)	Ref.
$[Re_2(CO)_8\{\mu\text{-InRe(CO)}_5\}_2]$, **12**[e,f]	In—Re(1)	2.738(1)	22
	In—Re(2)	2.754(1)	
	In—Re(2)	2.807(1)	
$[Re_4(CO)_{12}\{\mu_3\text{-InRe(CO)}_5\}_4]$, **13**[g]	In(1)—Re(1)	2.796(2)	24a
	In(1)—Re(1)	2.820(2)	
	In(1)—Re(1)	2.848(2)	
	In(1)—Re(2)	2.720(2)	
$[Re_2(CO)_4(PPh_3)_2(\mu\text{-I})_2$-	Ga—Re(1)	2.531(3)	25
$\{\mu\text{-GaRe(CO)}_4(PPh_3)\}]$, **14a**[h]	Ga—Re(2)	2.516(2)	
	Ga—Re(3)	2.494(2)	
$[Re_2(CO)_4(PPh_3)_2(\mu\text{-Br})_2$-	Ga—Re(1)	2.523(7)	26
$\{\mu\text{-GaRe(CO)}_4(PPh_3)\}]$, **14b**[h]	Ga—Re(2)	2.500(6)	
	Ga—Re(3)	2.485(7)	
$[\{Fe(CO)_4\}_2\{\mu\text{-InMn(CO)}_5\}_2]$, **16**[i,j]	In—Fe(1)	2.662(1)	27
	In—Fe(2)	2.663(1)	
	In—Mn(1)	2.635(1)	
$[\{Fe(CO)_4\}_2\{\mu\text{-TlFe(CO)}_4\}_2]^{2-}$, **17**[k,l]	Tl—Fe(1)	2.553(5)	30a
	Tl—Fe(2)	2.632(5)	
	Tl—Fe(2')	3.038(4)	
$[Tl_4Fe_8(CO)_{30}]^{4-}$, **18**[m]	Tl—Fe	2.530(7)–2.786(6)	30a
$[Tl_6Fe_{10}(CO)_{36}]^{6-}$, **19**[n]	Tl—Fe	2.540(4)–2.815(3)	30a
$[Tl_2Fe_6(CO)_{24}]^{2-}$, **20**[o]	Tl—Fe	2.746(2)–2.765(2)	30b
$[Tl\{Ru_6C(CO)_{16}\}_2]^-$, **22**	Tl—Ru	2.775(3)–2.875(2)	32
$[In\{Co(CO)_4\}_3]$, **25**	In—Co	2.597(3)	36
		2.595(3)	
		2.590(3)	
$[TlPt_3(CO)_3(PCy_3)_3]^+$, **30**	Tl—Pt	3.034(1)	41
		3.047(1)	

[a] See Fig. 3 for atom numbering scheme.
[b] Mn–Mn = 3.052(1) Å, Ga–Ga = 3.846(1) Å.
[c] Re–Re = 3.139(2) Å, Ga–Ga = 4.142(5) Å.
[d] Mn–Mn = 3.227(1) Å, In–In = 4.103(1) Å.
[e] Re(1) is the exocyclic rhenium.
[f] Re–Re = 3.232(1) Å, In–In = 4.525(2) Å.
[g] See Fig. 4 for atom numbering scheme.
[h] Re(3) is the exocyclic rhenium.
[i] See Fig. 5 for atom numbering scheme.
[j] Fe–Fe = 4.218(1) Å, In–In = 3.250(1) Å.
[k] See Fig. 6 for atom numbering scheme.
[l] Fe–Fe = 4.352(5) Å, Tl–Tl = 3.658(5) Å.
[m] Tl–Tl = 3.604(3), 3.859(2) Å.
[n] Tl–Tl = 3.706(1)–3.773(1) Å.
[o] Tl–Tl = 3.507(1) Å.

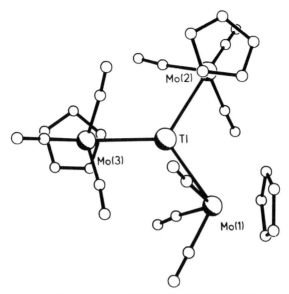

FIG. 2. Structure of $[Tl\{Mo(CO)_3(C_5H_5)\}_3]$ (5).

The thallium complex $[Tl\{Mo(CO)_3(C_5H_5)\}_3]$ (5) (see Fig. 2) has also been prepared from the reaction between TlC_5H_5 and $[Mo(CO)_6]$ (9) and structurally characterized (10). Although, in general, similar to 2 and 3, the conformations about the metal–metal bonds are different (as shown in the diagram) and there is a marked distortion at thallium toward pyramidality as well as a considerable spread in Tl—Mo bond distances. These structural distortions and variations are interesting and worthy of further study but we will not speculate further on their origins here. The chromium and tungsten analogues are also known, although no structural details are available (10a).

$$5$$

Two further points of interest are (1) the apparently facile heterolysis of the E—M bonds in donor solvents (8,8a) suggesting considerable ionic character in the metal–metal bonding, and (2) the comparison with related aluminum complexes. The former point is discussed in more detail in Section II,B whereas the latter concerns the nature of the donor site on the transition metal fragment. If the aforementioned complexes are considered as Lewis acid–base adducts, the basic site is clearly the transition metal center in **1–5**. In the aluminum complex $[Al\{W(CO)_3(C_5H_5)\}_3]\cdot 3$ THF (*11*), however, the interaction is between the aluminum and carbonyl oxygen atoms. This situation can be understood using the concept of hard and soft acids and bases, because Al^{3+} is a hard acid and prefers the harder basic site, i.e., carbonyl oxygen rather than metal center. These points have been discussed by Burlitch with regard to EPh_3 complexes of transition metal carbonylate anions (*12*).

B. *Complexes Involving Manganese and Rhenium*

Indium and thallium complexes containing the pentacarbonyl manganese and rhenium fragments $[E\{M(CO)_5\}_3]$ (**6**, E = In, M = Mn; **7**, E = In, M = Re; **8**, E = Tl, M = Mn) have been known for a number of years. No

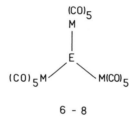

6 - 8

crystal structure data are available for any of these species, although these are assumed to contain trigonal–planar indium and thallium centers. Mays has reported the syntheses of **6** (*13,13a*) and **7** (*8a*) from reactions between the respective metal carbonylate anions and $InCl_3$, with additional synthetic routes to **7** involving reactions between $[HRe(CO)_5]$ and Me_3In or $[Hg\{Re(CO)_5\}_2]$ and powdered metallic indium (*8a*). The latter route is particularly interesting because it involves a direct metal exchange reaction, several examples of which are known for related zinc and cadmium complexes (*14*). Moreover, the indium complex **6** may also be prepared from the thallium analogue **8** and metallic indium (*13a*).

The thallium complex **8** is available by two general routes involving the use of either Tl(I) or Tl(III) salts. The former involves a disproportionation reaction [Eq. (1)] presumably via a

$$3Na[Mn(CO)_5] + 3TlNO_3 \xrightarrow{H_2O} [Tl\{Mn(CO)_5\}_3] + 2Tl + 3NaNO_3 \qquad (1)$$

Tl(I)–manganese intermediate (*15–17*), which has not been isolated, whereas the latter involves direct substitution at a Tl(III) center [Eq. (2)] (*18*).

$$TlCl_3 + 3Na[Mn(CO)_5] \rightarrow [Tl\{Mn(CO)_5\}_3] + 3NaCl \tag{2}$$

A number of reactivity studies have been performed on **6** and **8** and indicate a strongly polar (if not ionic) Mn—E bond; $Mn^{\delta-}$—$E^{\delta+}$ (E = In, Tl). Thus heterolytic bond dissociation occurs in polar ligating solvents such as MeCN or DMF, and halogens, hydrogen halides, and alkyl halides readily add across the metal–metal bond in a manner consistent with the polarity described above (*13,13a,18*). In the thallium example, however, the reactions are generally more complicated and result in Tl(I) salts [e.g., Eq. (3)], and metal exchange reactions are also more facile, e.g., the synthesis of **6** from **8** and indium metal. In general, therefore, the chemistry of **6** and **8** is consistent with predominantly ionic behavior.

$$[Tl\{Mn(CO)_5\}_3] + 3HCl \rightarrow TlCl + H_2 + [HMn(CO)_5] + 2[ClMn(CO)_5] \tag{3}$$

A further class of compounds involving the Group 7 metals and either gallium or indium has the general formula $[M_2(CO)_8\{\mu\text{-}EM(CO)_5\}_2]$ (**9–12**). Although originally thought to contain unsupported Ga—Ga or

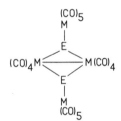

9 E=Ga, M=Mn; 10 E=Ga, M=Re;
11 E=In, M=Mn; 12 E=In, M=Re.

In—In bonds (*19*), later structural data obtained from X-ray crystallography established the presence of an approximately planar M_4E_2 framework with a central M_2E_2 rhombus, as shown in the diagram and in Fig. 3 for **11**. In the isomorphous manganese complexes **9** and **11** (*20,20a*), the angles at each gallium or indium center are bilaterally symmetrical [Ga, 76.86(2), 141.27(2), 141.30(2)°; In, 76.36(2), 141.07(2), 141.47(2)°], giving the molecule approximate D_{2h} symmetry, but a slight transoid distortion of the terminal $Mn(CO)_5$ fragments relative to the central Mn_2E_2 unit is observed that renders the gallium and indium centers slightly pyramidal and lowers the actual symmetry to C_{2h}. The gallium–rhenium complex **10** has also been structurally characterized (*21*) and is isomorphous with **9** and **11**, whereas

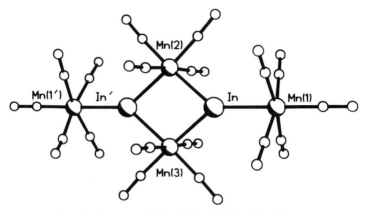

FIG. 3.   Structure of $[Mn_2(CO)_8\{\mu\text{-}InMn(CO)_5\}_2]$ (**11**).

the rhenium analogue **12**, although similar in its gross structural aspects to
9–11 (*22*), shows a marked asymmetry in the angles at indium [71.07(3),
137.32(5), 150.50(5)°], but the origin of this distortion is not clear.

All the above complexes, **9–12**, are isolated from reactions involving gallium or indium metal and $[Mn_2(CO)_{10}]$ or $[Re_2(CO)_{10}]$ in an autoclave,
although **11** has also been obtained by thermolysis of **6**. Some details on the
reactivity of **11** toward phosphonium halides (*23*) have been reported and
indicate formation of the halo-indium complexes $[PPh_4]_2[Mn_2(CO)_8\{\mu\text{-}InX_2\}_2]$. A higher nuclearity cluster has also been obtained in the reaction
of $[Re_2(CO)_{10}]$ with indium metal, viz. $[Re_4(CO)_{12}\{\mu_3\text{-}InRe(CO)_5\}_4]$, **13**
(*24,24a*). This consists of a central tetrahedral $Re_4(CO)_{12}$ core with each
face capped by an $InRe(CO)_5$ fragment as shown in Fig. 4.

14a, X=I; 14b, X=Br

15

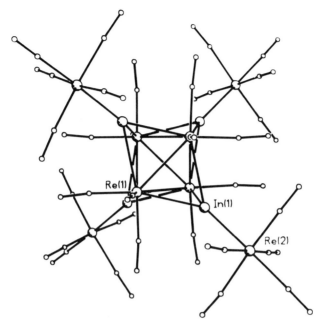

FIG. 4.   Structure of $[Re_4(CO)_{12}\{\mu_3\text{-}InRe(CO)_5\}_4]$ (13).

Also of interest are some gallium–rhenium complexes obtained from reactions between gallium trihalides and $[Re_2(CO)_8(PPh_3)_2]$. The first, derived
from GaI_3 or $GaBr_3$, comprises a planar Re_3Ga spiked-triangular unit with
two bridging halide ligands, viz. $[Re_2(CO)_4(PPh_3)_2(\mu\text{-}X)_2\{\mu\text{-}GaRe(CO)_4\text{-}
PPh_3\}]$ (X = Br, I), 14 (25,26), and the second, derived from $GaCl_3$, involves
a $GaRe(CO)_4PPh_3$ unit triply bridging an Re_3 triangle, $[Re_3(CO)_6(PPh_3)_3\text{-}
(\mu\,Cl)_3(\mu_3\text{-}GaRe(CO)_4PPh_3)]$, 15 (26). The reaction of 14 with CO has also
been reported to give $[Re_2(CO)_6(PPh_3)_2(\mu\text{-}I)(\mu\text{-}GaI_2)]$ (26).

Finally, we note a report of the synthesis and structure of a mixed transition metal–indium complex containing iron and manganese derived from
the reaction between $[Hg\{Mn(CO)_5\}_2]$, $[Fe(CO)_5]$, and indium metal, viz.
$[\{Fe(CO)_4\}_2\{\mu\text{-}InMn(CO)_5\}_2]$, 16 (27) (see Fig. 5). This complex is structurally similar to the manganese–indium complex 11 with each $Mn(CO)_4$
fragment replaced in 16 by an $Fe(CO)_4$ moiety. The increase in the electron
count by two has been assumed to account for the absence of a Fe—Fe bond
in 16 in contrast to the suggested presence of a corresponding Mn—Mn

$$\begin{array}{c}
(CO)_5 \\
Mn \\
| \\
In \\
\diagdown \diagup \\
(CO)_4Fe \qquad\qquad Fe(CO)_4 \\
\diagup \diagdown \\
In \\
| \\
Mn \\
(CO)_5
\end{array}$$

16

bond in **11**. The bond distance and angle data for the central M_2In_2 rings are shown in the diagrams, a comparison of which indicates that the Fe_2In_2 rhombus in **16** is significantly larger than that found for the Mn_2In_2 unit in **11**. Moreover, the Mn—Mn distance in **11** is long for such a bond [e.g., Mn—Mn = 2.9038(6) Å in $[Mn_2(CO)_{10}]$ (*28*)] and the acuteness of the internal ring angles at indium in **11**, taken as evidence consistent with the presence of a Mn—Mn bond (*20a*), is also observed in **16**. We suggest that the bonding in these interesting structures is best viewed with reference to the isolobal principle (*29*). Thus in **11**, $Mn(CO)_4$ is isolobal with CH where-

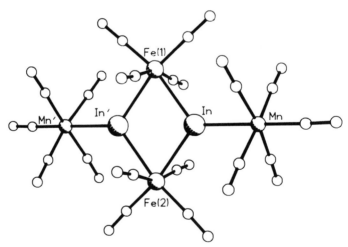

FIG. 5.   Structure of $[\{Fe(CO)_4\}_2\{\mu\text{-}InMn(CO)_5\}_2]$ (**16**).

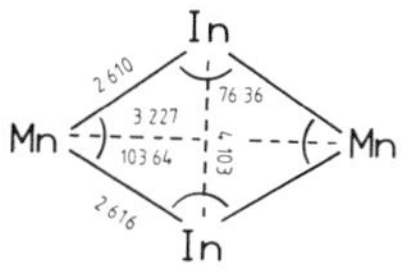

11

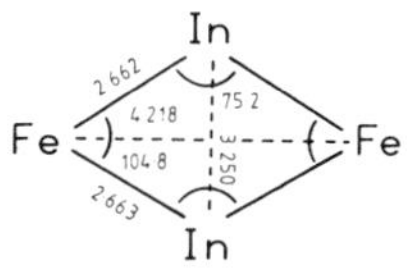

16

as $InMn(CO)_5$ can be viewed as isolobal with CH^+, which suggests a comparison between the central Mn_2In_2 unit of **11** and $C_4H_4^{2+}$. This would lead to a bonding description of the ring involving four σ-type orbitals and one delocalized π-type bonding orbital. In **16** the analogy would be with C_4H_4 (cyclobutadiene), with the additional two electrons residing in a π-type nonbonding orbital. Detailed theoretical studies on these two molecules would clearly be worthwhile.

C. Complexes involving Iron and Ruthenium

The most extensive series of compounds that fall within this section are the anionic iron carbonyl–thallium clusters recently described by Whitmire (30–30b). These are generally made by reactions between iron carbonylate anions and either Tl(I) or Tl(III) salts. Thus treatment of $K[HFe(CO)_4]$ with a variety of thallium salts followed by addition of $[Et_4N]Cl$ or Br affords the hexanuclear complex $[Et_4N]_2[\{Fe(CO)_4\}_2\{\mu\text{-TlFe(CO)}_4\}_2]$, **17**

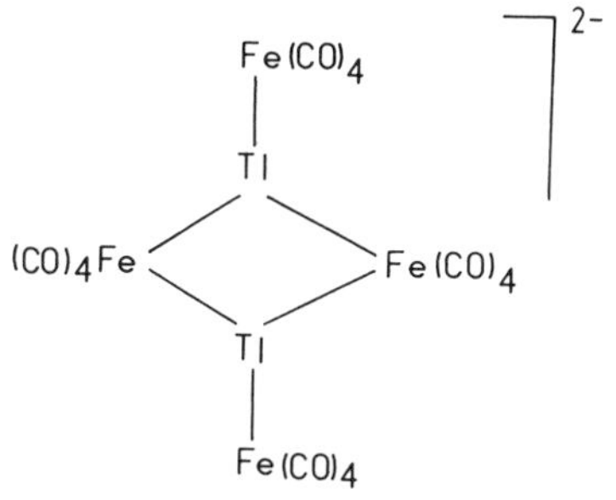

17

(see Fig. 6), which consists of a central Fe_2Tl_2 parallelogram with exocyclic $Fe(CO)_4$ groups bonded to each thallium atom. This complex is formally isoelectronic with the iron–manganese indium complex, **16**, which was described previously although in **17** the central metal ring is quite asymmetric. The

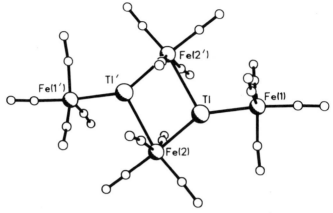

FIG. 6.   Structure of $[\{Fe(CO)_4\}_2\{\mu\text{-TlFe(CO)}_4\}_2]^{2-}$ **(17)**.

possible origin of this distortion relative to **16** (or **11**) provides another reason for carrying out theoretical molecular orbital studies.

$$
\begin{array}{c}
\text{Tl} \\
\text{Fe} \cdots \cdots \cdots \text{Fe} \\
\text{Tl}
\end{array}
$$

17

Solutions of **17** are prone to loss of CO and result in the larger clusters, $[Tl_4Fe_8(CO)_{30}]^{4-}$ **(18)** and $[Tl_6Fe_{10}(CO)_{36}]^{6-}$ **(19)** (see Figs. 7 and 8), both

18

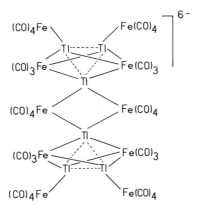

19

of which have been characterized by X-ray crystallography (*30a*). The bonding in these fascinating clusters is the subject of much interest and has been studied by Whitmire *et al.* with regard to factors such as electron deficiency, multicenter bonding, and Tl—Tl bonding for **19** by extended Hückel molecular orbital (EHMO) methods (*30*).

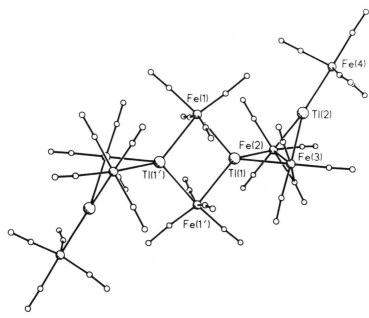

FIG. 7. Structure of $[Tl_4Fe_8(CO)_{30}]^{4-}$ **(18)**.

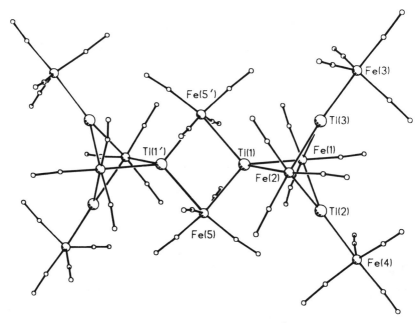

FIG. 8.   Structure of $[Tl_6Fe_{10}(CO)_{36}]^{6-}$ (**19**).

Subsequent reports by Whitmire (*30b*) have shown that a further anionic cluster, viz. $[Tl_2Fe_6(CO)_{24}]^{2-}$ (**20**) (see Fig. 9), is formed by oxidation of **17** and that reactions of **17** with a variety of bidentate nitrogeneous bases afford the anionic monothallium complexes **21a–d** (*31*).

20

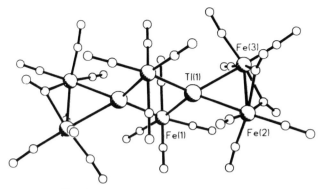

FIG. 9.   Structure of $[Tl_2Fe_6(CO)_{24}]^{2-}$ **(20)**.

Finally, we note that an anionic thallium–ruthenium cluster has been reported which contains a thallium atom bridging between two $Ru_6C(CO)_{16}$ clusters **(22)** *(32)* and that details of the synthesis of $[In\{Fe(CO)_2(C_5H_4R)\}_3]$ (R = H, Me) **23** *(33)* from $InCl_3$ and $Na[Fe(CO)_2(C_5H_4R)]$ have been described. Analytical and infrared data have also been presented for $[Tl\{Fe(CO)_2(NO)(PPh_3)\}_3]$ *(33a)*.

22

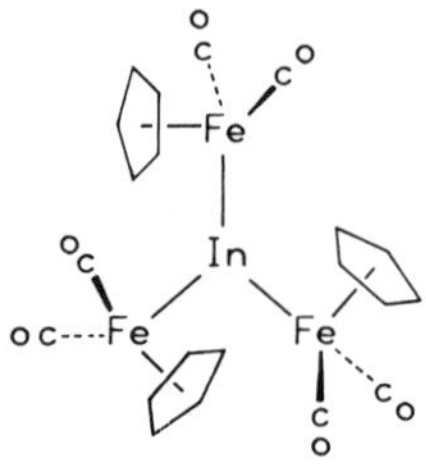

23

D. Complexes Involving Cobalt

Studies of complexes involving cobalt have a long history; a report by Hieber in 1942 (*34*) described the syntheses of $[E\{Co(CO)_4\}_3]$ (**24**, E = Ga; **25**, E = In; **26**, E = Tl) from the respective Group 13 metal, cobalt metal, and CO under high pressures. Later work by Graham (*35*) described the syntheses

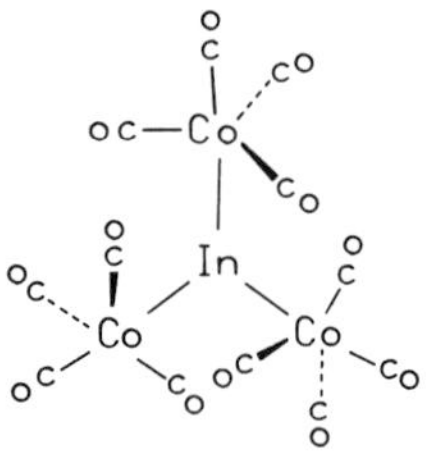

25

of **25** and **26** from $Na[Co(CO)_4]$ and $InBr_3$ and $TlCl_3$, respectively, and a crystal structure determination of **25** (see Fig. 10) has been carried out (*36*). This latter study shows that the indium center resides in a trigonal–planar coordination environment with angles close to idealized values and very similar In—Co bond lengths. Thus in **25** the metal–metal bond lengths span only a narrow range in contrast to the wide range seen in the related complexes **2**, **3**, and **5** described in Section II,A.

A tetracobalt anionic complex, viz. $[In\{Co(CO)_4\}_4]^-$ (**27**) (*37,37a*), has been briefly described together with the thallium analogue (**28**) (*37a*), both formed by addition of $[Co(CO)_4]^-$ to either **25** or **26**. No structural details have been reported although the indium and thallium centers are presumably tetrahedrally coordinated by the four cobalt atoms. Mention is also made (*37a*) of the facile heterolytic bond dissociation (In—Co or Tl—Co) observed in polar solvents. Little has been reported about the reactivity of these complexes, although a discussion on the use of **25** as a catalyst in the dimerization of norbornadiene has appeared (*38*).

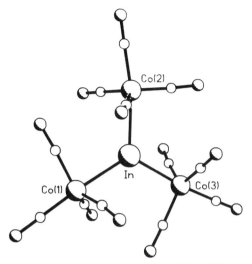

FIG. 10.   Structure of $[In\{Co(CO)_4\}_3]$ (25).

A final complex of interest is the Tl(I) species, $Tl[Co(CO)_4]$ (29) (39,40). This has been characterized by X-ray crystallography, but does not contain a Tl—Co bond. Further details on Tl(I) transition-metal complexes can be found in refs. (10a) and (33a), although we will not discuss them further here.

E. Complexes Involving Platinum

Only one complex defined by this heading is known, namely, the cationic thallium–triplatinum species $[TlPt_3(CO)_3(PCy_3)_3]^+$ (30) (41) (see Fig. 11) derived from the reaction between $TlPF_6$ and $[Pt_3(CO)_3(PCy_3)_3]$. Crystals of 30 were obtained from a similar reaction, but in the presence of $[Rh_2Cl_2-(\eta-C_8H_{12})_2]$, which produced the anion $[RhCl_2(\eta-C_8H_{12})]^-$, short contacts being observed between the thallium and chlorine atoms. The cluster, 30, is interesting in that it is only the third example of a 54-valence electron

$$Cy_3P—Pt \overset{O}{\underset{}{\Longrightarrow}} Pt—PCy_3$$

30

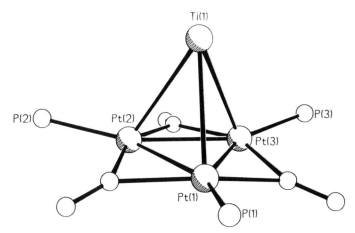

FIG. 11.   Structure of $[TlPt_3(CO)_3(PCy_3)_3]^+$ **(30)**.

tetrahedral cluster *(41)* and also because it contains thallium in an unusual trigonal–pyramidal coordination environment. In view of this unusual coordination and the fact that the Tl–Pt distances (average 3.038 Å)
are much longer than the Pt–Pt distances (average 2.667 Å), a description
based on a closely associated ion pair might be appropriate, i.e, $[Tl]^{3+}$-
$[Pt_3(PCy_3)_3(\mu\text{-}CO)_3]^{2-}$.

III

GERMANIUM, TIN, AND LEAD

A. Complexes Involving Chromium, Molybdenum, and Tungsten

The simplest class of complexes involving the Group 6 transition metals are
those represented by the general formula $E(ML_n)_2$, where ML_n represents a
17-electron fragment, and containing a two-coordinate and formally divalent
main-group element. Although there appear to be no germanium examples of
this type, early work by Harrison and Stobart *(42)* described the synthesis of
$[Sn\{W(CO)_3(C_5H_5)\}_2]$, **31**, from $[Sn(C_5H_4Me)_2]$ and $[WH(CO)_3(C_5H_5)]$.
These authors suggested, on the basis of tin-119 Mössbauer spectroscopy, that
31 underwent a rapid autopolymerization reaction resulting in a structurally
undefined complex of tin(IV) written as $\{Sn^{IV}[W(CO)_3(C_5H_5)]_2\}_n$. However,
subsequent work by Zuckerman *(43)* suggested that the product from the
aforementioned reaction was, in fact, $[SnH\{W(CO)_3(C_5H_5)\}_3]$ rather than

31. More recently, the molybdenum analogue of **31**, viz. [Sn{Mo(CO)$_3$-(C$_5$H$_5$)}$_2$], **32**, has been reported by Lappert and Power (*44*) from the reaction between [MoH(CO)$_3$(C$_5$H$_5$)] and [Sn{N(SiMe$_3$)$_2$}$_2$]. Compound **32** was characterized by spectroscopic and analytical methods, but no structural details have been presented for either **31** or **32**. Lappert *et al.* (*45*) have also reported an attempted preparation of the lead complex [Pb{Mo(CO)$_3$-(C$_5$H$_5$)}$_2$], **33**, which decomposes very readily to lead and [Mo$_2$(CO)$_6$-(C$_5$H$_5$)$_2$], together with details on more tractable complexes containing

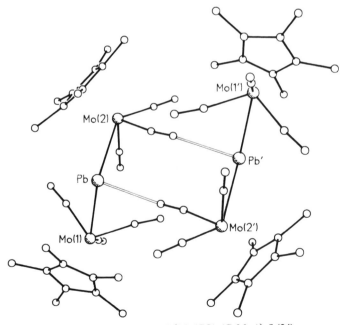

34

substituted cyclopentadienyl ligands. Specifically, the complexes [Pb{Mo-(CO)$_3$(C$_5$Me$_5$)}$_2$] (**34**) (see Fig. 12) and [Pb{Mo(CO)$_3$(C$_5$H$_3${SiMe$_3$}$_2$)}$_2$] **35**

FIG. 12.   Structure of [Pb{Mo(CO)$_3$(C$_5$Me$_5$)}$_2$] (**34**).

were described, the former having been characterized by X-ray crystallography and shown to exist as a dimer, involving isocarbonyl bridging to lead, in the solid state. In noncoordinating solvents, both **34** and **35** appear to exist as monomers, but in THF they form adducts, the nature of one of which, $[Pb(THF)\{Mo(CO)_3(C_5Me_5)\}_2]$, was established by X-ray crystallography (*45*).

Compounds of germanium and tin containing three transition metal fragments have been described by Huttner *et al.* (*46*) from the reaction between ECl_4 and $[W_2(CO)_{10}]^{2-}$. Both reactions result in similar complexes containing a trigonal–planar main-group element, $[E\{W(CO)_5\}\{W_2(CO)_{10}\}]$ (**36,**

$$\begin{array}{c} \text{(CO)}_5 \\ \text{W} \\ | \quad \diagdown \\ | \quad \rangle E =\!\!=\!\! W(CO)_5 \\ | \quad \diagup \\ \text{W} \\ \text{(CO)}_5 \end{array}$$

36, E=Ge; 37,E=Sn.

E = Ge; **37,** E = Sn) (see Fig. 13), and formally contain Ge(IV) and Sn(IV), respectively. Moreover, a formal double bond exists between the main-group element and the symmetry-unique tungsten atom, which factor is reflected in the X-ray-derived bond lengths (Table II). Reactions involving **37** with THF

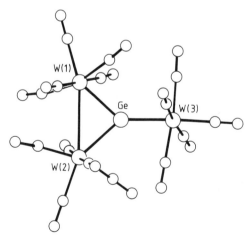

FIG. 13.   Structure of $[Ge\{W(CO)_5\}\{W_2(CO)_{10}\}]$ (**36**).

TABLE II

TRANSITION METAL BOND LENGTHS TO GERMANIUM, TIN AND LEAD

Complex	Bond	$r(M—E)$ (Å)	Ref.
$[Pb\{Mo(CO)_3(C_5Me_5)\}_2]$, **34**	Pb—Mo	2.935(1)	45
		2.989(1)	
$[Ge\{W(CO)_5\}\{W_2(CO)_{10}\}]$, **36**[a]	Ge—W(1)	2.589(1)	46
	Ge—W(2)	2.579(2)	
	Ge—W(3)	2.505(2)	
$[Sn\{W(CO)_5\}\{W_2(CO)_{10}\}]$, **37**[a]	Sn—W(1)	2.789(2)	46
	Sn—W(2)	2.776(2)	
	Sn—W(3)	2.702(2)	
$[Ge\{Mn(CO)_2(C_5H_4Me)\}_2]$, **39**	Ge—Mn	2.204(1)	49
$[Ge\{Mn(CO)_2(C_5Me_5)\}_2]$, **40**	Ge—Mn	2.18(2)	50
$[Pb\{Mn(CO)_2(C_5H_5)\}_2]$, **43**	Pb—Mn	2.463(1)	56
$[Ge\{Mn(CO)_2(C_5H_4Me)\}$-	Ge—Mn(1)	2.359(2)	49
$\{Mn_2(CO)_4(C_5H_4Me)_2\}]$, **45**[b]	Ge—Mn(2)	2.260(2)	
	Ge—Mn(3)	2.380(1)	
$[Sn\{Mn(CO)_2(C_5H_4Me)\}$-	Sn—Mn(1)	2.555(1)	54
$\{Mn_2(CO)_4(C_5H_4Me)_2\}]$, **47**[b]	Sn—Mn(2)	2.445(1)	
	Sn—Mn(3)	2.542(1)	
$[Pb\{Mn(CO)_2(C_5H_4Me)\}$-	Pb—Mn(1)	2.622(1)	56,58
$\{Mn_2(CO)_4(C_5H_4Me)_2\}]$, **48**	Pb—Mn(2)	2.490(1)	
	Pb—Mn(3)	2.611(1)	
$[Sn\{Fe(CO)_4\}_3]^{2-}$, **49**[c]	Sn—Fe	2.548(3)	61
		2.539(3)	
		2.556(3)	
		2.541(3)	
		2.552(3)	
		2.554(3)	
$[Pb\{Fe(CO)_4\}_3]^{2-}$, **50**[c]	Pb—Fe	2.627(4)	61
		2.619(5)	
		2.629(4)	
		2.624(4)	
		2.624(4)	
		2.622(5)	
$[Sn\{Fe_2(CO)_8\}_2]$, **54**	Sn—Fe (av)	2.54	68
$[Ge\{Fe_2(CO)_8\}_2]$, **55**	Ge—Fe	2.388(1)	69
		2.407(1)	
		2.374(2)	70
		2.438(2)	
$[Pb\{Fe_2(CO)_8\}_2]$, **56**	Pb—Fe	2.606(3)–	71
		2.635(3)	

(*continued*)

TABLE II (*continued*)

Complex	Bond	r(M—E) (Å)	Ref.
[Ge{Fe$_2$(CO)$_8$}-{FeMn(CO)$_6$(C$_5$H$_4$Me)}], **57**	Ge—Mn	2.431(1)	70
	Ge—Fe	2.394(1)	
		2.416(1)	
		2.427(1)	
[Pb{Fe$_2$(CO)$_8$}{Fe(CO)$_4$}$_2$]$^{2-}$, **58**^{d}	Pb—Fe(1)	2.651(5)	72
	Pb—Fe(2)	2.659(5)	
	Pb—Fe(3)	2.832(4)	
	Pb—Fe(4)	2.823(4)	
[Ge{Fe$_3$(CO)$_{10}$}{Fe(CO)$_4$}]$^{2-}$, **60**	Ge—Fe(1)	2.400(4)	71
	Ge—Fe(2)	2.394(3)	
	Ge—Fe(3)	2.389(3)	
	Ge—Fe(4)	2.327(3)^e	
[Sn{Fe$_2$(CO)$_8$(μ-SnMe$_2$)}$_2$], **61**^{f}	Sn—Fe	2.747(8)	73
[Fe$_2$(CO)$_7${μ-Ge(Fe$_2$(CO)$_8$)}$_2$], **63**	Ge—Fe	2.379(1)–	75
		2.452(1)	
[Fe$_2$(CO)$_7${μ-Sn(Fe$_2$(CO)$_8$)}$_2$], **64**	Sn—Fe	2.529(5)–	75
		2.570(4)	
[Fe$_3$(CO)$_{10}${μ-Ge(Fe$_2$(CO)$_8$)}$_2$], **65**	Ge—Fe	2.373(8)–	75
		2.415(8)	
[Fe$_3$(CO)$_9${μ_3-SnFe(CO)$_2$(C$_5$H$_5$)}$_2$], **66**	Sn—Fe(CO)$_3$ (av)	2.537	76
	Sn—Fe(CO)$_2$(C$_5$H$_5$) (av)	2.471	
[Ge{Co$_2$(CO)$_7$}$_2$], **69**	Ge—Co	2.339(2)–	78
		2.387(2)	
[Co$_3$(CO)$_9${μ-GeCo(CO)$_4$}], **70**^{g}	Ge—Co(1)	2.281(1)	81
	Ge—Co(2)	2.284(1)	
	Ge—Co(3)	2.277(1)	
	Ge—Co(4)	2.349(1)	
[Ge{Co$_2$(CO)$_7$}-{Co$_2$(CO)$_6$(μ-HgCo(CO)$_4$)}]$^-$, **73**	Ge—Co	2.317(2)–	86
		2.432(2)	
[Ge{Co$_2$(CO)$_7$}{Co$_3$(CO)$_9$}]$^{2-}$, **74**	Ge—Co	2.33(1)–	79
		2.51(1)	
[Ge$_2$Co$_7$(CO)$_{21}$]$^-$, **75**	Ge—Co	2.282(3)–	87
		2.332(2)	
[Ge{Co$_3$(CO)$_9$}{Co$_4$(CO)$_{11}$}]$^-$, **76**	Ge—Co	2.27(1)–	87
		3.16(1)	
[Co$_4$(CO)$_{11}${μ_4-GeCo(CO)$_4$}$_2$], **77**	Ge—Co	2.390(5)–	88
		2.434(5)	
[Ge$_3$Co$_8$(CO)$_{26}$], **78**	Ge—Co	2.338(2)–	89
		2.407(2)	
[Ni$_{10}$Ge(CO)$_{20}$]$^{2-}$, **79**	Ge—Ni	2.493(1)–	90
		2.691(1)	
[Ni$_{12}$Ge(CO)$_{22}$]$^{2-}$, **80**	Ge—Ni	2.470(1)	90

TABLE II (*continued*)

Complex	Bond	$r(M—E)$ (Å)	Ref.
$[Fe_2(CO)_7\{\mu\text{-}Ge(Co_2(CO)_7)\}_2]$, **83**	Ge—Co	2.309(5)–2.399(4)	*91*
	Ge—Fe	2.365(4)–2.437(4)	
$[Fe_3(CO)_9\{\mu_3\text{-}GeRe(CO)_5\}_2]$, **84**	Ge—Fe	2.356(4) 2.410(4)	*92*
	Ge—Re	2.533(3) 2.551(3)	
$[Fe_3(CO)_9\{\mu_3\text{-}SnRe(CO)_5\}_2]$, **85**	Sn—Fe	2.496(3)–2.592(3)	*93*
	Sn—Re	2.712(2) 2.740(2)	
$[Co_3(CO)_9\{\mu_3\text{-}GeFe(CO)_2(C_5H_5)\}]$, **86**	Ge—Co	2.308(1)–2.320(1)	*94*
	Ge—Fe	2.303(1)	
$[Co_2Mo(CO)_8(C_5H_5)$-$\{\mu_3\text{-}GeW(CO)_3(C_5H_5)\}]$, **90**	Ge—Co	2.298(6) 2.328(6)	*96*
	Ge—Mo	2.565(4)	
	Ge—W	2.589(4)	

[a] See Fig. 13 for numbering scheme.
[b] See Fig. 16 for numbering scheme.
[c] Two molecules per asymmetric unit.
[d] See Fig. 19 for numbering scheme.
[e] Fe(4) is the apical iron atom.
[f] Bond lengths are to the central tin atom.
[g] See Fig. 22 for numbering scheme.

and H_2Te have also been described (*47*) which result in $[Sn(THF)\{W(CO)_5\}$-$\{W_2(CO)_{10}\}]$ and $[Te_2\{W(CO)_5\}_3]$; the latter complex is described in more detail in Section V.

No homonuclear complexes containing more than three transition metal fragments have been described, although a number of mixed-metal species are known. One example has been reported from the reaction between $[SnCl_2$-$\{Fe(CO)_2(C_5H_5)\}_2]$ and $Na[Mo(CO)_3(C_5H_5)]$, which results in the rather unstable Sn(IV) complex, $[Sn\{Fe(CO)_2(C_5H_5)\}_2\{Mo(CO)_3(C_5H_5)\}_2]$ (**38**) (*48*). Other examples are described in later sections.

B. *Complexes Involving Manganese*

A number of interesting complexes are known to involve multiple bonding to manganese. These fall into two classes, the first comprising two manganese centers in a linear, two-coordinate geometry about the Group 14 element center, and the second with three manganese centers and trigonal–planar coordination.

The first example of the former was described by Weiss and Gäde (*49*) from the reaction of $K[Mn(GeH_3)(CO)_2(C_5H_4Me)]$ with acetic acid, which affords $[Ge\{Mn(CO)_2(C_5H_4Me)\}_2]$, **39**, the structure of which is shown in

39

Fig. 14. Complex **39** contains a linear, two-coordinate germanium atom with conformations about the Mn—Ge bonds such that the molecule adopts a centrosymmetric, anti structure. Work by Herrmann (*50*) describes the C_5Me_5

40

derivative $[Ge\{Mn(CO)_2(C_5Me_5)\}_2]$, **40** (see Fig. 15), obtained from the reaction between $[Mn(THF)(CO)_2(C_5Me_5)]$ and GeH_4/H_2SO_4. The structure of this complex was also described, being similar to **39** in containing a

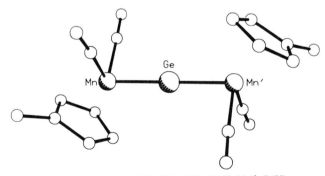

FIG. 14.   Structure of $[Ge\{Mn(CO)_2(C_5H_4Me)\}_2]$ (**39**).

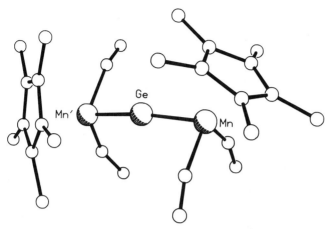

FIG. 15.   Structure of [Ge{Mn(CO)₂(C₅Me₅)}₂] (**40**).

linear, two-coordinate germanium but differing in adopting a gauche-type structure with respect to the Mn—Ge—Mn vector. The presumed multiple bonding in **39** and **40** is consistent with the short Ge—Mn bonds observed (**39**, Ge–Mn = 2.204(1) Å; **40**, Ge–Mn 2.18(2) Å), these being about 0.3 Å shorter than typical Ge—Mn single bond values (2.48–2.60 Å) (*50*). Moreover, theoretical studies on the bonding in these complexes have been reported by Fenske (*51*) and suggest that a possible analogy to allene is not accurate, because the Ge—Mn bonds contain partial triple-bond character. In this regard, the similarity to the isoelectronic chromium sulfur complex [S{Cr(CO)₂(C₅H₅)}₂] (*52*) is evident and these studies further suggest that rotation about the Ge—Mn bonds should be a low-energy process consistent with solution infrared evidence that **39** exists as a mixture of conformers (presumably the anti and gauche forms observed in the solid state structure of **39** and **40**, respectively). Brief mention of the synthesis of the C₅H₅ derivative [Ge{Mn(CO)₂(C₅H₅)}₂] (**41**) has been made by Herrmann (*53*), from the reaction between [Mn(THF)(CO)₂(C₅H₅)] and acidified Me₂Ge, although no structural details were reported.

The tin derivative [Sn{Mn(CO)₂(C₅Me₅)}₂] (**42**) has been prepared from the reaction between [Mn(THF)(CO)₂(C₅Me₅)] and SnH₄ (*54*), and the lead complexes [Pb{Mn(CO)₂(C₅H₄R)}₂] (**43**, R = H; **44**, R = Me) (*55,56*) have been obtained from the reaction between [Mn(THF)(CO)₂(C₅H₄R)] and PbCl₂. Complex **43** has been the subject of an X-ray structure determination that reveals a linear, two-coordinate lead atom and a gauche conformation analogous to **40**.

The second class of compound contains three manganese fragments and

45, E=Ge; 47, E=Sn; 48, E=Pb.

these are often obtained under the same reaction conditions employed in the synthesis of the previously described dimanganese compounds. The first to be reported and structurally characterized was $[Ge\{Mn(CO)_2(C_5H_4Me)\}\{Mn_2$-$(CO)_4(C_5H_4Me)_2\}]$ (**45**) (see Fig. 16) (*49*) in low yield from the reaction between $K[Mn(GeH_3)(CO)_2(C_5H_4Me)]$ and Hg^{2+}. The molecule contains a trigonal–planar germanium atom bridging a $Mn_2(CO)_4(C_5H_4Me)_2$ fragment and doubly bonded to a $Mn(CO)_2(C_5H_4Me)$ moiety. It is similar, and indeed formally isoelectronic, with the previously described tritungsten–germanium complex, **36**, with similar differences in Ge—Mn bond lengths (Table II), reflecting the degree of multiple Ge–Mn bonding involved.

Subsequent work by Herrmann *et al.* describes the C_5H_5 derivative of **45**, viz. **46**, from the reaction between $[Mn(THF)(CO)_2(C_5H_5)]$ and

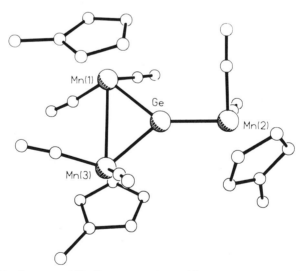

Fig. 16.   Structure of $[Ge\{Mn(CO)_2(C_5H_4Me)\}\{Mn_2(CO)_4(C_5H_4Me)_2\}]$ (**45**).

GeCl$_2$/dioxane/Zn, which proceeds via the intermediate chloro complex [Mn$_2$(CO)$_4$(C$_5$H$_5$)$_2$(μ-GeCl$_2$)] (*57*), together with the tin analogue of **45**, [Sn{Mn(CO)$_2$(C$_5$H$_4$Me)}{Mn$_2$(CO)$_4$(C$_5$H$_4$Me)$_2$}] (**47**) (*54*). The latter complex was characterized by X-ray crystallography and is similar in all respects to **45**. It is synthesized by a method similar to that for **45** using SnCl$_2$, but can also be obtained directly from [Mn(THF)(CO)$_2$(C$_5$H$_4$Me)] and SnH$_4$. This series of complexes is completed by the lead analogue [Pb{Mn(CO)$_2$(C$_5$H$_4$Me)}{Mn$_2$(CO)$_4$(C$_5$H$_4$Me)$_2$}] (**48**) (*56,58*), which was similarly prepared and characterized.

C. Complexes Involving Iron

Many complexes involving iron are known, mostly clusters containing iron carbonyl fragments, and reports for some come from the early years of metal cluster chemistry. Perhaps the simplest compound is {PbFe(CO)$_4$}$_n$, described originally in 1960 (*59*) and later in more detail by Marks in 1978 (*60*). The structure is unknown although it is postulated to be oligomeric on the basis of infrared spectroscopic data.

The tri-iron species [E{Fe(CO)$_4$}$_3$]$^{2-}$ (**49**, E = Sn; **50**, E = Pb) (*61*) (see Fig. 17) have been characterized by Whitmire and contain trigonal–planar

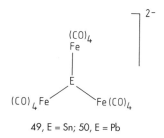

49, E = Sn; 50, E = Pb

tin and lead atoms bonded to three Fe(CO)$_4$ fragments. These molecules are isoelectronic and isostructural with the neutral tricobalt indium complex **25** described in Section II, and may, in principle, contain a degree of multiple bonding between the transition and main-group element centres. However, the X-ray-derived bond lengths for these molecules are not significantly shorter than single bond values, and EHMO calculations on the related trimolybdenum indium complex, **3**, suggest that π-bonding effects are small (*62*) in these systems. Moreover, the axial coordination of the tin or lead within the Fe(CO)$_4$ fragments (C_{3v} symmetry) are not those expected to maximize π-bonding.

Compounds containing four iron atoms were first described by Hieber (*63*) from the reaction between [NH$_4$]$_2$[SnCl$_6$] and [Fe(CO)$_3$(NO)]$^-$, which

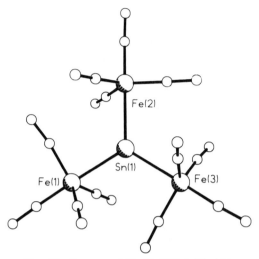

FIG. 17.   Structure of $[Sn\{Fe(CO)_4\}_3]^{2-}$ **(49)**.

affords $[Sn\{Fe(CO)_3(NO)\}_4]$, **51**, in high yield, and later work *(64)* reported details of the phosphine-substituted derivative $[Sn\{Fe(CO)_2(NO)(PPh_3)\}_4]$ **(52)**. No structural data were reported but the molecules are assumed to have a central tetrahedrally coordinated tin atom; this assumption is supported by infrared spectroscopy in work by Manning *(65)* describing the lead complex $[Pb\{Fe(CO)_2(NO)(P(OPh)_3)\}_4]$ **(53)**.

Another class of tetrairon complexes containing $Fe_2(CO)_8$ fragments has also been described. Stone *(66,67)* reported a low-yield synthesis of the spiro-cyclic tin complex $[Sn\{Fe_2(CO)_8\}_2]$, **54**, the structure of which was determined by X-ray crystallography *(68)*. As shown in Fig. 18, **54** contains a tetragonally distorted tetrahedral tin atom bonded by Sn—Fe bonds to two $Fe_2(CO)_8$ fragments. The original report by Stone *(67)* briefly mentioned the

54, E=Sn; 55, E=Ge; 56, E=Pb.

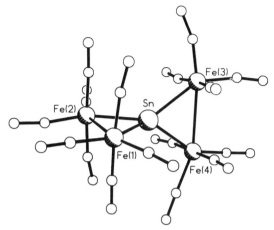

FIG. 18.   Structure of $[Sn\{Fe_2(CO)_8\}_2]$ (54).

germanium and lead analogues of **54**, $[E\{Fe_2(CO)_8\}_2]$, (**55**, E = Ge; **56**, E = Pb), but these were not fully characterized until many years later. The germanium complex, **55**, was independently prepared and structurally characterized by Batsanov (*69*) and Weiss (*70*). The former report described the synthesis of **55** from $[Fe_3(CO)_{12}]$ and $[Ge(CH=CH_2)_4]$ while in the latter, **55** was obtained from the reaction between the manganese germanium complex **39** and $[Fe_2(CO)_9]$. In addition, this latter synthesis also afforded the heteronuclear complex $[Ge\{Fe_2(CO)_8\}\{FeMn(CO)_6(C_5H_4Me)\}]$, **57**, the first example of a mixed transition metal germanium complex.

$$
\begin{array}{c}
\text{(structure 57)} \\
\mathbf{57}
\end{array}
$$

The lead analogue, **56**, was described by Whitmire (*71*) as resulting from the oxidation of $[Pb\{Fe_2(CO)_8\}\{Fe(CO)_4\}_2]^{2-}$ (see later) with $[Cu(CH_3CN)_4][BF_4]$, and was also structurally characterized. Both **55** and **56** have structures similar to **54**, with terminal carbonyls on iron in all cases.

As mentioned previously, **56** was prepared by oxidation of $[Pb\{Fe_2(CO)_8\}\{Fe(CO)_4\}_2]^{2-}$, **58**, itself prepared in high yield from lead(II) acetate or

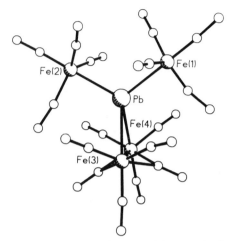

FIG. 19.   Structure of $[Pb\{Fe_2(CO)_8\}\{Fe(CO)_4\}_2]^{2-}$ **(58)**.

chloride and $[Fe(CO)_5]$ in methanolic KOH (*72*). The structure of **58** (see Fig. 19) is related to **56**, but as a consequence of the additional two electrons present in the former, one Fe—Fe bond found in **56** is absent in **58**. Also, the

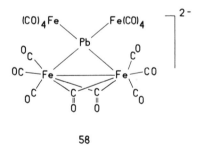

58

Fe—Fe bond present in **58** is bridged by two carbonyls, contrasting with the absence of bridging carbonyls in **56**. The tin analogue of **58**, $[Sn\{Fe_2(CO)_8\}-\{Fe(CO)_4\}_2]^{2-}$, **59**, has also been synthesized by a similar route (*71*), and oxidation by Cu(I) proceeds in a similar manner to the example for lead and provides another synthetic route to **54**. With germanium, however, the more condensed cluster species $[Ge\{Fe_3(CO)_{10}\}\{Fe(CO)_4\}]^{2-}$, **60**, is formed (*71*), although oxidation of this complex with Cu(I) affords **55**, analogous to the formation of **54** and **56** from **59** and **58**, respectively.

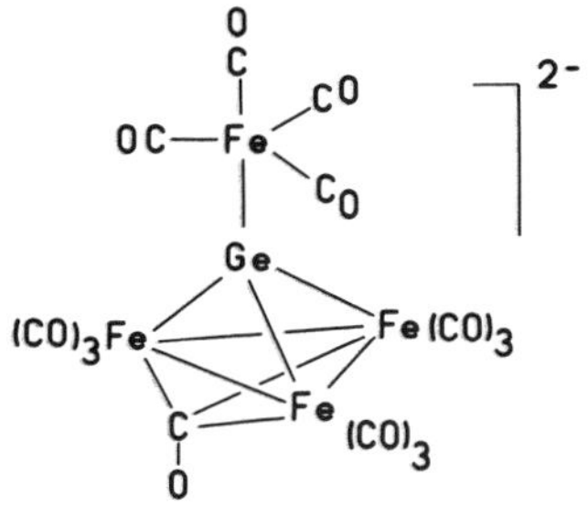

60

Another class of tetrairon tin compounds, described by Stone (67) and Fritchie (73), are derived from reactions between alkylchlorostannanes and iron carbonyl species. The latter report describes the structure of $[Sn\{Fe_2(CO)_8(\mu\text{-}SnMe_2)\}_2]$, 61, which contains a central tetrahedral tin atom bonded to two $Fe_2(CO)_8(\mu\text{-}SnMe_2)$ fragments.

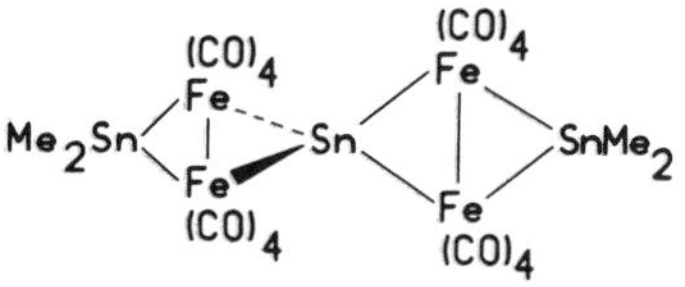

61

Finally, we note a number of higher nuclearity iron-containing clusters. Hieber (74) reported a complex formulated as $[Sn_2Fe_5(CO)_{20}]$, 62, although no structural details were presented. Hieber also described a complex formulated as $[PbFe_3(CO)_{12}]$, but later work by Whitmire (71) indicates that this compound is probably the tetrairon species, 56. Mackay and Nicholson (75) have described the synthesis and structures of three polynuclear species $[Fe_2(CO)_7\{\mu\text{-}E(Fe_2(CO)_8)\}_2]$ (63, E = Ge; 64, E = Sn) and $[Fe_3(CO)_{10}\{\mu\text{-}Ge(Fe_2(CO)_8)\}_2]$, 65, from reactions involving germanium or tin

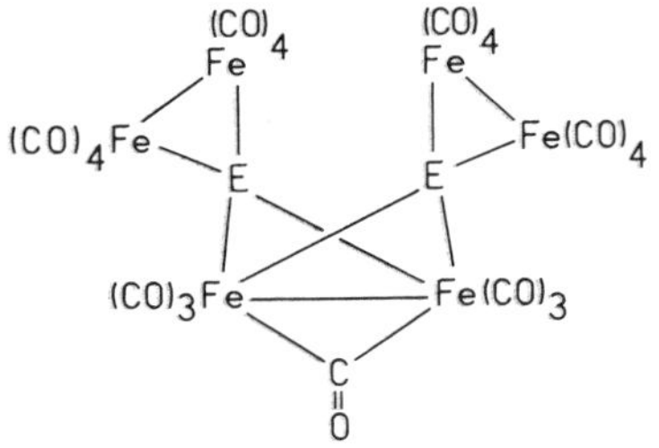

63, E=Ge; 64, E=Sn.

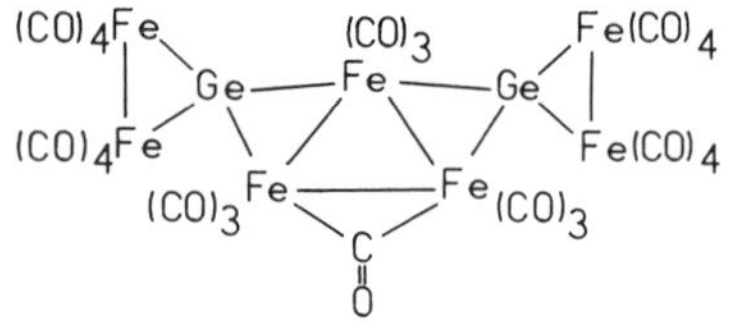

65

hydrides with $[Fe_2(CO)_9]$, together with some heteronuclear iron–cobalt complexes described later. In addition, the cluster, $[Fe_3(CO)_9\{\mu_3\text{-SnFe-}(CO)_2(C_5H_5)\}_2]$, **66**, has been reported by Wreford and Bau (*76*) from the

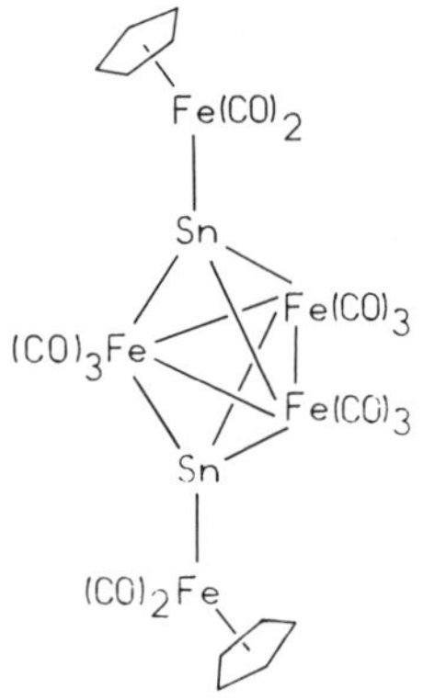

66

thermal decomposition of $[Fe_2(CO)_8\{\mu\text{-Sn}(\eta^1\text{-}C_5H_5)_2\}]$. The structure consists of a trigonal–bipyramidal Fe_3Sn_2 core with apical $Fe(CO)_2(C_5H_5)$ fragments as shown in Fig. 20.

D. Complexes Involving Cobalt

It is apparent from most of the examples previously described that the most common formal oxidation state found for the Group 14 element is E(IV) (E = Ge, Sn, Pb). Relatively few examples of divalent germanium, tin, or lead complexes have been described, and of these, many are not well characterized. Cobalt-containing compounds are no exception in this regard and there appears to be only one report in the literature that describes a species of this type, viz. $[Ge\{Co(CO)_4\}_2]$, **67**, although the precise structure of this complex is unknown (*77*). Two main synthetic routes are described, Eqs. (4) and (5), the starting complex in the latter reaction being

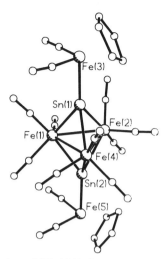

FIG. 20.   Structure of $[Fe_3(CO)_9\{\mu_3\text{-}SnFe(CO)_2(C_5H_5)\}_2]$ **(66)**.

initially derived from the reaction of **67** with 2,3-dimethylbutadiene. It is evident that **67** is readily oxidized to products containing germanium(IV)

$$GeF_2 + 2[Co(CO)_4(SiPh_3)] \longrightarrow 2Ph_3SiF + [Ge\{Co(CO)_4\}_2] \qquad (4)$$

$$\{(CO)_4Co\}_2Ge \overset{\Delta}{\longrightarrow} + [Ge\{Co(CO)_4\}_2] \qquad (5)$$

(77), and the propensity of germanium to form complexes in this oxidation state is illustrated by the larger number of complexes known for which this is true. Three distinct types of tetracobalt germanium compounds have been described, largely by Nicholson, all of which can be produced from reactions between GeI_4 and $[Co(CO)_4]^-$ or GeH_4 and $[Co_2(CO)_8]$. The fully open cluster species $[Ge\{Co(CO)_4\}_4]$ **(68)** has been mentioned (78,79), but no X-ray structure has been described and it appears to readily lose CO, affording more condensed clusters. Loss of two CO ligands affords $[Ge\{Co_2\text{-}(CO)_7\}_2]$, **69**, which is closely related to the tetrairon cluster **55** and contains

$$
\begin{array}{c}
Co(CO)_4 \\
| \\
(CO)_4Co - Ge - Co(CO)_4 \\
| \\
Co(CO)_4
\end{array}
$$

68

69

70

a distorted tetrahedral germanium at the center of a spirobicyclic $GeCo_4$ metal core (see Fig. 21). Further loss of another CO ligand affords the germylidyne cobalt cluster $[Co_3(CO)_9\{\mu_3\text{-}GeCo(CO)_4\}]$, **70**, (see Fig. 22), which has been described in detail by Schmid (*80,81*). This cluster contains a $GeCo_3$ tetrahedral core with an additional $Co(CO)_4$ group bonded to the germanium center and, as with **69**, is also related to a tetrairon cluster, namely the isoelectronic, dianionic species, **60**. It is also similar to the gallium–rhenium complex, **15**, although this contains two more electrons than **70**.

Several reports dealing with related tin and lead analogues exist that describe the open complexes $[E\{Co(CO)_4\}_4]$ (**71**, E = Sn; **72**, E = Pb). Bigorgne (*82*) outlines the synthesis of **71** from $[Co_2(CO)_8]$ and $SnCl_2$ and Graham describes a similar route but starting with $SnCl_4$ (*83*). Work by Schmid (*84*) reports the preparation of **71** from $[Co_2(CO)_8]$ and metallic tin, and of **72** by a similar method involving metallic lead. Various phosphine-substituted derivatives have also been described (*65,84*) but no X-ray struc-

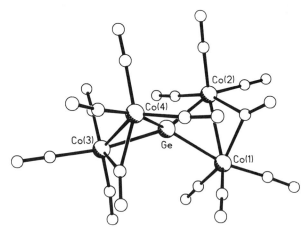

FIG. 21.   Structure of $[Ge\{Co_2(CO)_7\}_2]$ (**69**).

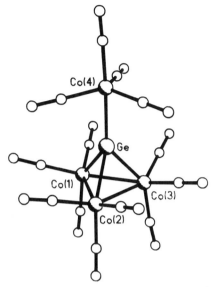

FIG. 22.   Structure of $[Co_3(CO)_9\{\mu_3\text{-GeCo(CO)}_4\}]$ (**70**).

tures have been reported for either **71, 72**, or any of the phosphine complexes. Complexes **68, 71**, and **72** are isoelectronic with the anionic cobalt indium complex, **27**.

It is interesting that no reports dealing with tin or lead analogues of **69** and **70** have appeared and it has been suggested (*85*) that a possible reason for this (particularly in the case of analogues of **70**) is the large size of the heavier main-group atoms, which imposes a severe steric strain in any closo-type cluster geometry. It is likely, in view of some bismuth chemistry (to be described later), that this view is no longer tenable, especially with regard to a possible lead derivative of **69** that would be very similar to the iron complex, **56**, reported by Whitmire (*71*).

Other examples of germanium–cobalt clusters include $[Ge\{Co_2(CO)_7\}\text{-}\{Co_2(CO)_6(\mu\text{-HgCo(CO)}_4)\}]^-$ (**73**) (*86*) and the higher nuclearity species $[Ge\{Co_2(CO)_7\}\{Co_3(CO)_9\}]^{2-}$ (**74**) (*79*), $[Ge_2Co_7(CO)_{21}]^-$ (**75**) (*87*), $[Ge\{Co_3(CO)_9\}\{Co_4(CO)_{11}\}]^-$ (**76**) (*87*), $[Co_4(CO)_{11}\{\mu_4\text{-GeCo(CO)}_4\}_2]$ (**77**) (*88*), and $[Ge_3Co_8(CO)_{26}]$ (**78**) (*89*). Cluster **78** can be viewed as an oligomer based on the structure type exhibited by **69**, and the possibility of synthesizing longer chains clearly exists. In general, these larger clusters are made from reactions between germanes, Ge_nH_{2n+2} or smaller CoGe complexes, and $[Co_2(CO)_8]$ or $[Co(CO)_4]^-$. There appear to be no corresponding examples incorporating tin or lead.

74

75

76

77

78

E. *Complexes Involving Nickel*

Examples of nickel complexes are rare and are confined to three reported nickel carbonyl clusters containing interstitial germanium or tin atoms (*90*). The reaction between $[Ni_6(CO)_{12}]^{2-}$ and $GeCl_4$ affords two species, $[Ni_{10}Ge(CO)_{20}]^{2-}$, **79**, and $[Ni_{12}Ge(CO)_{22}]^{2-}$, **80**. The former structure is based on a pentagonal antiprismatic framework of nickel atoms with an

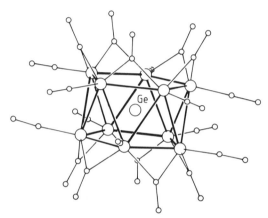

FIG. 23.   Structure of $[Ni_{10}Ge(CO)_{20}]^{2-}$ **(79)**.

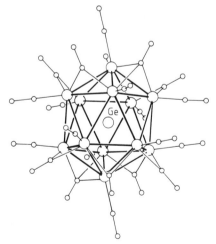

FIG. 24.   Structure of $[Ni_{12}Ge(CO)_{22}]^{2-}$ **(80)**.

interstitial germanium, whereas, in the latter, an icosahedral arrangement of nickel atoms is observed. A similar reaction involving $SnCl_2 \cdot 2H_2O$ affords the isostructural tin analogue of **80**, $[Ni_{12}Sn(CO)_{22}]^{2-}$ (**81**).

F. *Mixed Transition Metal Complexes*

There are a sufficiently large number of complexes, mainly of germanium, containing more than one type of transition metal, to warrant a separate discussion; some examples have already been mentioned (viz. **38** and **57**). Mackay and Nicholson (*89,91*) have described the reaction between $[Fe_2\text{-}(CO)_8(\mu\text{-}GeH_2)_2]$ and $[Co_2(CO)_8]$, which affords the mixed cobalt–iron clusters $[Fe_2(CO)_8\{\mu\text{-}Ge(Co_2(CO)_7)\}_2]$, **82**, and $[Fe_2(CO)_7\{\mu\text{-}Ge(Co_2\text{-}(CO)_7)\}_2]$, **83**, the latter having been characterized by X-ray diffraction. This is isoelectronic with the iron–germanium cluster, **63**, and both adopt very similar structures.

83

Two mixed iron–rhenium clusters have been reported by Haupt as resulting from high-temperature sealed-tube reactions involving $[MnRe(CO)_{10}]$, $[Fe(CO)_5]$, and either $SnCl_2$ or $Ge/GeCl_4$ (*92,93*). These are isostructural and comprise an E_2Fe_3 trigonal–bipyramidal core with apical $Re(CO)_5$ fragments $[Fe_3(CO)_9\{\mu_3\text{-}ERe(CO)_5\}_2]$ (**84**, E = Ge; **85**, E = Sn) (see Fig. 25). This structural form is also adopted by the closely related Sn_2Fe_5 cluster, **66**.

84, E=Ge; 85, E=Sn.

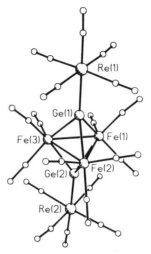

FIG. 25.   Structure of $[Fe_3(CO)_9\{\mu_3\text{-}GeRe(CO)_5\}_2]$ (84).

Finally, we note a number of clusters reported by Vahrenkamp (94–96), $[Co_3(CO)_9\{\mu_3\text{-}GeML_n\}]$ [86, $ML_n = Fe(CO)_2(C_5H_5)$; 87, $ML_n = Ni(CO)$-(C_5H_5); 88, $ML_n = M(CO)_3(C_5H_5)$, M = Cr, Mo, W; 89, $ML_n = Mn(CO)_5$]

and $[Co_2Mo(CO)_8(C_5H_5)\{\mu_3\text{-}GeW(CO)_3(C_5H_5)\}]$ (90). Clusters 86 and 90 were characterized by X-ray diffraction and are similar to the previously reported tetracobalt cluster, 70, described by Schmid.

G. Spectroscopic Studies

1. Mössbauer

[119]Sn Mössbauer spectroscopy has been used to a limited extent to characterize some of the tin complexes described herein. A table of isomer shifts is provided here, but it is clear that much more data are required before any trends might emerge.

2. NMR

Both tin and lead have suitable $I = \frac{1}{2}$ nuclei (^{119}Sn and ^{207}Pb), but there appear to be no chemical shift data available for either of these elements for any of the compounds described herein.

Compound	IS/mm sec^{-1}	QS/mm sec^{-1}	Ref.
31[a]	2.08(5)	2.05(10)	*42*
37	2.56(4)	4.60	*47*
42	2.65	4.60	*54*
47	2.43	3.96	*54*
81	2.5(1)	—	*90*

[a] The precise nature of this complex is ambiguous; see text.

IV

ANTIMONY AND BISMUTH

A wide variety of compounds exist incorporating antimony or bismuth atoms, and many of the bismuth examples have been reported in a recent review (*97*). There are some structural parallels with previously mentioned compounds involving elements of Groups 13 and 14, but many examples of quite different structural types are also found. In particular, there are several examples of complexes containing mutually bonded Sb_2 and Bi_2 fragments; the following sections will be further subdivided to take account of this factor. It is interesting to note the presence of main-group to main-group element bonds, because such compounds involving the Group 13 and 14 elements are conspicuous by their absence. Bond lengths are given in Table III.

A. Complexes Involving Chromium, Molybdenum, and Tungsten

1. Monoelement Complexes

Huttner has described the reaction between $[SbCl\{Cr(CO)_5\}_2]$ and $[Na]_2[M_2(CO)_{10}]$ (M = Cr, Mo, W) that affords the anionic complexes $[Sb\{Cr(CO)_5\}_2\{M(CO)_5\}]^-$ (**91**, M = Cr; **92**, M = Mo; **93**, M = W) (see Fig. 26) (*46*) containing a trigonal–planar antimony atom. These complexes are isoelectronic with the trimanganese–indium compound **6** (together with

TABLE III

TRANSITION METAL BOND LENGTHS TO ANTIMONY AND BISMUTH

Complex	Bond	$r(\text{M—E})$ (Å)	Ref.
[Sb{Cr(CO)$_5$}$_3$], **91**	Sb—Cr (av)	2.630(3)	46
[Sb$_2${μ-W(CO)$_5$}$_3$], **100**	Sb—W	2.991(3)– 3.040(3)	102
[Bi$_2${μ-W(CO)$_5$}$_3$], **101**	Bi—W	3.083(3)– 3.134(3)	103
[W$_2$(CO)$_8$(μ-η^2-Bi$_2$){μ-Bi(Me)W(CO)$_5$}], **102**	Bi—W	2.987(1)– 3.001(1)	105
[Mo$_2$(CO)$_4$(C$_5$H$_4$Me)$_2$(μ-η^2-Bi$_2$)], **103**	Bi—Mo	2.848(1) 2.985(1)	106
[W$_2$(CO)$_4$(C$_5$H$_4$Me)$_2$(μ-η^2-Bi$_2$)], **104**	Bi—W	2.852(2) 2.985(2)	107
[Bi{Mn(CO)$_5$}$_3$], **105**	Bi—Mn	2.884(1) 2.911(1) 2.916(1)	108
[Bi$_2${μ-Mn(CO)$_2$(C$_5$H$_4$Me)}$_3$], **107**	Bi—Mn	2.897– 2.950	110
[BiFe$_3$(CO)$_{10}$]$^-$, **108**	Bi—Fe	2.648(3) 2.650(2) 2.652(1)	111
[BiFe$_3$(CO)$_9$(μ-COMe)], **109**	Bi—Fe	2.644(1) 2.650(1) 2.657(1)	112
[Bi{Fe(CO)$_4$}$_4$]$^{3-}$, **110**	Bi—Fe	2.748(2)– 2.753(3)	113
[BiFe$_3$(CO)$_9$(μ-H)$_3$], **111**	Bi—Fe	2.638(1) 2.639(1) 2.641(1)	112
[BiRu$_3$(CO)$_9$(μ-H)$_3$], **112**	Bi—Ru	2.755(1) 2.756(1) 2.763(1)	114
[BiOs$_3$(CO)$_9$(μ-H)$_3$], **113**	Bi—Os	2.799(2) 2.800(2) 2.807(1)	114
[Fe$_3$(CO)$_9$(μ-CO){μ_3-SbFe(CO)$_4$}]$^-$, **114**	Sb—Fe	2.460(3)– 2.518(3)	115
[Fe$_3$(CO)$_9$(μ-H){μ_3-SbFe(CO)$_4$}]$^{2-}$, **115**[a]	Sb—Fe(1) Sb—Fe(2) Sb—Fe(3) Sb—Fe(4)	2.474(2) 2.460(2) 2.467(2) 2.521(2)	116
[Fe$_3$(CO)$_9$(μ-H)$_2${μ_3-SbFe(CO)$_4$}]$^-$, **116**[a]	Sb—Fe(1) Sb—Fe(2) Sb—Fe(3) Sb—Fe(4)	2.458(1) 2.509(1) 2.470(1) 2.507(1)	116

(*continued*)

TABLE III (*continued*)

Complex	Bond	r(M—E) (Å)	Ref.
$[Sb\{Fe(CO)_4\}_4]^{3-}$, **117**	Sb—Fe	2.663(1)– 2.670(1)	*116*
$[BiRu_5H(CO)_{18}]$, **118**	Bi—Ru	2.698(2)– 2.759(2)	*117*
$[Fe_2(CO)_8(\mu\text{-}Sb\{Fe(CO)_4\}\{Cr(CO)_5\})]^-$, **119**[b]	Sb—Fe	2.581(1) 2.649(1) 2.654(1)	*118*
	Sb—Cr	2.638(1)	
$[Fe_2(CO)_8(\mu\text{-}Bi\{Fe(CO)_4\}\{Cr(CO)_5\})]^-$, **120**[b]	Bi—Fe	2.651(2) 2.732(1) 2.742(2)	*118*
	Bi—Cr	2.718(2)	
$[Sb\{Fe(CO)_2(C_5H_5)\}_3]$, **121**	Sb—Fe	2.638(2)– 2.676(2)	*99*
$[Bi\{Fe(CO)_2(C_5H_4Me)\}_3]$, **122**	Bi—Fe	2.727(1) 2.737(1) 2.751(1)	*121*
$[BiFe_3(C_5H_4R)_3(\mu\text{-}CO)_3]$, **123** (R = Me)	Bi—Fe	2.582(1) 2.584(1)	*121*
$[BiFe_3(C_5H_4R)_3(\mu\text{-}CO)_3]$, **123** (R = H)	Bi—Fe	2.573(1) 2.574(1)	*122*
$[Bi_2Fe_2Co(CO)_{10}]^-$, **124**	Bi—Fe	2.669(5)– 2.693(5)	*123,124*
	Bi—Co	2.868(5) 2.894(5)	
$[Bi_2Os_4(CO)_{12}]$, **125**	Bi—Os	2.803(1) 2.923(1)	*114*
$[Bi_2Fe_3(CO)_9]$, **126**	Bi—Fe	2.617(2)– 2.643(2)	*125*
$[Bi_2Ru_3(CO)_9]$, **127**	Bi—Ru	2.725(2)– 2.781(2)	*126*
$[Bi_2Ru_4(CO)_{12}]$, **129**	Bi—Ru	2.833(1)– 2.839(1)	*126*
$[Bi_2Fe_4(CO)_{13}]^{2-}$, **132**	Bi—Fe	2.645(2)– 2.687(2)	*128*
$[Bi_4Fe_4(CO)_{13}]^{2-}$, **133**	Bi—Fe	2.699(6)– 2.753(6)	*130*
$[Fe_3(CO)_9\{\mu_3\text{-}SbFe(CO)_4\}_2]^{2-}$, **134**	Sb—Fe	2.504(1) 2.540(1)	*115*
$[Fe_2(CO)_6\{\mu\text{-}Sb(Fe_2(CO)_8)_2\}_2]$, **135**	Sb—Fe	2.494(1)– 2.567(1)	*131,132*
$[Bi\{Co(CO)_4\}_3]$, **136**	Bi—Co	2.767(2) 2.770(2) 2.760(2)	*133*

TABLE III (continued)

Complex	Bond	r(M—E) (Å)	Ref.
[BiIr$_3$(CO)$_9$], **138**	Bi—Ir	2.733(1)	*137*
		2.734(2)	
		2.736(2)	
[BiCo$_3$(CO)$_9$], **139**	Bi—Co	2.613(6)	*138*
[Bi{Co(CO)$_4$}$_4$]$^-$, **140**	Bi—Co	2.866(2)–	*140*
		2.969(2)	
		2.876(2)–	*141*
		2.939(2)	
[Sb{Co(CO)$_3$(PPh$_3$)}$_4$]$^+$, **142**	Sb—Co	2.593(3)–	*142*
		2.602(2)	
[SbRh$_{12}$(CO)$_{27}$]$^{3-}$, **143**	Sb—Rh	2.712(1)	*143*
		2.922(1)	
[Bi$_2$Co$_4$(CO)$_{11}$]$^-$, **144**	Bi—Co	2.704(1)–	*139*
		2.792(3)	
[Sb$_2$Co$_4$(CO)$_{11}$]$^-$, **145**	Sb—Co	2.605(2)–	*144*
		2.692(2)	
[Sb$_2$Co$_4$(CO)$_{11}$]$^{2-}$, **147**	Sb—Co	2.616(4)–	*144*
		2.676(4)	
[Sb$_4$Co$_4$(CO)$_{12}$], **150**	Sb—Co	2.612(3)	*145*
		2.616(3)	
[Bi$_4$Co$_4$(CO)$_{12}$], **151**	Bi—Co	2.742(2)	*146*
		2.751(2)	

[a] Fe(4) is the exotetrahedral iron atom.

[b] The shorter Sb–Fe and Bi–Fe distances are to the exocyclic iron atom.

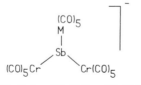

91, M=Cr; 92, M=Mo; 93, M=W.

the related rhenium and thallium derivatives) and, more generally, to all complexes of the type E(ML$_n$)$_3$, where E = Ga, In, Tl, and ML$_n$ represents a 17-electron transition metal fragment. An isoelectronic relationship is also apparent with the dianionic iron–tin and lead complexes **49** and **50**.

Trigonal–planar coordination is rare for antimony and the possibility of multiple bonding between antimony and chromium must be considered. However, as has already been pointed out, the presence of multiple bonding

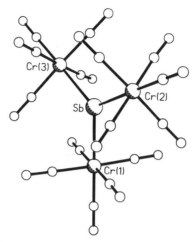

FIG. 26.   Structure of $[Sb\{Cr(CO)_5\}_3]^-$ (**91**).

in related indium and tin complexes is questionable and this is an area wherein detailed theoretical work on electronic structure would clearly be of value. We note also the preparation of the mixed metal complexes $[Sb\{Cr(CO)_5\}_2\{Mn(CO)_5\}]$ (**94**) and $[Sb\{Cr(CO)_5\}_2\{Mo(CO)_3(C_5H_5)\}]$ (**95**) from $[SbCl\{Cr(CO)_5\}_2]$ and $[Mn(CO)_5]^-$ and $[Mo(CO)_3(C_5H_5)]^-$, respectively (*46*).

One other class of compound is known, namely, the trimetallastibine species $[Sb\{M(CO)_3(C_5H_5)\}_3]$ (**96**, M = Mo; **97**, M = W). Compound **96** was originally reported by Malisch (*98*) and subsequently by Norman (*99*) along with the related tungsten complex. The bismuth analogues $[Bi\{M(CO)_3(C_5H_5)\}_3]$ (**98**, M = Mo; **99**, M = W) have also been described (*100*)

96, E=Sb, M=Mo; 97, E=Sb, M=W;
98, E=Bi, M=Mo; 99, E=Bi, M=W.

and all are assumed to contain a trigonal–pyramidal antimony or bismuth center. No single-crystal X-ray crystallographic results have yet been reported, but extended X-ray absorption fine structure (EXAFS) work supports this structural assignment (*101*).

2. Dielement Complexes

Sb$_2$ and Bi$_2$ are isoelectronic with N$_2$ and contain a formal triple bond between the elements. Although unstable in condensed phases (they are predominant species in the vapor phase of these elements), they can be stabilized by coordination to a transition metal center as demonstrated by Huttner with the synthesis of [E$_2${μ-W(CO)$_5$}$_3$] (**100**, E = Sb; **101**, E = Bi) from ECl$_3$ and

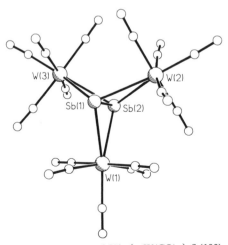

100, E=Sb; 101, E=Bi.

[Na]$_2$[W$_2$(CO)$_{10}$] (*102,103*) (see Fig. 27). Both molecules contain an E$_2$ unit coordinated to three W(CO)$_5$ fragments such that the metal core has approximate D_{3h} symmetry. The bonding in these compounds is interesting because it involves the E$_2$ unit as a formal six-electron donor in which donation of electrons occurs through the π and σ E—E bonding orbitals rather than through the lone pairs colinear with the E—E bond (as is usually found in N$_2$ complexes). This is something of an oversimplification and these molecules are also candidates for detailed theoretical studies, although we note that a description of the bonding in the analogous diarsenic compound has appeared (*104*). The Sb—Sb and Bi—Bi bond lengths are given in Table IV and are considerably shorter than single-bond values.

FIG. 27. Structure of [Sb$_2${μ-W(CO)$_5$}$_3$] (**100**).

TABLE IV

Sb—Sb AND Bi—Bi BOND LENGTHS IN COMPLEXES
CONTAINING Sb$_2$ AND Bi$_2$

Complex	Bond	$r(E—E)$ (Å)	Ref.
100	Sb—Sb	2.663	*102*
101	Bi—Bi	2.818(3)	*103*
102	Bi—Bi	2.796(1)	*105*
103	Bi—Bi	2.838(1)	*106*
104	Bi—Bi	2.845(3)	*107*
107	Bi—Bi	2.813	*110*
124	Bi—Bi	3.092(2)	*123,124*
125	Bi—Bi	3.017(2)	*114*
144	Bi—Bi	3.088(1)	*139*
145	Sb—Sb	2.911(1)	*144*
147	Sb—Sb	2.882(2)	*144*
133[a]	Bi—Bi		
	Bi(1)—Bi(2)	3.140(2)	*130*
	Bi(1)—Bi(3)	3.162(2)	
	Bi(1)—Bi(4)	3.168(2)	
	Bi(2)—Bi(3)	3.453(2)	
	Bi(2)—Bi(4)	3.473(2)	
	Bi(3)—Bi(4)	3.473(2)	

[a] See Fig. 36 for numbering scheme.

Dibismuth complexes are also known in which this unit acts as a four-electron donor. The complex $[W_2(CO)_8(\mu\text{-}\eta^2\text{-}Bi_2)\{\mu\text{-}Bi(Me)W(CO)_5\}]$ (**102**) has been described by Cowley (*105*) and $[Mo_2(CO)_4(C_5H_4Me)_2(\mu\text{-}\eta^2\text{-}Bi_2)]$- (**103**) by Norman (*106*). The latter complex (see Fig. 28) is obtained by either

102   103

photolysis or thermolysis of the C_5H_4Me derivative of **98** and is closely related to the analogous P_2 and As_2 species (*2*). The tungsten complex $[W_2\text{-}(CO)_4(C_5H_4Me)_2(\mu\text{-}\eta^2\text{-}Bi_2)]$ (**104**) can also be made by this route (*107*), and EHMO studies on these compounds are in progress (*62*).

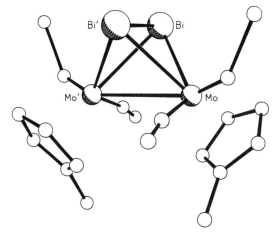

FIG. 28.   Structure of $[Mo_2(CO)_4(C_5H_4Me)_2(\mu\text{-}\eta^2\text{-}Bi_2)]$ **(103)**.

B. Complexes Involving Manganese and Rhenium

1. Monoelement Complexes

The trimetallabismuthine complex $[Bi\{Mn(CO)_5\}_3]$, **105**, has been prepared from the reaction between $BiCl_3$ and $3[Mn(CO)_5]^-$ and was structurally characterized by Wallis and Schmidbaur (*108*) (see Fig. 29). The bismuth

$$
\begin{array}{c}
\diagup\!\text{Bi}\diagdown \\
(CO)_5Mn \big| Mn(CO)_5 \\
Mn(CO)_5 \\
105
\end{array}
$$

adopts a trigonal–pyramidal geometry consistent with the presence of a stereochemically active lone pair. The rhenium analogue $[Bi\{Re(CO)_5\}_3]$, **106**, has also been prepared (*109*).

2. Dielement Complexes

The only example in this class is the manganese–bismuth complex, $[Bi\{\mu\text{-}Mn(CO)_2(C_5H_4Me)\}_3]$ **(107)** (see Fig. 30) prepared by Huttner from $BiCl_3$ and $[Mn_2(CO)_4(C_5H_4Me)_2(\mu\text{-}H)]^-$ (*110*). This is structurally and electronically very similar to the tritungsten complexes **100** and **101** and a similar bonding description is applicable.

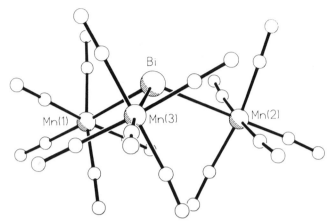

FIG. 29.   Structure of $[Bi\{Mn(CO)_5\}_3]$ **(105).**

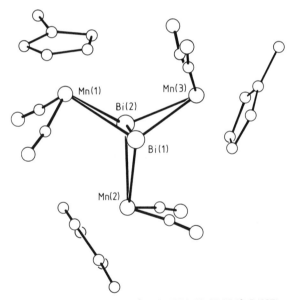

FIG. 30.   Structure of $[Bi_2\{\mu\text{-}Mn(CO)_2(C_5H_4Me)\}_3]$ **(107).**

107

C. Complexes Involving Iron, Ruthenium, and Osmium

1. Monoelement Complexes

A considerable range of iron carbonyl clusters is now known to contain antimony or bismuth, and later developments have seen this extended to ruthenium and osmium complexes. Whitmire (*111*) has described the reaction between $NaBiO_3$ and methanolic $[Fe(CO)_5]$, which affords the anionic triiron cluster $[BiFe_3(CO)_{10}]^-$ (**108**), which contains a tetrahedral $BiFe_3$ core. Cluster **108** is an example of a closo tetrahedral cluster with three first-row

108   109

transition elements and a large main-group element, which at one time were considered unlikely to exist for steric reasons (*85*), but, as subsequent examples will show, are now quite well known. Methylation of **108** can be effected with $CH_3SO_3CF_3$, which affords the neutral species $[BiFe_3(CO)_9(\mu\text{-COMe})]$ (**109**) (*112*).

In the report that described the synthesis of **108** (*111*), Whitmire also observed that the reaction between $NaBiO_3$ and methanolic $[Fe(CO)_5]$ in the presence of hydroxide afforded the trianionic species $[Bi\{Fe(CO)_4\}_4]^{3-}$, **110**, which was subsequently characterized by X-ray crystallography (*113*). Cluster **110** contains a central bismuth atom tetrahedrally coordinated by four $Fe(CO)_4$ fragments and is isoelectronic with the anionic tetracobalt–indium complex, **27**, and the neutral tetracobalt–germanium, –tin, and –lead complexes, **68, 71,** and **72**. Oxidation of **110** affords **108** (*111*), while acidification (*112*) yields the hydride-containing cluster $[BiFe_3(CO)_9(\mu\text{-H})_3]$, **111**, which also contains a tetrahedral $BiFe_3$ core. Johnson and Lewis (*114*) have

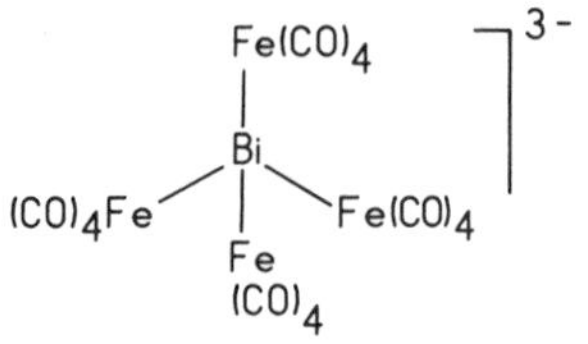

110

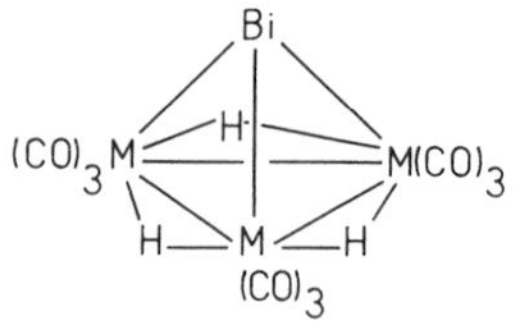

111, M=Fe; 112, M=Ru; 113, M=Os.

isolated the ruthenium analogue of **111** from the reaction between $NaBiO_3$ or $Bi(NO_3)_3$ and $[Ru_3(CO)_{12}]$ and the osmium derivative using $NaBiO_3$ and $[Os_3(CO)_{12}]$.

Many clusters containing four transition metal atoms have also been isolated, in addition to **110**, which has already been described. Whitmire has reported the reaction between $SbCl_3$ and $[Et_4N]_2[Fe_4(CO)_{13}]$, which affords the cluster $[Fe_3(CO)_9(\mu\text{-}CO)\{\mu_3\text{-}SbFe(CO)_4\}]^-$ (**114**) (*115*), which contains an iron triangle capped by a $SbFe(CO)_4$ fragment. Cluster **114** is isoelectronic with the dianionic iron–germanium cluster **60**, although the two

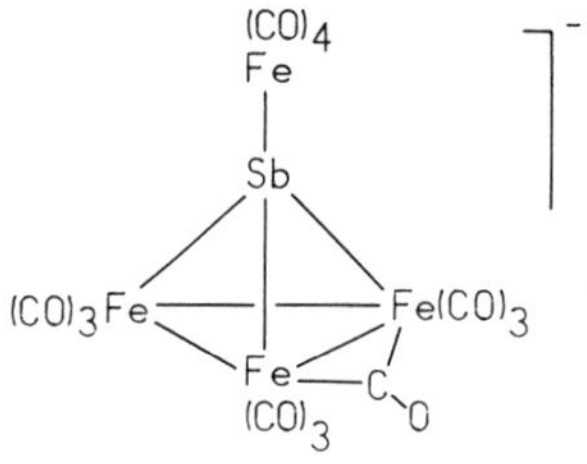

114

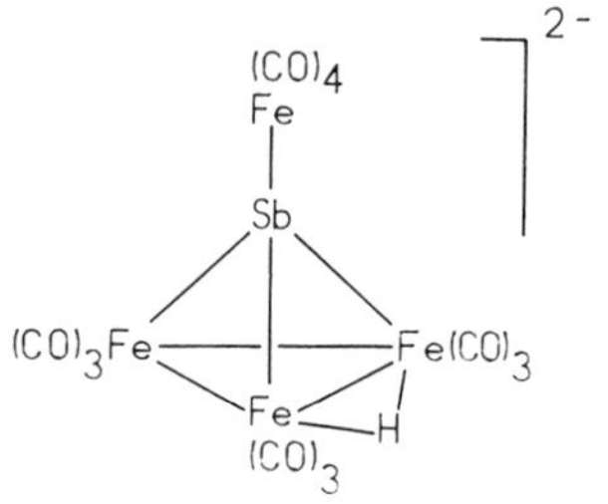

115

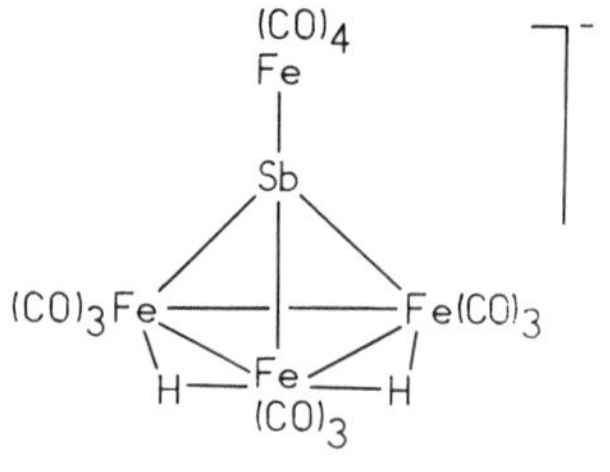

116

structures differ slightly in having a μ_2-carbonyl in **114** and a μ_3-carbonyl in **60**. A similar isoelectronic relationship with the tetracobalt–germanium cluster **70** is also apparent. The related clusters $[Fe_3(CO)_9(\mu\text{-}H)\{\mu_3\text{-}SbFe(CO)_4\}]^{2-}$ (**115**) and $[Fe_3(CO)_9(\mu\text{-}H)_2\{\mu_3\text{-}SbFe(CO)_4\}]^-$ (**116**) (*116*) have also been described as resulting from sequential protonation of $[Sb\{Fe(CO)_4\}_4]^{3-}$ (**117**). This latter complex is synthesized from $SbCl_3$ or $SbCl_5$ and $[Na]_2[Fe(CO)_4]$ or $[Fe(CO)_5]/KOH/MeOH$ and is analogous to the bismuth complex **110**.

A spirocyclic pentaruthenium–bismuth cluster has been described by Johnson and Lewis (*117*) from the reaction between $Bi(NO_3)_3 \cdot 5H_2O$ and $Na[Ru_3H(CO)_{11}]$, viz. $[BiRu_5H(CO)_{18}]$ (**118**), and the heteronuclear iron–chromium complexes $[Fe_2(CO)_8(\mu\text{-}E\{Fe(CO)_4\}\{Cr(CO)_5\})]^-$ (**119**, E = Sb;

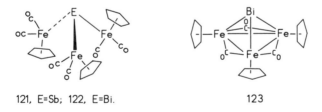

118

119, E=Sb;
120, E=Bi.

120, E = Bi) have been reported by Whitmire (*118*). These latter complexes were synthesized from the reaction between $[ECl\{Fe(CO)_4\}_3]^{2-}$ and $[Cr(CO)_5THF]$.

Also relevant to this discussion are a number of cyclopentadienyl iron–antimony and –bismuth complexes. Malisch has reported the synthesis of $[Sb\{Fe(CO)_2(C_5H_5)\}_3]$, **121**, from the reaction between SbF_3 and $Na[Fe(CO)_2(C_5H_5)]$ (*119*) and this complex has been structurally characterized by Norman (*99*). This is another example of a trimetallastibine analogous to **96** and **97** in which the antimony adopts a trigonal–pyramidal geometry. The bismuth complex $[Bi\{Fe(CO)_2(C_5H_5)\}_3]$, **122** (Fig. 31) has been described by Schmidbaur (*120*) and the methylcyclopentadienyl derivative, by Norman (*121*), and both groups have reported that **122** (C_5H_5 and C_5H_4Me) photolytically loses CO to give the *closo*-triiron–bismuth clusters $[BiFe_3(C_5H_4R)_3(\mu\text{-}CO)_3]$ (R = H, Me), **123** (*121,122*) (Fig. 32), which contain, as with other $BiFe_3$ clusters, an almost regular $BiFe_3$ tetrahedron.

121, E=Sb; 122, E=Bi.   123

2. Dielement Complexes

a. Mutually Bonded Elements. Two complexes of this type have been described. Whitmire has reported the reaction between $[Bi_2Fe_3(CO)_9]$ (see later) and $[Co(CO)_4]^-$, which affords the anionic complex $[Bi_2Fe_2Co(CO)_{10}]^-$, **124** (*123,124*) (see Fig. 33), containing a six-electron donor

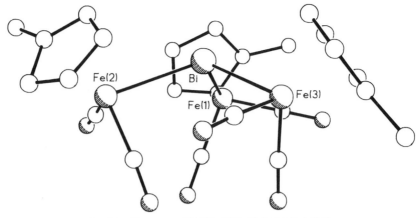

FIG. 31.   Structure of $[Bi\{Fe(CO)_2(C_5H_4Me)\}_3]$ **(122)**.

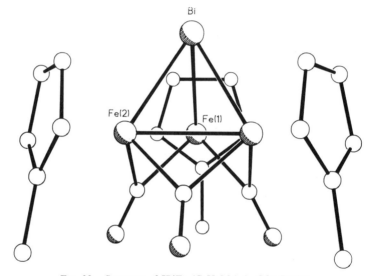

FIG. 32.   Structure of $[BiFe_3(C_5H_4Me)_3(\mu\text{-}CO)_3]$ **(123)**.

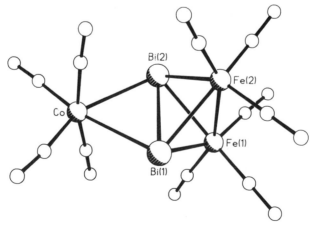

FIG. 33.   Structure of $[Bi_2Fe_2Co(CO)_{10}]^-$ **(124)**.

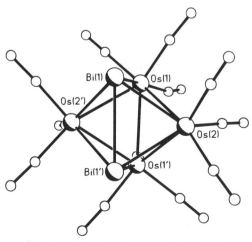

dibismuth ligand; Johnson and Lewis (*114*) have isolated the tetraosmium cluster [Bi$_2$Os$_4$(CO)$_{12}$], **125** (see Fig. 34), as one product from the reaction between NaBiO$_3$ and [Os$_3$(CO)$_{12}$].

b. Mutually Nonbonded Elements. One of the first clusters to be isolated in this class was [Bi$_2$Fe$_3$(CO)$_9$], **126** (*125*), as one product from the reaction between [HFe(CO)$_4$]$^-$ and NaBiO$_3$ followed by protonation with

126, M=Fe; 127, M=Ru; 128, M=Os.

H$_2$SO$_4$. This cluster has a trigonal–bipyramidal core geometry with the two bismuth atoms in apical positions; in other work, Johnson and Lewis have described the synthesis and structure of the ruthenium analogue, **127**, which

FIG. 34.   Structure of [Bi$_2$Os$_4$(CO)$_{12}$] **(125)**.

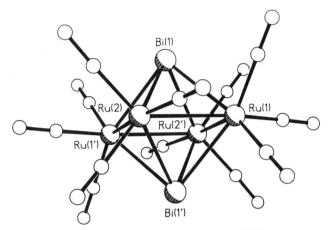

Fig. 35.   Structure of $[Bi_2Ru_4(CO)_{12}]$ (**129**).

is prepared from $Bi(NO_3)_3 \cdot 5H_2O$ and $[Ru_3H(CO)_{11}]^-$ (*114,126*), and the synthesis of the osmium derivative, **128**, as a third product from the reaction between $[Os_3(CO)_{12}]$ and $NaBiO_3$ (cf. the syntheses of **113** and **125**). These workers have also described the preparation and structure of $[Bi_2Ru_4\text{-}(CO)_{12}]$, (**129**) (*114,126*) (see Fig. 35) from $[Ru_3(CO)_{12}]$ and $Bi(NO_3)_3 \cdot 5H_2O$

$$\begin{array}{c} Bi \\ (CO)_3Ru{-}{-}{-}Ru(CO)_3 \\ (CO)_3Ru{-}{-}{-}Ru(CO)_3 \\ Bi \end{array}$$

129

in KOH/MeOH. The structure of **129** is particularly interesting when compared with the osmium analogue, **125**. Both comprise an octahedron containing two bismuth atoms and four ruthenium or osmium atoms. In the osmium complex (**125**), these are adjacent (1,2-isomer), while in **129** they are on opposite vertices (1,6-isomer). There is no evidence for mixtures of isomers in either system (*114*), and it has been suggested (*126*) that a possible explanation for the occurrence of different isomeric forms for different elements is thermodynamic in origin. Thus, in **125**, the number of Os–Os interactions is maximized in the 1,2-isomer (5 vs. 4), which factor is probably important in view of the increased strength of metal–metal bonds for the 5*d* transition metals. A similar situation has been observed in related ruthenium and osmium carbonyl clusters containing sulfur atoms (*127*).

Further examples of Bi_2 clusters in this class are the anionic and dianionic complexes $[Bi_2Fe_3(CO)_9]^{-,2-}$ (130 and 131), reported by Whitmire (*128*) as resulting from sequential one-electron reduction of 126 with sodium amalgam, and the dibismuth–tetrairon dianion $[Bi_2Fe_4(CO)_{13}]^{2-}$ (132) obtained from

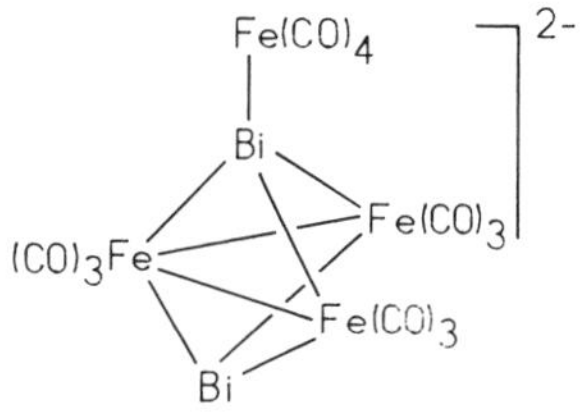

132

the reaction between 126 and $[Fe(CO)_4]^{2-}$ (*128*). The structure of 132 may be described as an open triangle of iron atoms triply bridged by one bismuth atom and one $BiFe(CO)_4$ fragment. Cluster 132 is also obtained from the reaction between 110 and $BiCl_3$, which also affords 108 and the novel tetrabismuth complex $[Bi_4Fe_4(CO)_{13}]^{2-}$ (133). This latter complex has been

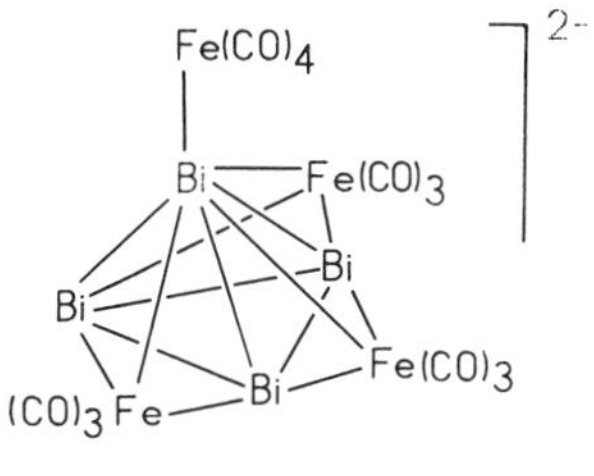

133

structurally characterized and studied by means of EHMO calculations (*129,130*). The structure comprises a tetrahedral Bi_4 core in which three faces are capped by $Fe(CO)_3$ fragments and one of the bismuth atoms is bonded to an exopolyhedral $Fe(CO)_4$ group, such that the overall metal core geometry has approximate C_{3v} symmetry. Cluster 133 (see Fig. 36) has been described as a Zintl metal carbonylate, of which there are relatively few examples. Moreover, the theoretical studies indicate a significant degree of Bi–Bi bonding within the tetrahedral Bi_4 unit. It will be clear from the foregoing discussion that a wide range of structure types exist for iron carbonyl–bismuth clusters. The analogous chemistry with ruthenium and osmium is less well developed, although with the exception of 118, 125, and 129, all

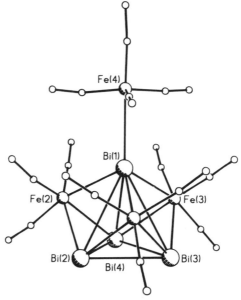

FIG. 36.   Structure of $[Bi_4Fe_4(CO)_{13}]^{2-}$ **(133)**.

the known compounds are analogous to iron–containing species. The synthetic interactions between these clusters are also complex and are most easily summarized as shown in Scheme 1, which is based on that given by Whitmire (*128*).

Only two diantimony–iron carbonyl clusters are known. Whitmire (*115*) has described the synthesis of $[Fe_3(CO)_9\{\mu_3\text{-}SbFe(CO)_4\}_2]^{2-}$ **(134)** from the reaction between $NaSbO_3$ and methanolic $[Fe(CO)_5]$. The structure is similar to that of **132** except that in **132** only one bismuth carries an $Fe(CO)_4$

134

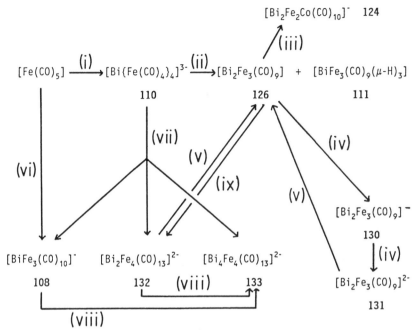

SCHEME 1. (i) KOH; NaBiO$_3$, (ii) H$^+$, (iii) [Co(CO)$_4$]$^-$, (iv) Na/Hg (v)Cu$^+$, (vi) NaBiO$_3$, (vii) BiCl$_3$, (viii) CO, (ix) [Fe(CO)$_4$]$^{2-}$.

group, whereas in **134** both antimony atoms are bonded to Fe(CO)$_4$ fragments.

Finally, we note that both Whitmire (*131*) and Cowley (*132*) have reported the synthesis and structure of [Fe$_2$(CO)$_6${μ-Sb(Fe$_2$(CO)$_8$)$_2$}$_2$] (**135**) from the reaction between SbCl$_3$ and [Fe$_2$(CO)$_8$]$^{2-}$. This compound is

$$
\begin{array}{c}
(CO)_4 \quad (CO)_4 \\
Fe \quad\quad Fe \\
\end{array}
$$

(CO)$_4$Fe——Sb   Sb——Fe(CO)$_4$

(CO)$_3$Fe——Fe(CO)$_3$

135

isoelectronic with and structurally similar to the iron–germanium and iron–tin clusters **63** and **64** and the mixed iron–cobalt–germanium complex **83**.

D. *Complexes Involving Cobalt, Rhodium, and Iridium*

1. *Monoelement Complexes*

The tricobalt–bismuthine complex [Bi{Co(CO)$_4$}$_3$], **136**, has been described by Schmid (*133*) and can be synthesised by two routes involving either the reaction between BiCl$_3$ and [Co(CO)$_4$]$^-$ or between bismuth metal and

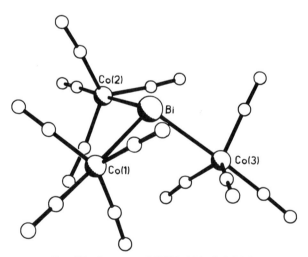

136

[Co$_2$(CO)$_8$]. The structure (*133,134*) (Fig. 37) reveals a trigonal–pyramidal bismuth atom coordinated to three Co(CO)$_4$ fragments and thus **136** may be described as a trimetallabismuthine, analogous to the previously mentioned complexes, **98, 99, 105, 106,** and **122**. The phosphine derivative [Bi{Co(CO)$_3$-(PPh$_3$)}$_3$], **137**, has also been described as resulting either from CO substitution by PPh$_3$ in **136** (*133*) or from the reaction between BiCl$_3$ and [Co(CO)$_3$(PPh$_3$)]$^-$ (*135,136*). Schmid has also reported spectroscopic data on AsPh$_3$ and SbPh$_3$ derivatives (*133*).

F$_{IG}$. 37. Structure of [Bi{Co(CO)$_4$}$_3$] (**136**).

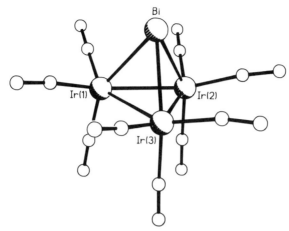

FIG. 38.   Structure of $[BiIr_3(CO)_9]$ (**138**).

There appear to be no reports of an analogous reaction involving [Rh-$(CO)_4]^-$, but Schmid has described the reaction between $BiCl_3$ and $Na[Ir-(CO)_4]$, which affords the tetrahedral cluster $[BiIr_3(CO)_9]$, **138** (*137*). Complex **138** (see Fig. 38) contains only terminal carbonyls, and an iridium

$$(CO)_3Ir \diagram Ir(CO)_3$$

138

complex analogous to **136** was not isolated. Moreover, initial attempts to generate the cobalt analogue of **138** were unsuccessful, which led to a suggestion (as previously discussed) that tetrahedral clusters involving three first-row transition elements and a large main-group element were too sterically strained to exist. Such cluster formation presents no difficulties with the larger iridium atom and the propensity of third-row transition elements to form metal–metal bonds is consistent with these observations.

As was described in the preceding section, however, a range of complexes is now known to contain a $BiFe_3$ tetrahedral core; subsequent work by Whitmire (*138*) and Martinengo (*139*) demonstrated that the cobalt analogue of **138** $[BiCo_3(CO)_9]$ (**139**) can be prepared by heating either **136** (*138*) or the anionic complex $[Bi\{Co(CO)_4\}_4]^-$ (**140**) (*139*) (see later). Complex **139** (see Fig. 39) is similar to **138**, but contains three bridging carbonyls and is therefore analogous to the isoelectronic iron complex **123**.

$$
\begin{array}{c}
\text{Bi} \\
\text{(CO)}_2\text{Co} = \overset{\overset{\text{O}}{\underset{\|}{\text{C}}}}{=} \text{Co(CO)}_2 \\
\underset{\text{O}}{\text{C}} - \text{Co} - \underset{\text{O}}{\text{C}} \\
\text{(CO)}_2
\end{array}
$$

139

The anionic tetracobalt–bismuth compound $[Bi\{Co(CO)_4\}_4]^-$ (**140**) can be prepared from the reaction between $BiCl_3$ and $4[Co(CO)_4]^-$ (*139*) or

$$
\begin{array}{c}
\text{Co(CO)}_4 \\
| \\
\text{Bi} \\
\text{(CO)}_4\text{Co} \diagdown \diagdown \text{Co(CO)}_4 \\
\text{Co(CO)}_4
\end{array} \Bigg]^-
$$

140

from the reaction between **136** and $[Co(CO)_4]^-$ (*140,141*); its structure has been established as the $[NMe_4]^+$ salt (*140*) or the $[Co(C_5H_5)_2]^+$ salt (*141*). Compound **140** is structurally similar to the iron complex **110** but contains two more valence electrons around the bismuth center, i.e., 10 rather than eight. The presence of a lone pair on the bismuth does not markedly distort

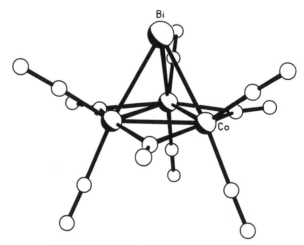

Fig. 39.   Structure of $[BiCo_3(CO)_9]$ (**139**).

the tetrahedral geometry and may, therefore, be described as stereochemically inactive, but the observed Bi—Co bond lengths (Table III) are long and this has been assumed to result from the presence of the extra pair of electrons. The symmetric structure of this species may also be the reason for its observed paramagnetism reported by Whitmire (*141*).

Corresponding eight-electron cationic species are not known for bismuth, but a range of antimony complexes (metallastibonium ions) has been described by Cullen (*135*), viz. $[Sb\{Fe(CO)_2(NO)(PPh_3)\}_4]^+$ (**141**) and $[Sb\{Co-(CO)_3(PPh_3)\}_4]^+$ (**142**), the latter having been structurally characterized as its BPh_4^- salt (*142*). These are isoelectronic with the iron complex **110** and related species.

$$\begin{array}{c} \text{(CO)}_3\text{PPh}_3 \\ \text{Co} \\ | \\ \text{Sb} \\ \text{PPh}_3\text{(CO)}_3\text{Co} \diagdown \quad \diagup \text{Co(CO)}_3\text{PPh}_3 \\ \text{Co} \\ \text{(CO)}_3\text{PPh}_3 \end{array} \Bigg]^+$$

142

Few complexes are known containing interstitial main-group atoms that also fall within the scope of this review (see **79–81**), but one example containing antimony has been reported by Vidal (*143*), in which this element resides in the center of an icosahedron of rhodium atoms, $[SbRh_{12}(CO)_{27}]^{3-}$ (**143**) (Fig. 40).

2. Dielement Complexes

One class of compound is known that falls within this category. Martinengo has reported that prolonged heating of solutions of **140** affords the paramagnetic anionic cluster $[Bi_2Co_4(CO)_{11}]^-$, **144** (*139*), which contains a Bi_2 unit bonded to four cobalt atoms as shown in Fig. 41. Whitmire has described the

$$\begin{array}{c} \text{(CO)}_3\text{Co} \diagdown \quad \diagup \text{Co(CO)}_3 \\ \text{E} \\ \text{(CO)}_2\text{Co} \quad \text{Co(CO)}_2 \\ \text{C} \\ \| \\ \text{O} \end{array} \Bigg]^-$$

144, E=Bi; 145, E=Sb.

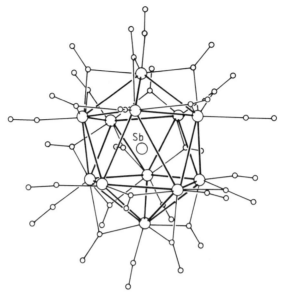

FIG. 40.   Structure of $[SbRh_{12}(CO)_{27}]^{3-}$ **(143)**.

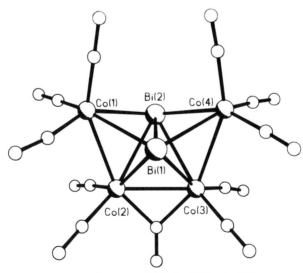

FIG. 41.   Structure of $[Bi_2Co_4(CO)_{11}]^-$ **(144)**.

preparation of the antimony analogue $[Sb_2Co_4(CO)_{11}]^-$, **145** (*144*) from the reaction between $SbCl_3$ and $3[Co(CO)_4]^-$, which appears to proceed via the unstable metallastibine complex $[Sb\{Co(CO)_4\}_3]$, **146** (it is interesting that **146** appears to be unstable in solution in view of the stability of the bismuth complex **136**). Compound **145** is isostructural with **144** and both of these complexes have the same structure as the diamagnetic dianion $[Sb_2Co_4$-$(CO)_{11}]^{2-}$, **147**, which is also formed in the above reaction. The dianion **147** can also be synthesized by reduction of **145**, either electrochemically or with sodium naphthalenide, and, in reverse, **145** can be prepared from **147** electrochemically or by oxidation with Cu^+.

Preliminary results from our own laboratory indicate that the reaction between $SbCl_3$ and $3[Co(CO)_3(PPh_3)]^-$ also proceeds via an unstable metallastibine intermediate, $[Sb\{Co(CO)_3(PPh_3)_3]$, which subsequently rearranges to the neutral dicobalt–diantimony complex $[Sb_2Co_2(CO)_4$-$(PPh_3)_2]$, **148** (*136*). There is also evidence for the bismuth analogue $[Bi_2Co_2$-$(CO)_4(PPh_3)_2]$, **149** (*136*), although neither have yet been structurally characterized.

3. Tetraelement Complexes

One type of this class of compound exists, namely, the cubic cluster species $[E_4Co_4(CO)_{12}]$ (**150**, E = Sb; **151**, E = Bi) (see Fig. 42). The antimony derivative was reported by Dahl (*145*) as resulting from a high-temperature and

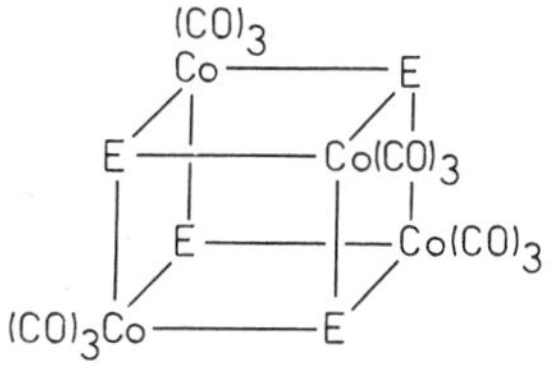

150, E=Sb; 151, E=Bi.

pressure reaction between $Co(OAc)_2\cdot 4H_2O$ and $SbCl_3$ in methanol under CO and H_2, and the bismuth complex was recently described by Martinengo (*146*) as resulting from prolonged pyrolysis of **136**. Both complexes contain cobalt and antimony or bismuth atoms at the vertices of a cube and are formal dimers of the as-yet unknown complexes $[E_2Co_2(CO)_6]$. As previously stated, however, there is evidence for the phosphine derivatives **148** and **149**, and this area of chemistry together with related work involving phosphorus and arsenic is discussed in Ref. *136*.

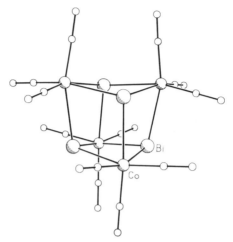

FIG. 42.   Structure of $[Bi_4Co_4(CO)_{12}]$ (**151**).

V

TELLURIUM

A. *Complexes Involving Zirconium and Vanadium*

The scarcity of compounds involving the heavier main-group elements and the early transition metals, apparent in previous discussions, is also apparent in the chemistry of tellurium. Thiele (*147*) has described the preparation of the zirconium complex $[\{Zr(C_5Me_5)_2(\mu\text{-Te})\}_2]$, **152**, from a divinyl complex

152

$[Zr(CH{=}CH_2)_2(C_5Me_5)_2]$ and elemental tellurium. The complex is assumed to contain tellurium atoms in a bent, two-coordinate geometry, which is common for this element in its divalent state.

A vanadium complex, $[V_2(CO)_6(dppe)_2(\mu\text{-Te})]$ (**153**), has been described by Weiss (*148*) resulting from the reaction between $Na[V(CO)_4(dppe)]$ and

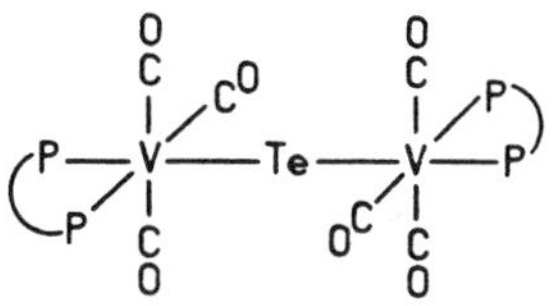

153

acidified Na_2TeO_3, which is assumed to contain a linear, two-coordinate tellurium atom by analogy with the crystallographically characterized sulfur analogue.

Each 15-electron $V(CO)_4(dppe)$ fragment requires three electrons to satisfy the effective atomic number rule, which clearly requires a degree of multiple bonding between the tellurium and vanadium centers. Two alternative descriptions, **A** and **B**, were proposed as representations for the bonding.

$$V\equiv Te\equiv V \qquad V\rightleftharpoons Te \Rightarrow V$$

A   **B**

B. Complexes Involving Chromium, Molybdenum, and Tungsten

Herrmann and Ziegler (*149*) have reported the synthesis of the monotellurium complexes $[M_2(CO)_6(C_5R_5)_2(\mu\text{-}Te)]$ (**154**, M = Cr, R = H; **155**, M = Cr, R = Me; **156**, M = Mo, R = H; **157**, M = W, R = H; **158**, M = W, R = Me.) from the reactions between $[M(CO)_3(C_5R_5)]^-$ and K_2TeO_3. The

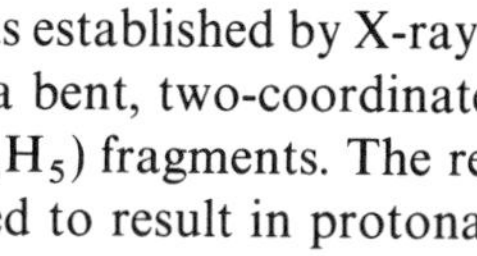

154

structure of **154** (Fig. 43) was established by X-ray crystallography and shows a single tellurium atom in a bent, two-coordinate geometry ($\angle Te = 117.2°$) bonded to two $Cr(CO)_3(C_5H_5)$ fragments. The reactivity of **154** toward H^+ and Me^+ has been described to result in protonation or methylation of the tellurium center, respectively (*150*), but attempts at decarbonylation were unsuccessful and did not afford complexes of the form $[(C_5R_5)(CO)_2\text{-}M\equiv Te\equiv M(CO)_2(C_5R_5)]$, analogous to **153**, for any of the derivatives (**154–158**) studied. A possible route to this type of complex involving the

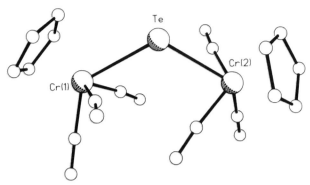

FIG. 43.   Structure of $[Cr_2(CO)_6(C_5H_5)_2(\mu\text{-}Te)]$ **(154)**.

reaction between tellurium metal and $[W_2(CO)_4(C_5Me_5)_2]$ has been examined by Ziegler *(151)*, but this affords the ditellurium–ditungsten complex $[W_2(CO)_4(C_5Me_5)_2(\mu\text{-}Te)_2]$ **(159)**, protonation of which affords the Te_2H_2 complex $[W_2(CO)_4(C_5Me_5)_2(\mu\text{-}Te_2H_2)]^{2+}$.

159

A tritungsten–ditellurium complex $[Te_2\{W(CO)_5\}_3]$, **160**, has been reported by Huttner *(47)* as resulting from the reaction between the tungsten–tin complex, **37**, and H_2Te. This complex contains a Te_2 moiety acting as a

160

six-electron donor to three $W(CO)_5$ fragments and is structurally similar to many six-electron donor diphosphene and distibene complexes *(152)*.

A comparison of **160** and **100** is also instructive. The former contains two more electrons than the latter (Te_2 vs. Sb_2) and this results in quite different structures being adopted, with the tellurium structure more "open" than the antimony structure (i.e., less M—E bonds).

Hybrid metal carbonyl Zintl-type complexes are still rare (one example, **133**, having already been mentioned), but three known examples are relevant to this discussion. Kolis has reported the reaction between $[Te_4]^{2+}$ and $[W(CO)_6]$, which affords the $[Te_3]^{2+}$ complex $[W(CO)_4(Te_3)]^{2+}$, **161**

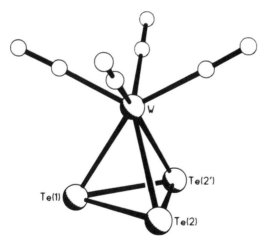

161   162, M=Cr; 163 M=W.

(Fig. 44) (*153*), and also the reaction between $[M(CO)_6]$ and $[Te_4]^{2-}$, which yields the $[Te_4]^{2-}$ complexes $[M(CO)_4(Te_4)]^{2-}$ (Fig. 45) (**162**, M = Cr; **163**, M = W) (*154*).

FIG. 44.   Structure of $[W(CO)_4(Te_3)]^{2+}$ (**161**).

C. Complexes Involving Manganese and Rhenium

Tellurium exhibits a wide range of coordination and bonding modes to manganese- and rhenium-containing fragments. One of the simplest is in the

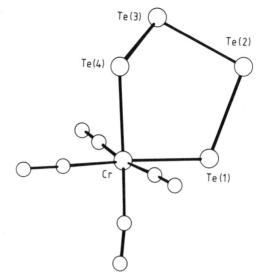

FIG. 45.   Structure of $[Cr(CO)_4(Te_4)]^{2-}$ **(162)**.

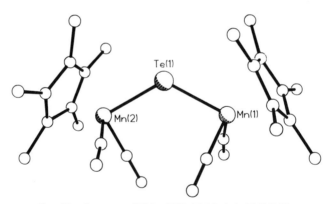

FIG. 46.   Structure of $[Mn_2(CO)_4(C_5Me_5)_2(\mu\text{-Te})]$ **(164)**.

complex $[Mn_2(CO)_4(C_5Me_5)_2(\mu\text{-Te})]$, **164** (Fig. 46), described by Herrmann (*155*), which results from the reaction between $[Mn(CO)_2(THF)(C_5Me_5)]$ and acidified Al_2Te_3. This complex contains a bent, two-coordinate tellurium atom $[\angle Te = 123.8(1)°]$ similar to that found in **154**, but electron counting requires the tellurium to act as a four-electron donor and it is therefore formally doubly bonded to each manganese center. Complex **164** is similar to

164

the tin complex, **42**, but contains two more electrons. These are apparently localized on the tellurium center, resulting in a bent geometry as opposed to the linear coordination observed for tin in **42**. A similar reaction involving [Mn(CO)$_2$(THF)(C$_5$H$_5$)] has been reported by Herberhold (*156*) to result in the trimanganese–tellurium complex [Te{Mn(CO)$_2$(C$_5$H$_5$)}$_3$], **165**, containing a trigonal–planar tellurium atom acting as a six-electron donor to three

165

manganese centers. The increased coordination around the tellurium is probably a result of the smaller size of the manganese fragment compared with that in **164**. Both compounds, however, contain formal Te=Mn double bonds, which is evident in the short bond lengths observed (Table VI) (*156a*). Complex **165** is related to the tin complex, **47**, which contains two electrons less. This factor is reflected in a structural difference such that in **47**, a single Mn—Mn bond is present, whereas in **165**, there are no Mn—Mn bonding interactions. An interesting relationship between **165** and the triiron–indium complex, **23**, is also apparent, because the two complexes are isoelectronic. Formal electron counting procedures require Mn=Te double and Fe—In single bonds, but the exact nature of the bonding is probably less clear-cut. Theoretical studies of these molecules would be of value, although simple electronegativity arguments would be consistent with a greater degree of π bonding in the tellurium example.

Herrmann has explored the corresponding chemistry with rhenium, which is rather more diverse (*157*). Treatment of [Re(CO)$_2$(THF)(C$_5$Me$_5$)] (**166**) with TeH$_2$ affords four products, as shown in Scheme 2. The monotellurium complex **170** is similar to the manganese complex **164** but differs in having a closed Re$_2$Te triangle with formal Re—Te single bonds and a Re—Re bond.

TABLE V

TRANSITION METAL BOND LENGTHS TO TELLURIUM

Complex	Bond	r(M—E) (Å)	Ref.
$[Cr_2(CO)_6(C_5H_5)_2(\mu\text{-Te})]$, **154**	Te—Cr	2.814(1)	*149*
		2.799(1)	
$[Te_2\{W(CO)_5\}_3]$, **160**	Te—W	2.881(3)	*47*
		2.739(2)	
$[W(CO)_4(Te_3)]^{2+}$, **161**	Te—W	2.802(1)	*153*
		2.802(1)	
		2.817(1)	
$[Cr(CO)_4(Te_4)]^{2-}$, **162**	Te—Cr	2.753(2)	*154*
		2.699(2)	
$[W(CO)_4(Te_4)]^{2-}$, **163**	Te—W	2.856(1)	*154*
		2.819(1)	
$[Mn_2(CO)_4(C_5Me_5)_2(\mu\text{-Te})]$, **164**	Te—Mn	2.459(2)	*155*
$[Te\{Mn(CO)_2(C_5H_5)\}_3]$, **165**	Te—Mn	2.469(2)	*156*
		2.474(2)	
		2.512(2)	
$[Re_2(CO)_4(C_5Me_5)_2(\mu\text{-Te})]$, **170**	Te—Re	2.679(1)	*157*
$[Te_2\{Re(CO)_2(C_5Me_5)\}_2]$, **168**	Te—Re	2.632(1)	*157*
		2.806(1)	
$[Te_2\{Re(CO)_2(C_5Me_5)\}\{Mn(CO)_2(C_5H_5)\}_2]$, **171**	Te—Re	2.807(1)	*157*
		2.827(1)	
	Te—Mn	2.520(1)	
		2.530(1)	
$[Mn_2(CO)_6(PEt_3)_4(\mu\text{-Te}_2)]$, **172**	Te—Mn	2.717(1)	*161*
$[Te\{Re(CO)_5\}_3]^+$, **173**	Te—Re (av)	2.842(1)	*162*
$[Te_2Fe_3(CO)_9]$, **177**	Te—Fe	2.524(2)–	*166*
		2.550(2)	
$[Te_2Fe_3(CO)_9(PPh_3)]$, **180**	Te—Fe	2.565(1)–	*169*
		2.668(1)	
$[TeOs_3(CO)_9(\mu\text{-H})_2]$, **181**	Te—Os	2.654(5)	*171*
		2.674(5)	
$[Te_2Co_4(CO)_{10}]$, **184**	Te—Co	2.54(1)	*175*
$[Co_2(PMe_3)_6(\mu\text{-Te})_2]$, **187**	Te—Co	2.528(2)	*178*
		2.584(2)	
$[Ni\{PhP(CH_2CH_2PPh_2)_2\}(Te_2)]$, **190**	Te—Ni	2.555(2)	*179*
		2.557(2)	
$[Ni_2\{HC(CH_2PPh_2)_3\}_2(\mu\text{-Te}_2)]$, **191**	Te—Ni	2.576(1)	*180*
		2.596(1)	
$[Ni_2(C_5H_5)_2\{\mu\text{-TeNi}(PPh_3)(C_5H_5)\}_2]$, **193**	Te—Ni	2.457(1)	*181*
		2.519(1)	
$[Pt_2(PPh_3)_4(\mu\text{-Te})_2]$, **194**	Te—Pt	2.612(1)	*182*
		2.628(1)	
$[Te_2FeMo_2(CO)_7(C_5H_5)_2]$, **195**[a]	Te—Fe	2.572(1)	*183*
		2.588(1)	
	Te—Mo	2.772(1)–	
		2.832(1)	

TABLE V (*continued*)

Complex	Bond	r(M—E) (Å)	Ref.
[Te₂Fe₂Mo₂(CO)₇(C₅H₅)₂], **198**	Te—Fe	2.472(2)	*183*
		2.468(2)	
	Te—Mo	2.673(1)–	
		2.684(1)	
[Te₂FeMo₂(CO)₃(C₂H₂)(C₅H₅)₂], **199**	Te—Fe	2.562(2)	*184*
		2.567(2)	
	Te—Mo	2.649(1)–	
		2.677(1)	
[TeFe₂Mn(CO)₈(C₅Me₅)], **201**	Te—Fe	2.472(1)	*186*
		2.476(1)	
	Te—Mn	2.518(1)	
[TeFeCo₂(CO)₉], **204**[b]	Te—Fe/Co	2.466	*176*
[Te₃Co₂Mo₂(CO)₄(C₅H₅)₂], **216**	Te—Co	2.451(2)–	*156a*
		2.493(1)	
	Te—Mo	2.649(1)–	
		2.710(1)	

[a] Four molecules per asymmetric unit.
[b] Iron and cobalt centers disordered.

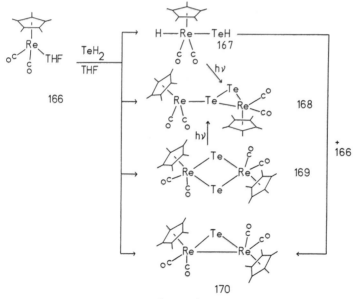

SCHEME 2

In **164**, no Mn—Mn bond is present and formal double bonding exists between the Mn and Te atoms. The propensity of the $5d$ transition metals to form M—M bonds may be cited as a reason for these differences, and the situation is similar to the valence isomerism observed in phosphinidene (RP) and sulfenium (SR$^+$) complexes (and their respective heavier congeners) (*158–160*), in which two types of structures termed "open" (**C**) and "closed" (**D**) are observed (ML_n = 16-electron transition metal fragment).

The structure of **169**, although not crystallographically characterized, is assumed to be as drawn in Scheme 2 with bent, two-coordinate tellurium atoms, but photolysis or thermolysis converts this complex into the isomeric species **168**, which contains a Te$_2$ ligand acting as a four-electron donor. If **168** is assumed to contain a doubly bonded ditellurium moiety, Te=Te, electron donation occurs from the π bond and one Te lone pair, as illustrated in **E**.

As with the structure of **160**, which contains Te$_2$ as a six-electron donor, comparisons can be made with the coordination chemistry of diphosphenes (RP = PR) (*152*). Scheme 2 illustrates that **170** can also be formed from **167** and **166**, but a similar reaction between **167** and [Mn(CO)$_2$(THF)(C$_5$H$_5$)] affords the cluster complex **171**, which contains another example of a Te$_2$ ligand acting as a six-electron donor. Complex **171** is thus very similar to **160**.

171

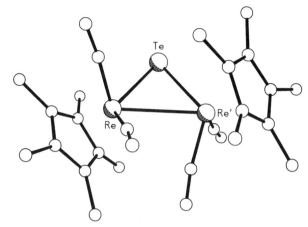

FIG. 47. Structure of [Re$_2$(CO)$_4$(C$_5$Me$_5$)$_2$(μ-Te)] (170).

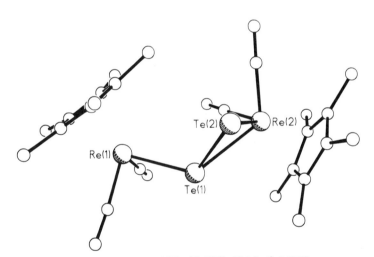

FIG. 48. Structure of [Te$_2${Re(CO)$_2$(C$_5$Me$_5$)}$_2$] (168).

Another example of a complex containing a Te$_2$ ligand has been reported by Steigerwald (161), viz. [Mn$_2$(CO)$_6$(PEt$_3$)$_4$(μ-Te$_2$)] (172). However, whereas the previous examples (160, 168, and 171) contain Te$_2$ as a four- or six-electron donor based formally on Te=Te, 172 is best described as a dimetalla–ditelluride with each tellurium atom contributing one electron to its adjacent manganese center. The absence of any formal Te—Te multiple bonding is reflected in the Te—Te bond length (Table VI). The synthesis of

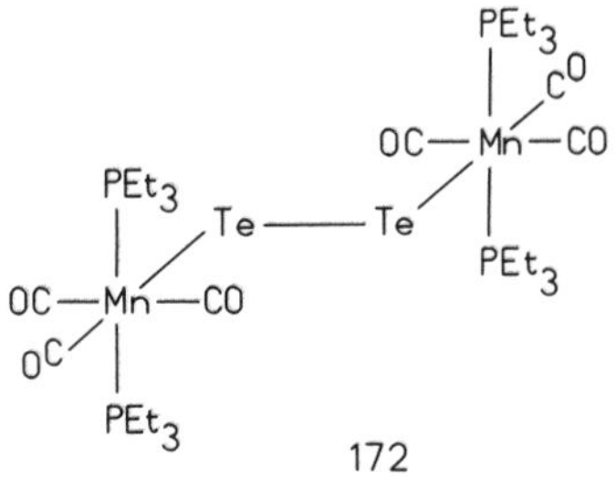

172

172 involves the reaction between $[Mn_2(CO)_{10}]$ and the phosphine telluride, Et_3PTe, which is a synthetically useful source of tellurium atoms. Moreover, **172** is noteworthy in being a low-temperature precursor to the magnetic semiconductor material, MnTe, uncontaminated by Mn, Te, or $MnTe_2$. This reaction illustrates the enormous potential for materials synthesis offered by the types of compounds discussed in this review.

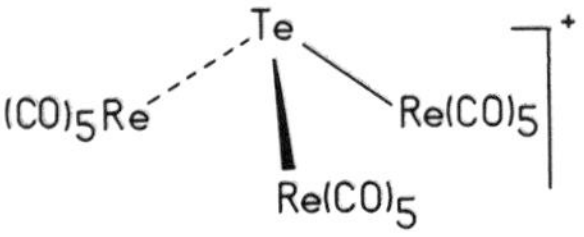

173

A final example is the cationic trirhenium–tellurium complex $[Te\{Re(CO)_5\}_3]^+$ (**173**) synthesized by Beck *et al.* (*162*) from the reaction between

TABLE VI

Te—Te Bond Lengths in Complexes
Containing Te$_n$

Complex	r(Te—Te) (Å)	Ref.
160	2.686(4)	*47*
161	2.718(1)	*153*
	2.736(1)	
162	2.705(1)	*154*
	2.717(1)	
	2.750(1)	
163	2.703(1)	*154*
	2.719(1)	
	2.764(2)	
168	2.703(1)	*157*
171	2.732(1)	*157*
172	2.763(1)	*161*
190	2.668(1)	*179*
191	2.802(1)	*180*

$3[Re(CO)_5]^+$ and $[Te]^{2-}$. Complex **173** contains a trigonal–pyramidal tellurium atom and is isoelectronic with the trimanganese–bismuth complex **105**.

D. Complexes Involving Iron, Ruthenium, and Osmium

Steigerwald has described the preparation of $[Fe_2(CO)_2(PEt_3)_2(C_5H_5)_2$-$(\mu\text{-}Te_n)]$ (**174**, $n = 1$; **175**, $n = 2$) (*163*), which are useful precursors for FeTe and FeTe$_2$, from reactions between $[Fe_2(CO)_4(C_5H_5)_2]$ and one or two equivalents of Et_3PTe. Similarities exist between **174** and **154** and between **175** and **172**, with **175** being a further example of a dimetalla–ditelluride complex.

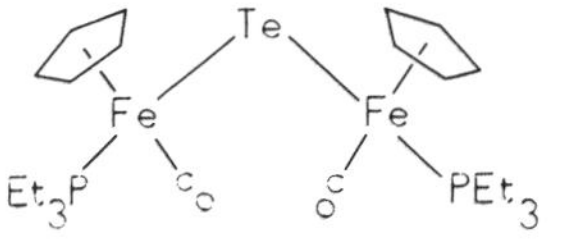

174 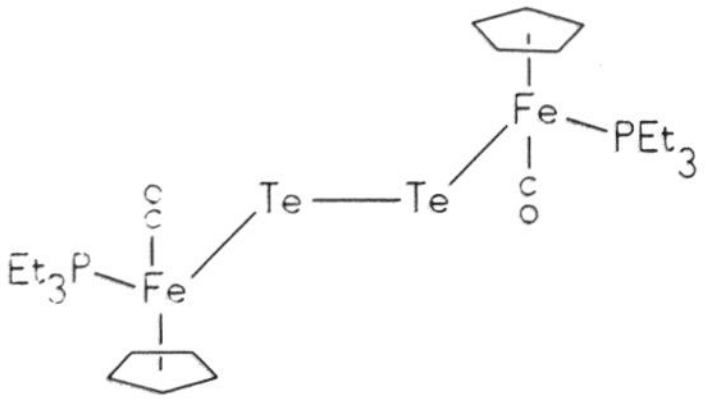   175

Another example of a diiron–ditellurium complex, $[Te_2Fe_2(CO)_6]$ (**176**), has been isolated by Rauchfuss (*164*) as one of the products from the reaction between $[Fe(CO)_5]/[OH]^-$ and $[TeO_3]^{2-}$. Complex **176** provides another

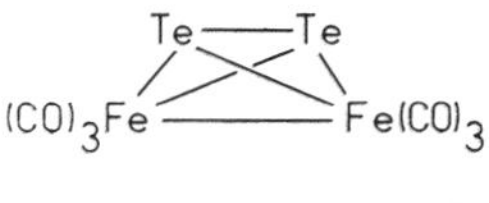

176

example of a Te$_2$ ligand acting as a six-electron donor. However, **176** is not very stable and the major product from the above reaction is $[Te_2Fe_3(CO)_9]$ (**177**), which was originally reported by Hieber (*165*) and has been structurally

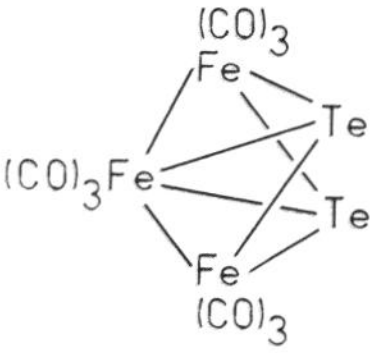

177

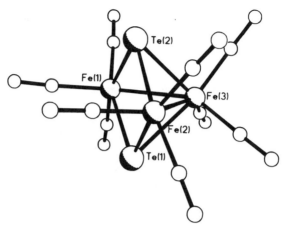

FIG. 49.   Structure of [Te$_2$Fe$_3$(CO)$_9$] (**177**).

characterized by Schumann (*166*) (see Fig. 49). The structure is based on an open triangle of iron atoms triply bridged by the two telluriums and is similar to the isoelectronic dianionic antimony and bismuth complexes, **134** and **132** [if the apical Fe(CO)$_4$ fragments are ignored] or, more exactly, to [Bi$_2$Fe$_3$-(CO)$_9$]$^{2-}$ (**131**), although the latter complex has not been structurally characterized. The ruthenium and osmium analogues [Te$_2$M$_3$(CO)$_9$] (**178**, M = Ru; **179**, M = Os) have also been reported (*167,168*).

A considerable amount of work on the reactivity of **177** has been reported, particularly with Lewis bases, which form adducts of the general formula [Te$_2$Fe$_3$(CO)$_9$L] (L = CO, PR$_3$, AsR$_3$, Bu^{t}NC) (*169,170*) or substitution products resulting from CO loss. The latter complexes presumably retain the core structure of **177**, but adduct formation results in the loss of one Fe—Fe bond (or cluster vertex) as revealed in the X-ray-derived structure of the PPh$_3$ adduct, [Te$_2$Fe$_3$(CO)$_9$(PPh$_3$)], **180** (*169*).

$$\text{PPh}_3(\text{CO})_3\text{Fe} \underset{\text{Te}}{\overset{\text{Te}}{\diamond}} \begin{array}{c} \text{Fe(CO)}_3 \\ | \\ \text{Fe(CO)}_3 \end{array}$$

180

Many heterometallic species have also been derived from **177** and these will be described later. A triosmium–tellurium cluster has been reported by Vahrenkamp and Shapley (*171*) from the reaction between [Os$_3$(CO)$_{10}$-(MeCN)$_2$] and H$_2$Te, viz. [TeOs$_3$(CO)$_9$(μ-H)$_2$] (**181**), which is related to the

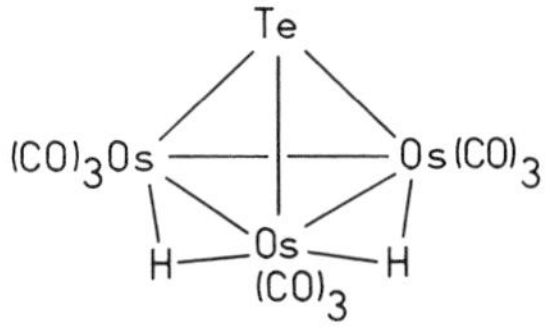

181

trihydrido bismuth clusters (111–113). Higher nuclearity clusters have also been reported, though not yet structurally characterized. Prolonged heating of 177 under high pressures of CO affords $[Te_4Fe_4(CO)_{12}]$, 182, which is assumed to have a cubic structure (172) analogous to the isoelectronic antimony and bismuth complexes, 150 and 151, while the reaction between 177 and $[Ru_3(CO)_{12}]$ in refluxing benzene affords $[Te_2Ru_4(CO)_{12}]$, 183 (173), which is assumed to have an octahedral structure.

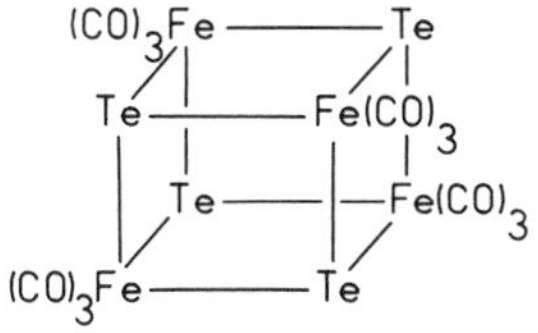

182

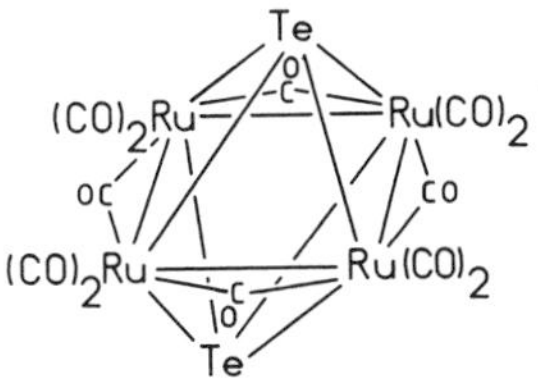

183

E. Complexes Involving Cobalt and Rhodium

The tetracobalt–ditellurium cluster $[Te_2Co_4(CO)_{10}]$, 184, was originally described by Hieber (174) and has been crystallographically characterized by

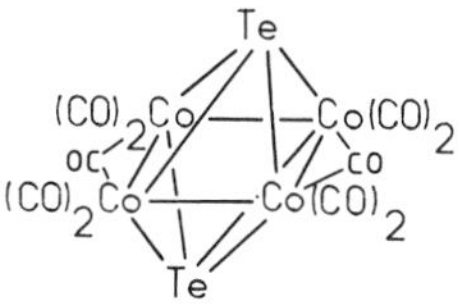

184

Dahl (175). The basic structure is defined by an octahedral core with *trans*-tellurium atoms and is isoelectronic with the ruthenium complex 183. Both clusters may also be compared with the ruthenium and osmium bismuth complexes, 125 and 129, which, although octahedral, contain two less electrons. A complex containing an additional carbonyl group, viz. $[Te_2Co_4(CO)_{11}]$, 185, has been described by Rauchfuss (172) as resulting from the reaction between **177** and $[Co_2(CO)_8]$ and also by Hieber (*174*) and Dahl (*176*), using Et_2Te and Ph_2Te as sources of tellurium, and $[Co_2(CO)_8]$.

Other examples are $[Rh_2(CO)_2(C_5Me_5)_2(\mu\text{-}Te)]$ (**186**), reported by Herrmann (*177*) from the reaction between $[Rh_2(C_5Me_5)_2(\mu\text{-}CO)_2]$ and elemental tellurium, which is assumed to contain a bent, two-coordinate tellurium analogous to that found in **164**. Also known is the phosphine cobalt complex $[Co_2(PMe_3)_6(\mu\text{-}Te)_2]$ (**187**), prepared by Klein (*178*) from the reaction between the Zintl anion $[SnTe_4]^{4-}$ and $[CoCl(PMe_3)_3]$. The tellurium

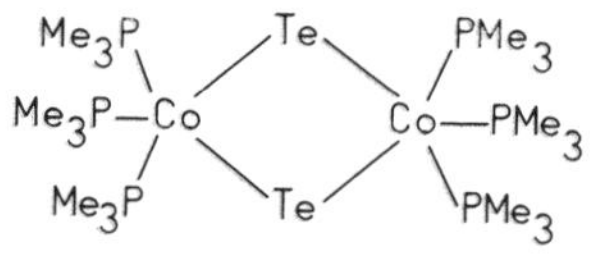

187

atoms are bent, and two coordinate in a manner analogous to the proposed structure of **159**. However, the Co–Co separation is long [4.093(2) Å] and clearly nonbonding, but the Te–Te distance is also long [3.062(2) Å] when compared with other known values (Table VI). Overall six-electron donation from the pair of tellurium atoms is required by electron-counting procedures, and the resonance structures shown here have been suggested to account for the bonding.

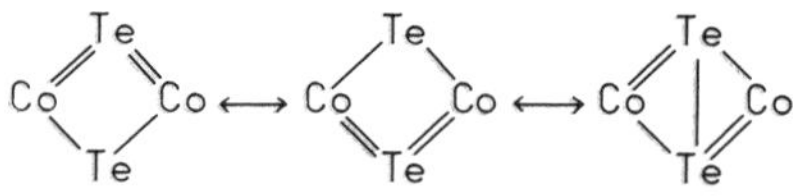

Addition of CO to **187** affords $[Co_2(PMe_3)_4(CO)_4(\mu\text{-}Te_2)]$, **188**, although the precise structure of this complex is not known.

F. Complexes Involving Nickel and Platinum

Stoppioni has reported the reaction between $Ni(ClO_4)_2 \cdot 6H_2O$ and $[Te]^{2-}$ in the presence of tridentate phosphines, which affords the complexes [Ni-(L)(Te_2)] [**189**, L = $HC(CH_2PPh_2)_3$; **190**, L = $PhP(CH_2CH_2PPh_2)_2$] (*179*)

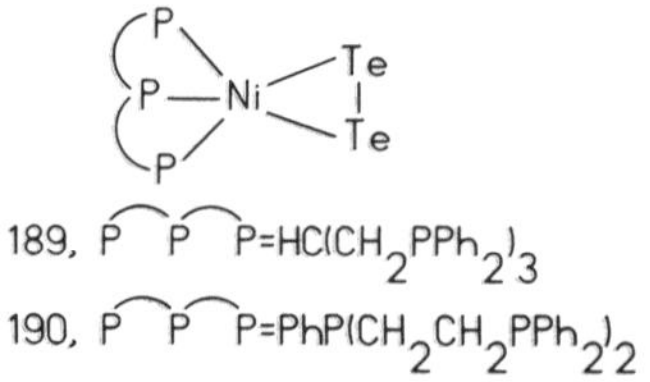

189, P⌢P P=HC(CH_2PPh_2)_3

190, P⌢P P=PhP(CH_2CH_2PPh_2)_2

containing a side-on bonded two-electron donor ditelluride ligand. Addition of another nickel fragment is also possible (*180*), and in the case where L = $HC(CH_2PPh_2)_3$, an X-ray structure has revealed a planar Ni_2Te_2 core

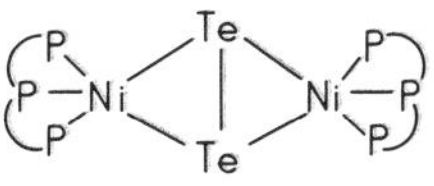

191, P͡ P͡ P = $HC(CH_2PPh_2)_3$

as shown in the diagram. The structure is similar to that found for **187**, although the Te–Te distance is much shorter in **191** and this complex may be viewed as containing a π-bound bridging Te_2 ligand. The cobalt analogue of **191** was also reported (*180*), although no structural details were presented, and also described (*179*) is a dicationic complex $[Ni_2\{PhP(CH_2\text{-}CH_2PPh_2)_2\}_2(\mu\text{-}Te_2)]^{2+}$, **192**, which is isoelectronic with the neutral cobalt species.

Fenske (*181*) has reported an interesting reaction between $[NiCl(PPh_3)\text{-}(C_5H_5)]$ and $Te(SiMe_3)_2$ that affords a nickel–tellurium complex $[Ni_2\text{-}(C_5H_5)_2\{\mu\text{-}TeNi(PPh_3)(C_5H_5)\}_2]$, **193**, containing trigonal–pyramidal tellurium atoms. Finally, we note the platinum complex $[Pt_2(PPh_3)_4(\mu\text{-}Te_2)]$, **194**, described by Haushalter (*182*), which is derived from the reaction between $[Pt(PPh_3)_4]$ and $[Hg_4Te_{12}]^{4-}$.

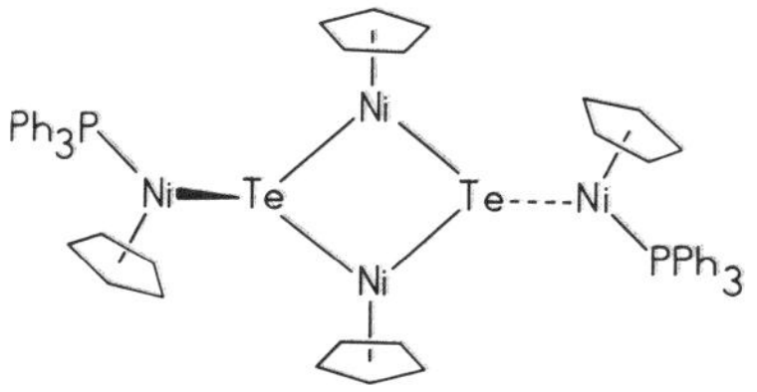

193

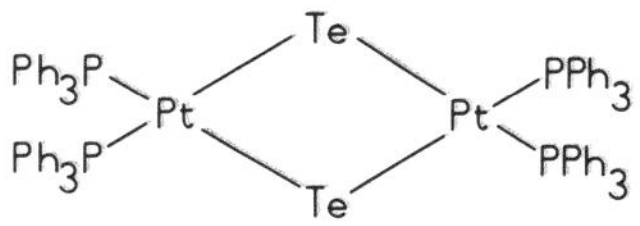

194

G. *Mixed Transition Metal Complexes*

Many complexes are known for tellurium that fall within this category, and further subdivision based on the metals present is warranted.

1. *Iron–Molybdenum Complexes*

Complex **177** reacts with $[Mo_2(CO)_6(C_5H_5)_2]$ under CO to afford $[Te_2\text{-}FeMo_2(CO)_7(C_5H_5)_2]$, **195** (*183*) (see Fig. 50), which contains two tellurium atoms separated by 3.13 Å. The structure is similar to that found in **180** but

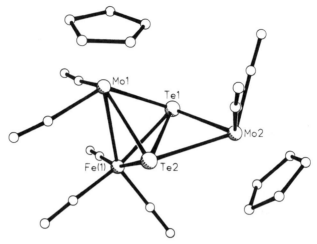

FIG. 50.   Structure of $[Te_2FeMo_2(CO)_7(C_5H_5)_2]$ **(195)**.

can also be compared with the dibismuth cluster **124** with which it is isoelectronic. Thermolysis of **195** affords three products, **196–198**, of which **198** (Fig. 51) was characterized crystallographically, but labeling studies indicated that the $(C_5H_5)Mo_2Fe$ core remained intact during these rearrangements and that initial activation occurred by CO loss. Attempts to trap a possible intermediate resulted in the alkyne adduct $[Te_2FeMo_2(CO)_3(C_2H_2)$-

195

196

197

198

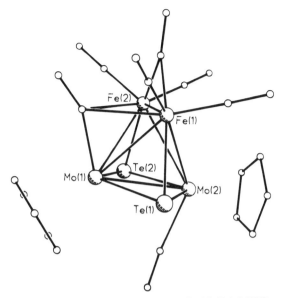

FIG. 51.   Structure of [Te$_2$Fe$_2$Mo$_2$(CO)$_7$(C$_5$H$_5$)$_2$] (198).

(C$_5$H$_5$)$_2$], 199 (*184*) containing a closo trigonal–bipyramidal Te$_2$FeMo$_2$ core. The reaction of 195 with Br$_2$ and CO has also been reported to afford

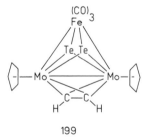

199

the crystallographically characterized Te$_2$Br complex, [FeMo(CO)$_5$(C$_5$H$_5$)-(μ-Te$_2$Br)], 200 (*185*), which ionizes in solution to give the cationic complex [FeMo(CO)$_5$(C$_5$H$_5$)(μ-Te$_2$)]$^+$, 200a.

200

2. Iron–Manganese Complexes

Herrmann (*186*) has reported that the reaction of **164** with $[Fe_2(CO)_9]$ affords the tetrahedral cluster $[TeFe_2Mn(CO)_8(C_5Me_5)]$, **201**, which is isoelectronic with **196**.

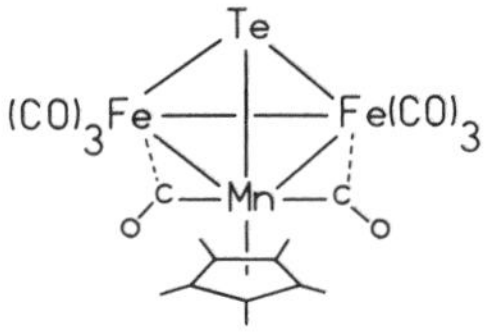

201

3. Iron–Ruthenium and Iron–Osmium Complexes

Two complexes of this type have been described by Mathur (*168*) to be derived from the reaction between **176** and $[M_3(CO)_{11}(MeCN)]$ (M = Ru, Os), which affords $[Te_2Fe_2M_3(CO)_{17}]$ (**202**, M = Ru; **203**, M = Os). Although not characterized crystallographically, the structures are assumed to be as shown in the diagram.

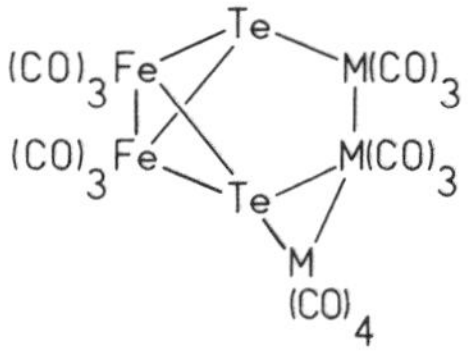

202, M=Ru; 203, M=Os.

4. Iron–Cobalt and Iron–Rhodium Complexes

Dahl (176) has described the preparation of $[TeFeCo_2(CO)_9]$, **204**, from the reaction between $[Co_2(CO)_8]$, $[Fe_2(CO)_9]$, and Et_2Te. The structure is tetrahedral and isoelectronic with **196** and **201**. Rauchfuss has reported

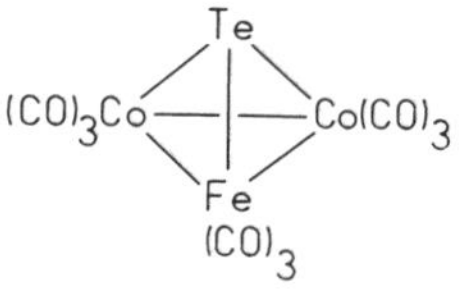

204

a number of FeCo clusters (*172*) derived from reactions of **177**. Reaction with [Co(CO)$_2$(C$_5$H$_5$)] affords **205**, which undergoes reversible decarbonylation to give two isomers, **206**, and **207**. The reaction between **177** and [Co$_2$(CO)$_8$]

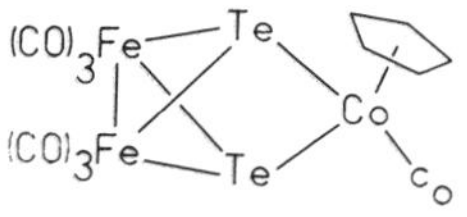

205

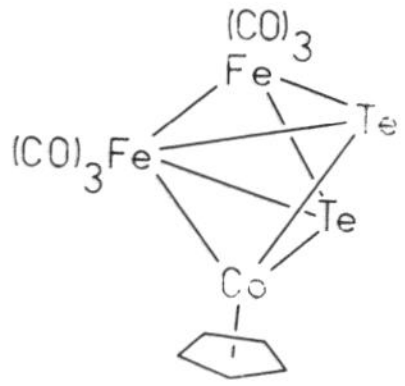

206

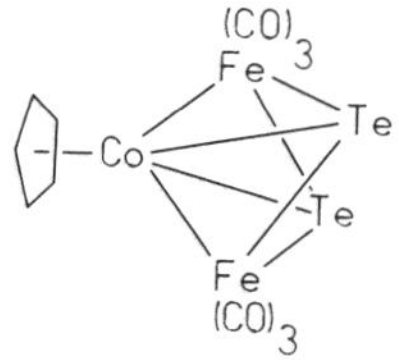

207

affords, as previously mentioned, **185**, under conditions of 150°C and 2000 psi of CO for 24 hr in CH$_2$Cl$_2$. However, at 180°C and 1700 psi of CO for 4 hr in hexane, this reaction provides an alternative route to **204**. The rhodium analogues of **205–207** have been described by Rauchfuss (*187*) together with some PPh$_3$ derivatives (see Scheme 3).

5. Iron Complexes Containing Nickel, Palladium, and Platinum

The reaction between **176** and [M(PPh$_3$)$_2$(C$_2$H$_4$)] affords [Te$_2$Fe$_2$M-(CO)$_6$(PPh$_3$)$_2$] (**213**, M = Pd; **214**, M = Pt) (*188,189*). These clusters are both

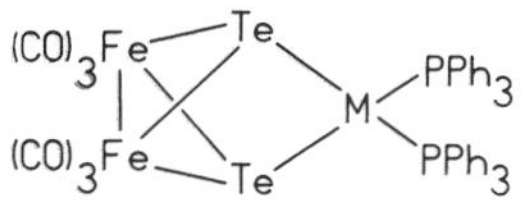

213, M=Pd; 214, M=Pt.

similar to the cobalt and rhodium complexes **205** and **208** and to the various adducts of **177** typified by **180**. Mathur (*190*) has also reported the synthesis of **213** and **214** from the reaction between **177** and [M(PPh$_3$)$_4$], together with the nickel analogue [Te$_2$Fe$_2$Ni(CO)$_6$(PPh$_3$)$_2$], **215**.

177 + [Rh(CO)$_2$(C$_5$H$_5$)] $\longrightarrow$

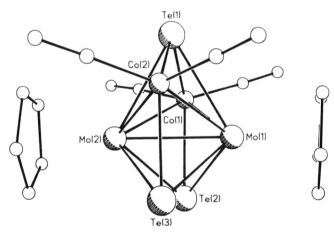

SCHEME 3

FIG. 52.   Structure of [Te$_3$Co$_2$Mo$_2$(CO)$_4$(C$_5$H$_5$)$_2$] (**216**).

6. *Cobalt–Molybdenum Complexes*

One example of this class of compound (**216**) is derived from the reaction between **185** and $[Mo_2(CO)_6(C_5H_5)_2]$, which has the structure shown in Fig. 52.

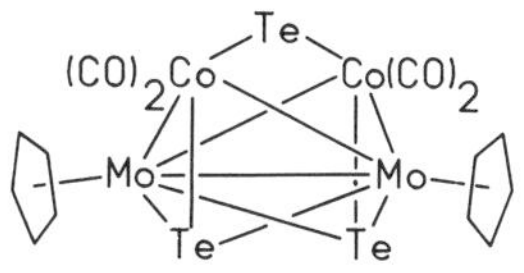

216

H. *Spectroscopic Studies*

1. ^{125}Te *NMR*

^{125}Te is a useful nucleus for NMR studies, and chemical shift data are available for many of the complexes described in this section, as shown in the following table.

Compound	Chemical shift[a]	Couplings (Hz)	Ref.
154	-1006		*150*
156	-1291		*150*
158	-1098		*150*
168	$-200.3, 434.4$		*157*
169	-510.0		*157*
170	613.7		*157*
171	848.0		*157*
175	$-863.1^{b,c}, -868.3$	$^2J(P, Te), 120$	*163*
176	-733^d		*164*
177	1123^d		*164*
180	$-887, -938^d$	$^2J(P, Te), 42$	*187*
195	-1092^d		*183*
199	931^d		*184*
200a	-839^d		*185*
$[Te_2Fe_3(CO)_8(PPh_3)]$	1062^d		*187*
205	-825^d		*187*
206 **207**	$1103^d, 1087$		*187*
208	-973^d	$^1J(Rh, Te), 93$	*187*

(*continued*)

Compound	Chemical shift[a]	Couplings (Hz)	Ref.
209 } 210 }	1081[d], 1109		187
211 } 212 }	−925[d], −838	[1]J(Rh, Te), 100; [2]J(P, Te), 21	187
214	−861[d]		187

[a] Chemical shift values are relative to Me_2Te, 0.0 ppm (15% in toluene, external) unless otherwise stated; positive values to high frequency.
[b] Relative to Ph_2Te.
[c] Two diastereomers in solution.
[d] Relative to neat Me_2Te.

It is evident from the above table that a considerable spread of chemical shift values is observed in tellurium–transition metal complexes, but the factors that determine the chemical shift are still poorly understood and data are not available for all known structural types. The most extensive compilations of data have been provided by Rauchfuss (*187*) and Herrmann (*191*), with the point being made in the former reference that chemical shifts are extremely sensitive to changes in cluster geometry. In principle, ^{125}Te NMR spectroscopy is a valuable method for studying tellurium–transition metal clusters in solution, but it is clear that more data are required before unambiguous structural assignments can be inferred.

2. ^{125}Te Mössbauer

The ^{125}Te nucleus is amenable to study by Mössbauer spectroscopy as well as by NMR, but far fewer studies dealing with the former have been reported. We are aware of one report by Jones (*192*) that reports isomer shift and quadrupole splitting data for **153, 164, 165,** and **154.** The isomer shifts are found to be essentially the same regardless of whether the tellurium acts as a two-, four-, or six-electron donor, and the small values observed for the quadrupole splittings in **153, 164,** and **165** are indicative of tellurium–metal multiple bonding.

REFERENCES

1. T. P. Fehlner, *Comments Inorg. Chem.* **7,** 307 (1988).
2. O. J. Scherer, *Angew. Chem., Int. Ed. Engl.* **24,** 924 (1985); *Comments Inorg. Chem.* **6,** 1 (1987).
3. M. Di Vaira, P. Stoppioni, and M. Peruzzini, *Polyhedron* **6,** 351 (1987).
4. W. A. Herrmann, *Angew. Chem., Int. Ed. Engl.* **25,** 56 (1986).
5. K. H. Whitmire, *J. Coord. Chem. B.* **17,** 95 (1988).
6. J. N. St. Denis, W. Butler, M. D. Glick, and J. P. Oliver, *J. Organomet. Chem.* **129,** 1 (1977).

7. A. J. Conway, P. B. Hitchcock, and J. D. Smith, *J. Chem. Soc., Dalton Trans.*, 1945 (1975).
8. L. M. Clarkson, W. Clegg, N. C. Norman, A. J. Tucker, and P. M. Webster, *Inorg. Chem.* **27**, 2653 (1988).
8a. A. T. T. Hsieh and M. J. Mays, *J. Organomet. Chem.* **37**, 9 (1972).
9. R. B. King, *Inorg. Chem.* **9**, 1936 (1970).
10. J. Rajaram and J. A. Ibers, *Inorg. Chem.* **12**, 1313 (1973).
10a. J. M. Burlitch and T. W. Theyson, *J. Chem. Soc., Dalton Trans.* 828 (1974).
11. R. B. Peterson, J. J. Stezowski, C. Wan, J. M. Burlitch, and R. E. Hughes, *J. Am. Chem. Soc.* **93**, 3532 (1971).
12. J. M. Burlitch, M. E. Leonowicz, R. B. Peterson, and R. E. Hughes, *Inorg. Chem.* **18**, 1097 (1979).
13. A. T. T. Hsieh and M. J. Mays, *J. Chem. Soc., Chem. Commun.*, 1234 (1971).
13a. A. T. T. Hsieh and M. J. Mays, *J. Chem. Soc., Dalton Trans.*, 516 (1972).
14. J. M. Burlitch and A. Ferrari, *Inorg. Chem.* **9**, 563 (1970).
15. A. T. T. Hsieh and M. J. Mays, *J. Organomet. Chem.* **38**, 243 (1972).
16. A. T. T. Hsieh and M. J. Mays, *Inorg. Synth.* **16**, 61 (1976).
17. H.-J. Haupt and F. Neumann, *J. Organomet. Chem.* **33**, C56 (1971).
18. A. T. T. Hsieh and M. J. Mays, *J. Organomet. Chem.* **22**, 29 (1970).
19. H.-J. Haupt and F. Neumann, *Z. Anorg. Allg. Chem.* **394**, 67 (1972).
20. H.-J. Haupt and F. Neumann, *J. Organomet. Chem.* **74**, 185 (1974).
20a. H. Preut and H.-J. Haupt, *Chem. Ber.* **107**, 2860 (1974).
21. H.-J. Haupt, U. Flörke, and H. Preut, *Acta Crystallogr., Sect. C* **C42**, 665 (1986).
22. H. Preut and H.-J. Haupt, *Chem. Ber.* **108**, 1447 (1975).
23. F. Neumann and H.-J. Haupt, *J. Organomet. Chem.* **84**, 329 (1975).
24. H.-J. Haupt, F. Neumann, and H. Preut, *J. Organomet. Chem.* **99**, 439 (1975).
24a. H. Preut and H.-J. Haupt, *Acta Crystallogr., Sect. B* **B35**, 1205 (1979).
25. U. Flörke, P. Balsaa, and H.-J. Haupt, *Acta Crystallogr., Sect. C* **C42**, 275 (1986).
26. H.-J. Haupt, P. Balsaa, and U. Flörke, *Z. Anorg. Allg. Chem.* **557**, 69 (1988).
27. H. Preut and H.-J. Haupt, *Acta Crystallogr., Sect. B* **B35**, 2191 (1979).
28. M. R. Churchill, K. N. Amoh, and H. J. Wasserman, *Inorg. Chem.* **20**, 1609 (1981).
29. R. Hoffmann, *Angew. Chem., Int. Ed. Engl.* **27**, 711 (1982).
30. K. H. Whitmire, R. R. Ryan, H. J. Wasserman, T. A. Albright, and S.-K. Kang, *J. Am. Chem. Soc.* **108**, 6831 (1986).
30a. K. H. Whitmire, J. M. Cassidy, A. L. Rheingold, and R. R. Ryan, *Inorg. Chem.* **27**, 1347 (1988).
30b. J. M. Cassidy and K. H. Whitmire, *Inorg. Chem.* **28**, 1432 (1989).
31. J. M. Cassidy and K. H. Whitmire, *Inorg. Chem.* **28**, 1435 (1989).
32. G. B. Ansell, M. A. Modrick, and J. S. Bradley, *Acta Crystallogr., Sect. C* **C40**, 1315 (1984).
33. N. C. Norman and P. M. Webster, *Z. Naturforsch. B: Chem. Sci.* **44B**, 91 (1989).
33a. S. E. Pederson and W. R. Robinson, *Inorg. Chem.* **14**, 2365 (1975).
34. W. Hieber and U. Teller, *Z. Anorg. Allg. Chem.* **249**, 43 (1942).
35. D. J. Patmore and W. A. G. Graham, *Inorg. Chem.* **5**, 1536 (1966).
36. W. R. Robinson and D. P. Schussler, *Inorg. Chem.* **12**, 848 (1973).
37. J. M. Burlitch, R. B. Peterson, H. L. Conder, and W. R. Robinson, *J. Am. Chem. Soc.* **92**, 1783 (1970).
37a. W.R. Robinson and D. P. Schussler. *J. Organomet. Chem.* **30**, C5 (1971).
38. G. N. Schrauzer, B. N. Bastian, and G. A. Fosselius, *J. Am. Chem. Soc.* **88**, 4890 (1966).
39. D. P. Schussler, W. R. Robinson, and W. F. Edgell, *Inorg. Chem.* **13**, 153 (1974).
40. P. Klüfers, *Z. Kristallogr.* **165**, 217 (1983).
41. O. J. Ezomo, D. M. P. Mingos, and I. D. Williams, *J. Chem. Soc., Chem. Commun.*, 924 (1987).
42. P. G. Harrison and S. R. Stobart, *J. Chem. Soc., Dalton Trans.*, 940 (1973).

43. T. S. Dory, J. J. Zuckerman, C. D. Hoff, and J. W. Connolly, *J. Chem. Soc., Chem. Commun.,* 521 (1981).

44. M. F. Lappert and P. P. Power, *J. Chem. Soc., Dalton Trans.,* 51 (1985).

45. P. B. Hitchcock, M. F. Lappert, and M. J. Michalczyk, *J. Chem. Soc., Dalton Trans.,* 2635 (1987).

46. G. Huttner, U. Weber, B. Sigwarth, O. Scheidsteger, H. Lang, and L. Zsolnai, *J. Organomet. Chem.* **282,** 331 (1985).

47. O. Scheidsteger, G. Huttner, K. Dehnicke, and J. Pebler, *Angew. Chem., Int. Ed. Engl.* **24,** 428 (1985).

48. S. V. Dighe and M. Orchin, *J. Am. Chem. Soc.* **87,** 1146 (1965).

49. W. Gäde and E. Weiss, *J. Organomet. Chem.* **213,** 451 (1981).

50. J. D. Korp, I. Bernal, R. Hörlein, R. Serrano, and W. A. Herrmann, *Chem. Ber.* **118,** 340 (1985).

51. N. M. Kostić and R. F. Fenske, *J. Organomet. Chem.* **233,** 337 (1982).

52. T. J. Greenhough, B. W. S. Kolthammer, P. Legzdins, and J. Trotter, *Inorg. Chem.* **18,** 3543 (1979).

53. W. A. Herrmann, J. Weichmann, R. Serrano, K. Blechschmitt, H. Pfisterer, and M. L. Ziegler, *Angew. Chem., Int. Ed. Engl.* **22,** 314 (1983).

54. W. A. Herrmann, H.-J. Kneuper, and E. Herdtweck, *Chem. Ber.* **122,** 437 (1989).

55. W. A. Herrmann, H.-J. Kneuper, and E. Herdtweck, *Angew. Chem., Int. Ed. Engl.* **24,** 1062 (1985).

56. W. A. Herrmann, H.-J. Kneuper, and E. Herdtweck, *Chem. Ber.* **122,** 445 (1989).

57. W. A. Herrmann, H.-J. Kneuper, and E. Herdtweck, *Chem. Ber.* **122,** 433 (1989).

58. H.-J. Kneuper, E. Herdtweck, and W. A. Herrmann, *J. Am. Chem. Soc.* **109,** 2508 (1987).

59. P. Krumholz and S. Bril, *Proc. Chem. Soc., London,* 116 (1960).

60. R. D. Ernst and T. J. Marks, *Inorg. Chem.* **17,** 1477 (1978).

61. J. M. Cassidy and K. H. Whitmire, *Inorg. Chem.* **28,** 2494 (1989).

62. T. B. Marder and N. C. Norman, unpublished work.

63. W. Hieber and H. Beutner, *Z. Anorg. Allg. Chem.* **320,** 101 (1963).

64. W. Hieber and W. Klingshirn, *Z. Anorg. Allg. Chem.* **323,** 292 (1963).

65. P. Hackett and A. R. Manning, *Polyhedron* **1,** 45 (1982).

66. J. D. Cotton, J. Duckworth, S. A. R. Knox, P. F. Lindley, I. Paul, F. G. A. Stone, and P. Woodward, *J. Chem. Soc., Chem. Commun.,* 253 (1966)

67. J. D. Cotton, S. A. R. Knox, I. Paul, and F. G. A. Stone, *J. Chem. Soc. A,* 264 (1967).

68. P. F. Lindley and P. Woodward, *J. Chem. Soc. A,* 382 (1967).

69. A. S. Batsanov, L. V. Rybin, M. I. Rybinskaya, Yu. T. Struchkov, I. M. Salimgareeva, and N. G. Bogatova, *J. Organomet. Chem.* **249,** 319 (1983).

70. D. Melzer and E. Weiss, *J. Organomet. Chem.* **255,** 335 (1983).

71. K. H. Whitmire, C. B. Lagrone, M. R. Churchill, J. C. Fettinger, and B. H. Robinson, *Inorg. Chem.* **26,** 3491 (1987).

72. C. B. Lagrone, K. H. Whitmire, M. R. Churchill, and J. C. Fettinger, *Inorg. Chem.* **25,** 2080 (1986).

73. R. M. Sweet, C. J. Fritchie, and R. A. Schunn, *Inorg. Chem.* **6,** 749 (1967).

74. W. Hieber, J. Gruber, and F. Lux, *Z. Anorg. Allg. Chem.* **300,** 275 (1959).

75. S. G. Anema, K. M. Mackay, and B. K. Nicholson, *Inorg. Chem.* **28,** 3158 (1989).

76. T. J. McNeese, S. S. Wreford, D. L. Tipton, and R. Bau, *J. Chem. Soc., Chem. Commun.,* 390 (1977).

77. A. Castel, P. Riviere, J. Satgé, J. J. E. Moreau, and R. J. P. Corriu, *Organometallics* **2,** 1498 (1983).

78. R. F. Gerlach, K. M. Mackay, B. K. Nicholson, and W. T. Robinson, *J. Chem. Soc., Dalton Trans.,* 80 (1981).

79. R. A. Croft, D. N. Duffy, and B. K. Nicholson, *J. Chem. Soc., Dalton Trans.*, 1023 (1982).
80. G. Schmid and G. Etzrodt, *J. Organomet. Chem.* **137**, 367 (1977).
81. R. Boese and G. Schmid, *J. Chem. Soc., Chem. Commun.*, 349 (1979).
82. M. Bigorgne and A. Quintin, *C. R. Hebd. Seances Acad. Sci, Ser. C* **264**, 2055 (1967).
83. D. J. Patmore and W. A. G. Graham, *Inorg. Chem.* **7**, 771 (1968).
84. G. Schmid and G. Etzrodt, *J. Organomet. Chem.* **131**, 477 (1977).
85. G. Schmid, *Angew. Chem., Int. Ed. Engl.* **17**, 392 (1978).
86. D. N. Duffy, K. M. Mackay, B. K. Nicholson, and W. T. Robinson, *J. Chem. Soc., Dalton Trans.*, 381 (1981).
87. D. N. Duffy, K. M. Mackay, B. K. Nicholson, and R. A. Thomson, *J. Chem. Soc., Dalton Trans.*, 1029 (1982).
88. S. P. Foster, K. M. Mackay, and B. K. Nicholson, *Inorg. Chem.* **24**, 909 (1985).
89. S. G. Anema, K. M. Mackay, L. C. McLeod, B. K. Nicholson, and J. M. Whittaker, *Angew. Chem., Int. Ed. Engl.* **25**, 759 (1986).
90. A. Ceriotti, F. Demartin, B. T. Heaton, P. Ingallina, G. Longoni, M. Manassero, M. Marchionna, and N. Masciocchi, *J. Chem. Soc., Chem. Commun.*, 786 (1989).
91. S. G. Anema, J. A. Audett, K. M. Mackay, and B. K. Nicholson, *J. Chem. Soc., Dalton Trans.*, 2629 (1988).
92. H.-J. Haupt and U. Flörke, *Acta Crystallogr., Sect. C* **C44**, 472 (1988).
93. H.-J. Haupt, A. Götze, and U. Flörke, *Z. Anorg. Allg. Chem.* **557**, 82 (1988).
94. P. Gusbeth and H. Vahrenkamp, *Chem. Ber.* **118**, 1746 (1985).
95. P. Gusbeth and H. Vahrenkamp, *Chem. Ber.* **118**, 1758 (1985).
96. P. Gusbeth and H. Vahrenkamp, *Chem. Ber.* **118**, 1770 (1985).
97. N. C. Norman, *Chem. Soc. Rev.* **17**, 269 (1988).
98. W. Malisch and P. Panster, *Z. Naturforsch. B: Anorg. Chem., Org. Chem.* **30B**, 229 (1975).
99. A. M. Barr, M. D. Kerlogue, N. C. Norman, P. M. Webster, and L. J. Farrugia, *Polyhedron* **8**, 2495 (1989).
100. W. Clegg, N. A. Compton, R. J. Errington, N. C. Norman, A. J. Tucker, and M. J. Winter, *J. Chem. Soc., Dalton Trans.*, 2941 (1988).
101. N. A. Compton, R. J. Errington, N. C. Norman, and A. G. Orpen, in preparation.
102. G. Huttner, U. Weber, B. Sigwarth, and O. Scheidsteger, *Angew. Chem., Int. Ed. Engl.* **21**, 215 (1982).
103. G. Huttner, U. Weber, and L. Zsolnai, *Z. Naturforsch. B: Anorg. Chem., Org. Chem*, **37B**, 707 (1982).
104. B. Sigwarth, L. Zsolnai, H. Berke, and G. Huttner, *J. Organomet. Chem.* **226**, C5 (1982).
105. A. M. Arif, A. H. Cowley, N. C. Norman, and M. Pakulski, *Inorg. Chem.* **25**, 4836 (1986).
106. W. Clegg, N. A. Compton, R. J. Errington, and N. C. Norman. *Polyhedron* **7**, 2239 (1988).
107. W. Clegg, N. A. Compton, R. J. Errington, and N. C. Norman, unpublished results.
108. J. M. Wallis, G. Müller, and H. Schmidbaur, *Inorg. Chem.* **26**, 458 (1987).
109. N. A. Compton, R. J. Errington, G. A. Fischer, N. C. Norman, P. M. Webster, and A. G. Orpen, *J. Chem. Soc., Dalton Trans.* (submitted for publication).
110. K. Plössl, G. Huttner, and L. Zsolnai, *Angew. Chem., Int. Ed. Engl.* **28**, 446 (1989).
111. K. H. Whitmire, C. B. Lagrone, M. R. Churchill, J. C. Fettinger, and L. V. Biondi, *Inorg. Chem.* **23**, 4227 (1984).
112. K. H. Whitmire, C. B. Lagrone, and A. L. Rheingold, *Inorg. Chem.* **25**, 2472 (1986).
113. M. R. Churchill, J. C. Fettinger, K. H. Whitmire, and C. B. Lagrone, *J. Organomet. Chem.* **303**, 99 (1986).
114. H. G. Ang, C. M. Hay, B. F. G. Johnson, J. Lewis, P. R. Raithby, and A. J. Whitton, *J. Organomet. Chem.* **330**, C5 (1987).
115. K. H. Whitmire, J. S. Leigh, S. Luo, M. Shieh, M. D. Fabiano, and A. L. Rheingold, *New J. Chem.* **12**, 397 (1988).
116. S. Luo and K. H. Whitmire, *Inorg. Chem.* **28**, 1424 (1989).

117. B. F. G. Johnson, J. Lewis, P. R. Raithby, and A. J. Whitton, *J. Chem. Soc., Chem. Commun.*, 401 (1988).
118. K. H. Whitmire, M. Shieh, and J. Cassidy, *Inorg. Chem.* **28**, 3164 (1989).
119. W. Malisch and P. Panster, *Angew. Chem., Int. Ed. Engl.* **15**, 618 (1976).
120. J. M. Wallis, G. Müller, and H. Schmidbaur, *J. Organomet. Chem.* **325**, 159 (1987).
121. W. Clegg, N. A. Compton, R. J. Errington, and N. C. Norman, *J. Chem. Soc., Dalton Trans.*, 1671 (1988).
122. J. M. Wallis, G. Müller, J. Riede, and H. Schmidbaur, *J. Organomet. Chem.* **369**, 165 (1989).
123. K. H. Whitmire, K. S. Raghuveer, M. R. Churchill, J. C. Fettinger, and R. F. See, *J. Am. Chem. Soc.* **108**, 2778 (1986).
124. K. H. Whitmire, M. Shieh, C. B. Lagrone, B. H. Robinson, M. R. Churchill, J. C. Fettinger, and R. F. See, *Inorg. Chem.* **26**, 2798 (1987).
125. M. R. Churchill, J. C. Fettinger, and K. H. Whitmire, *J. Organomet. Chem.* **284**, 13 (1985).
126. C. M. Hay, B. F. G. Johnson, J. Lewis, P. R. Raithby, and A. J. Whitton, *J. Chem. Soc., Dalton Trans.*, 2091 (1988).
127. R. D. Adams, I. T. Horvath, P. Mathur, B. E. Segmuller, and L.-W. Yang, *Organometallics* **2**, 1078 (1983).
128. K. H. Whitmire, M. Shieh, C. B. Lagrone, B. H. Robinson, M. R. Churchill, J. C. Fettinger, and R. F. See, *Inorg. Chem.* **26**, 2798 (1987).
129. K. H. Whitmire, M. R. Churchill, and J. C. Fettinger, *J. Am. Chem. Soc.* **107**, 1056 (1985).
130. K. H. Whitmire, T. A. Albright, S.-K. Kang, M. R. Churchill, and J. C. Fettinger, *Inorg. Chem.* **25**, 2799 (1986).
131. A. L. Rheingold, S. J. Geib, M. Shieh, and K. H. Whitmire, *Inorg. Chem.* **26**, 463 (1987).
132. A. M. Arif, A. H. Cowley, and M. Pakulski, *J. Chem. Soc., Chem. Commun.*, 622 (1987).
133. G. Etzrodt, R. Boese, and G. Schmid, *Chem. Ber.* **112**, 2574 (1979).
134. P. Klüfers, *Z. Kristallogr.* **156**, 74 (1981).
135. W. R. Cullen, D. J. Patmore, and J. R. Sams, *Inorg. Chem.* **12**, 867 (1973).
136. W. Clegg, N. A. Compton, R. J. Errington, D. C. R. Hockless, N. C. Norman, M. Ramshaw, and P. M. Webster. *J. Chem. Soc., Dalton Trans.* (in press).
137. W. Kruppa, D. Bläser, R. Boese, and G. Schmid, *Z. Naturforsch. B: Anorg. Chem., Org. Chem.* **37B**, 209 (1982).
138. K. H. Whitmire, J. S. Leigh, and M. E. Gross. *J. Chem. Soc., Chem. Commun.*, 926 (1987).
139. S. Martinengo and G. Ciani, *J. Chem. Soc., Chem. Commun.*, 1589 (1987).
140. S. Martinengo, A. Fumagalli, G. Ciani, and M. Moret, *J. Organomet. Chem.* **347**, 413 (1988).
141. J. S. Leigh and K. H. Whitmire, *Angew. Chem., Int. Ed. Engl.* **27**, 396 (1988).
142. R. E. Cobbledick and F. W. B. Einstein, *Acta Crystallogr., Sect. B* **B35**, 2041 (1979).
143. J. L. Vidal and J. M. Troup, *J. Organomet. Chem.* **213**, 351 (1981).
144. J. S. Leigh, K. H. Whitmire, K. A. Yee, and T. A. Albright, *J. Am. Chem. Soc.* **111**, 2726 (1989).
145. A. S. Foust and L. F. Dahl, *J. Am. Chem. Soc.* **92**, 7337 (1970).
146. G. Ciani, M. Moret, A. Fumagalli, and S. Martinengo, *J. Organomet. Chem.* **362**, 291 (1989).
147. R. Beckhaus and K.-H. Thiele, *Z. Anorg. Allg. Chem.* **573**, 195 (1989).
148. J. Schiemann, P. Hübener, and E. Weiss, *Angew Chem., Int. Ed. Engl.* **22**, 980 (1983).
149. W. A. Herrmann, J. Rohrmann, M. L. Ziegler, and T. Zahn, *J. Organomet. Chem.* **273**, 221 (1984).
150. W. A. Herrmann, J. Rohrmann, and C. Hecht, *J. Organomet. Chem.* **290**, 53 (1985).
151. K. Endrich, E. Guggolz, O. Serhadle, M. L. Ziegler, and R. P. Korswagen, *J. Organomet. Chem.* **349**, 323 (1988).
152. A. H. Cowley and N. C. Norman, *Prog. Inorg. Chem.* **34**, 1 (1986).
153. R. Faggiani, R. J. Gillespie, C. Campana, and J. W. Kolis, *J. Chem. Soc., Chem. Commun.*, 485 (1987).

154. W. A. Flomer, S. C. O'Neal, J. W. Kolis, D. Jeter, and A. W. Cordes, *Inorg. Chem.* **27**, 969 (1988).
155. W. A. Herrmann, C. Hecht, M. L. Ziegler, and B. Balbach, *J. Chem. Soc., Chem. Commun.*, 686 (1984).
156. M. Herberhold, D. Reiner, and D. Neugelbauer, *Angew. Chem., Int. Ed. Engl.* **22**, 59 (1983).
156a. A. L. Rheingold, *Acta Crystallogr., Sect. C* **C43**, 585 (1987).
157. W. A. Herrmann, C. Hecht, E. Herdtweck, and H.-J. Kneuper, *Angew. Chem., Int. Ed. Engl.* **26**, 132 (1987).
158. A. M. Arif, A. H. Cowley, M. Pakulski, M-A. Pearsall, W. Clegg, N. C. Norman, and A. G. Orpen, *J. Chem. Soc., Dalton Trans.*, 2713 (1988).
159. G. Huttner and K. Evertz, *Acc. Chem. Res.* **19**, 406 (1986).
160. H. Braunwarth, F. Ettel, and G. Huttner, *J. Organomet. Chem.* **355**, 281 (1988).
161. M. L. Steigerwald and C. E. Rice, *J. Am. Chem. Soc.* **110**, 4228 (1988).
162. W. Beck, W. Sacher, and U. Nagel, *Angew. Chem., Int. Ed. Engl.* **25**, 270 (1986).
163. M. L. Steigerwald, *Chem. Mater.* **1**, 52 (1989).
164. D. A. Lesch and T. B. Rauchfuss, *Inorg. Chem.* **20**, 3583 (1981).
165. W. Hieber and J. Gruber, *Z. Anorg. Allg. Chem.* **296**, 91 (1958).
166. H. Schumann, M. Magerstädt, and J. Pickardt, *J. Organomet. Chem.* **240**, 407 (1982).
167. B. F. G. Johnson, J. Lewis, P. G. Lodge, P. R. Raithby, K. Henrick, and M. McPartlin, *J. Chem. Soc., Chem. Commun.*, 719 (1979).
168. P. Mathur, I. J. Mavunkal, and V. Rugmini, *Inorg. Chem.* **28**, 3616 (1989).
169. D. A. Lesch and T. B. Rauchfuss, *Organometallics* **1**, 499 (1982).
170. G. Cetini, P. L. Stanghellini, R. Rossetti, and O. Gambino, *J. Organomet. Chem.* **15**, 373 (1968).
171. H.-T. Schacht, A. K. Powell, H. Vahrenkamp, M. Koike, H.-J. Kneuper, and J. R. Shapley, *J. Organomet. Chem.* **368**, 269 (1989).
172. L. E. Bogan, D. A. Lesch, and T. B. Rauchfuss, *J. Organomet. Chem.* **250**, 429 (1983).
173. P. Mathur, and B. H. S. Thimmappa, *J. Organomet. Chem.* **365**, 363 (1989).
174. W. Hieber and T. Kruck, *Chem. Ber.* **95**, 2027 (1962).
175. R.C. Ryan and L. F. Dahl, *J. Am. Chem. Soc.* **97**, 6904 (1975).
176. C. E. Strouse and L. F. Dahl, *J. Am. Chem. Soc.* **93**, 6032 (1971).
177. W. A. Herrmann, C. Bauer, and J. Weichmann, *J. Organomet. Chem.* **243**, C21 (1983).
178. H.-F. Klein, M. Gass, U. Koch, B. Eisenmann, and H. Schäfer, *Z. Naturforsch., B: Chem. Sci,* **43B**, 830 (1988).
179. M. Di Vaira, M. Peruzzini, and P. Stoppioni, *Angew. Chem., Int. Ed. Engl.* **26**, 916 (1987).
180. M. Di Vaira, M. Peruzzini, and P. Stoppioni, *J. Chem. Soc., Chem. Commun.*, 374 (1986).
181. D. Fenske, A. Hollnagel, and K. Merzweiler, *Angew. Chem., Int. Ed. Engl.* **27**, 965 (1988).
182. R. D. Adams, T. A. Wolfe, B. W. Eichhorn, and R. C. Haushalter, *Polyhedron* **8**, 701 (1989).
183. L. E. Bogan, T. B. Rauchfuss, and A. L. Rheingold, *J. Am. Chem. Soc.* **107**, 3843 (1985).
184. L. E. Bogan, G. R. Clark, and T. B. Rauchfuss, *Inorg. Chem.* **25**, 4050 (1986).
185. L. E. Bogan, T. B. Rauchfuss, and A. L. Rheingold, *Inorg. Chem.* **24**, 3722 (1985).
186. W. A. Herrmann, C. Hecht, M. L. Ziegler, and T. Zahn, *J. Organomet. Chem.* **273**, 323 (1984).
187. D. A. Lesch and T. B. Rauchfuss, *Inorg. Chem.* **22**, 1854 (1983).
188. D. A. Lesch and T. B. Rauchfuss, *J. Organomet. Chem.* **199**, C6 (1980).
189. V. W. Day, D. A. Lesch, and T. B. Rauchfuss, *J. Am. Chem. Soc.* **104**, 1290 (1982).
190. P. Mathur, I. J. Mavunkal, and V. Rugmini, *J. Organomet. Chem.* **367**, 243 (1989).
191. W. A. Herrmann and H.-J. Kneuper, *J. Organomet. Chem.* **348**, 193 (1988).
192. C. H. W. Jones and R. D. Sharma, *Organometallics* **5**, 1194 (1986).

NOTE ADDED IN PROOF. Since the completion of this manuscript a number of reports have appeared on compounds relevant to this topic. These are listed here briefly according to main group element type which corresponds to the section heading numbers used in this review.

SECTION II. Haupt and Flörke have reported the crystal structure of $[In\{Re_2(CO)_6(\mu\text{-}I)\}\text{-}\{Re_2(CO)_6(\mu\text{-}I)_2\}]$, which contains a spiro μ_4-In bound to four rhenium atoms. This complex was obtained as one product from the reaction between InI_3 and $[Re_2(CO)_{10}]$ at 437 K. [H. -J. Haupt and U. Flörke, *Acta Crystallogr., Sect. C.* **C45**, 1718 (1989)].

SECTION III. Spectroscopic and chemical evidence has been obtained for $[Ge\{Mn(CO)_5\}_2]$ [D. Lei and M. J. Hampden Smith, *J. Chem. Soc., Chem. Commun.,* 1211 (1989)], and X-ray crystal structures have been reported for $[Sn\{Co(CO)_4\}_4]$ (**71**) and $[Pb\{Co(CO)_4\}_4]$ (**72**) [A. Cabrera, H. Samain, A. Mortreux, F. Petit, and A. J. Welch, *Organometallics* **9**, 959 (1990); J. S. Leigh and K. H. Whitmire, *Acta Crystallogr., Sect. C.* **C46**, 732 (1990)], which contain tetrahedrally coordinated tin and lead atoms. Crystal structure data has been presented for $[Sn\{Fe_2(CO)_8\}\{Fe_3(CO)_{11}\}]$, which contains a tetrahedral, spiro, η^4-tin atom bridging the Fe—Fe bond in the $Fe_2(CO)_8$ fragment and one Fe—Fe vector in the triangular $Fe_3(CO)_{11}$ unit [S. G. Anema, K. M. Mackay and B. K. Nicholson, *J. Organomet. Chem.* **372**, 25 (1989)]. Structural data on the mixed metal complexes $[Sn\{Fe(CO)_2(C_5H_5)\}_2\{Fe(CO)_4\}]$ and $[Sn\{Fe(CO)_2(C_5H_5)\}_2\{Cr(CO)_5\}]$ reveal tin atoms that are in a trigonal planar coordination environment, although weak axial isocarbonyl type contacts are also present [P. B. Hitchcock, M. F. Lappert and M. J. McGeary, *Organometallics* **9**, 884 (1990)].

SECTION IV. Whitmire has reported crystal structure data on the iron–antimony complexes $[Fe_2(CO)_8\{\mu\text{-}Sb(Fe(CO)_4)_2\}]^-$ and $[Sb_2Fe_6(CO)_{20}]^{2-}$ [S. Luo and K. H. Whitmire, *J. Organomet. Chem.* **376**, 297 (1989)]. The former is structurally similar to **58**, whereas the latter comprises a 1,6-Sb_2Fe_4 octahedral core with each antimony further bonded to an exopolyhedral $Fe(CO)_4$ fragment. Zeigler has described the syntheses and structures of $[Fe_3(CO)_9(\mu_3\text{-}CH)\text{-}(\mu_3\text{-}E)]$ (E = Sb, Bi) [C. Caballero, B. Nuber, and M. L. Zeigler, *J. Organomet. Chem.* **386**, 209 (1990)], and structural data have also been reported for the nickel-antimony cluster $[Ni_{13}Sb_2(CO)_{24}]^{2-}$ [V. G. Albano, F. Demartin, M. C. Iapalucci, G. Longoni, A. Sironi, and V. Zanotti, *J. Chem. Soc., Chem. Commun.,* 547 (1990)], which consists of 1,12-Sb_2Ni_{10} icosahedron with two exopolyhedral $Ni(CO)_3$ attached to the antimony atoms.

SECTION V. The crystal structure of **152** has been reported [A. Zalkin and D. J. Berg, *Acta Crystallogr., Sect. C.* **C44**, 1488 (1988)] [Te—Te = 2.769(1) Å]. Several iron–tellurium complexes have also been described: $[Fe_2(CO)_6(\mu\text{-}Te)(\mu\text{-}\eta^1\text{-}Te_2)]$ [B. W. Eichhorn, R. C. Haushalter, and J. S. Merola, *Inorg. Chem.* **29**, 728 (1990)]; $[Fe_3(CO)_9(\mu_3\text{-}PPr^i)(\mu_3\text{-}Te)]$ [D. Buchholz, G. Huttner, L. Zsolnai, and W. Imhof, *J. Organomet. Chem.* **377**, 25 (1989)]. The dianionic complex $[Fe_2(CO)_6(\mu\text{-}Te)_2]^{2-}$ has been prepared, and reactions with $[MCl_2(PPh_3)_2]$ (M = Ni, Pd, Pt), $[TiCl_2(C_5H_5)_2]$, and R_2SnCl_2 have been examined, which afford a range of heteronuclear tellurium clusters [P. Mathur and V. D. Reddy, *J. Organomet. Chem.* **385**, 363 (1990)]. Also, a new synthesis of **179** has been described [P. Mathur and D. Chakrabarty, *J. Organomet. Chem.* **373**, 129 (1989)]. Stoppioni *et al.* have characterized the dirhodium–ditelluride complexes $[Rh_2\{X\text{-}(CH_2CH_2PPh_2)_3\}_2(\mu\text{-}Te_2)](X = N, P)$ (Te—Te = 2.691(1) Å, Rh—Te—Te = 111.2, 112.1°) [M. Di Vaira, M. Peruzzini, and P. Stoppioni, *Inorg. Chem.* **28**, 4614 (1989)], and Steigerwald has described the nickel–tellurium–phosphine clusters, $[Te_6Ni_9(PEt_3)_8]$ and $[Te_{18}Ni_{20}(PEt_3)_{12}]$ [J. G. Brennen, T. Siegrist, S. M. Stuczynski, and M. L. Steigerwald, *J. Am. Chem. Soc.* **111**, 9240 (1989)].

ADVANCES IN ORGANOMETALLIC CHEMISTRY, VOL. 31

Organo-Transition Metal Compounds Containing Perfluorinated Ligands

RUSSELL P. HUGHES

Department of Chemistry
Dartmouth College
Hanover, New Hampshire 03755

I

INTRODUCTION

The chemistry of transition metal compounds containing hydrocarbon ligands continues to be intensively explored. Investigation of the fundamental interactions between organic molecules and transition metal compounds has led to the discovery of novel types of structure, dynamic behavior, and chemical reactivity, so that there is now considerable predictive capability in this area, and a number of chemical processes utilizing homogeneous catalysis by transition metal complexes are currently operated by the U.S. chemical industry (*1*). Studies of the chemistry of hydrocarbon ligands bound to transition metals clearly have had a significant impact upon chemical thought and practice. This review discusses the effect of α-fluorine substituents on the structure and reactivity of organic ligands bound to transition metal centers, with particular emphasis on recent chemistry of octafluoro-cycloocta-1,3,5,7-tetraene (OFCOT).

II

FLUOROCARBONS VS. HYDROCARBONS

An excellent review of the properties of fluorinated organic molecules, replete with numerical data, has been published by Smart (*2*). Of all elements, fluorine forms the strongest bonds to carbon and has the unique ability to replace hydrogen wherever it occurs in organic molecules, while still maintaining a stable structure. It is clear that perfluorinated organic molecules have a chemistry as versatile and wide-ranging as that of their hydrocarbon counterparts. The van der Waals radius of fluorine is only slightly larger than that of hydrogen, and C—F bond lengths are on the order of 1.3 vs. 1.0 Å for C—H, making substitution of fluorine for hydrogen in most organic molecules possible without introducing prohibitively severe steric effects. In addition, the

183

powerful electron-withdrawing effect of fluorine, the excellent energy match between its lone pair orbitals and π-orbitals on adjacent carbon centers, which leads to a strong π-donor effect for F in competition with its σ-withdrawing effect, and the significantly larger bond dissociation energies (about 20–30 kcal/mol) for C—F bonds compared to their C—H analogues combine to produce dramatic changes in the chemical and physical properties of fluorinated organic molecules compared to their hydrocarbon analogues (2). Furthermore, the high electronegativity of fluorine leads to a preference to form bonds to carbon orbitals of low electronegativity (i.e., with minimal s character). This, combined with destabilizing π interactions of fluorine lone pairs with adjacent filled π orbitals on carbon atoms, generates a significant preference for F to reside on sp^3 rather than sp^2 or sp carbon centers (2). Therefore, the properties of molecules with fluorine "skins" on a carbon skeleton do not simply parallel those of analogues containing hydrogen "skins."

III

ORGANOMETALLIC COMPOUNDS WITH PERFLUORINATED LIGANDS

The previous comments also apply to organometallic compounds containing perfluorinated organic ligands. For the purpose of this article, such ligands will usually be defined as those containing fluorine directly bound to ligated carbon atoms, and in most cases discussion will be restricted to perfluorinated cases. Only with F in this α position can the opposing σ and π effects referred to previously both come into play and influence the metal–carbon interaction. Of course, this definition excludes the many transition metal compounds containing ligands in which F occupies a β or γ position relative to the ligated carbon. For example, the extensively studied chemistries of hexafluoro-2-butyne and hexafluoroacetone (3) and more recent studies of ligands such as trifluoromethylisocyanide (4) fall into this latter category. In such ligands the CF_3 group is powerfully electron withdrawing, without the option of π donation, and also has considerable steric demand compared to F or CH_3 (2). Pentafluorophenyl ligands are also excluded by this definition, although they are becoming extensively utilized in transition metal chemistry. Some aspects of the chemistry of pentafluorophenyl complexes are the subject of a recent review (5).

Transition metal complexes containing fluorinated ligands date back to the initial explosive development of organometallic chemistry in the period from 1950 to 1970, and some early reviews contain a great deal of useful and relevant information (3,6,7). Early work clearly established that these compounds do indeed exhibit dramatically different structural, bonding, and chemical reactivity patterns, together with enhanced thermal stability, compared to those hydrocarbon analogues that exist. This article is not intended to be a comprehensive review of all organotransition metal compounds containing perfluorinated ligands, but rather a comparative discussion of the effects of the presence of fluorine on the α-carbon of an organic ligand on the structures and chemistry of such complexes. Illustrative highlights for some representative compounds are presented below, and more comprehensive details on fluorocarbon complexes of the individual transition metals can be found in the appropriate chapters of "Comprehensive Organometallic Chemistry" (8). The most common kinds of compounds are those with metal–carbon σ bonds and metal–olefin bonds. These are discussed in somewhat more detail than are the much rarer enyl, diene, and arene complexes, and the bulk of this review covers extensions of these ideas to the coordination chemistry and reactions of OFCOT and some of its valence isomers that have been studied at Dartmouth over the past few years.

A. Perfluoroalkyl Complexes

It is now well accepted that compounds containing metal–perfluoroalkyl (M—R_f) bonds are in some ways less reactive and are invariably more thermally stable than analogues containing hydrocarbon alkyl (M—R) linkages. For the purposes of this review it is appropriate to consider evidence and arguments concerning the underlying reasons for these general phenomenological observations, together with some representative examples. Early (6,7) and subsequent (9) synthetic approaches to molecules of this general type have been the subject of extensive reviews. The physical properties and chemical reactivity of transition metal compounds containing trifluoromethyl ligands have also been exhaustively surveyed (10).

1. Relative Thermodynamic Stabilities, Bond Lengths, and
   Bond Strengths in M—R_f and M—R Complexes

The only quantitative numerical data that allow a comparison of the relative thermodynamic stabilities of M—R and M—R_f σ bonds are the bond dissociation energies $D([CF_3—Mn(CO)_5]) = 172 \pm 7$ kJ/mol and $D([CH_3—Mn(CO)_5]) = 153 \pm 5$ kJ/mol (11). The value for the CF_3 complex is identical within experimental error to that found for the corresponding Mn—$C(sp^2)$ bond in the phenyl derivative (11). A relative strengthening of the Mn—CF_3 bond compared to the Mn—CH_3 bond is implied by the fact that the difference (19 kJ/mol) between the values for the CF_3 and CH_3 bonds to manganese exceeds that (6 kJ/mol) between $D(H$—$CF_3)$ and $D(H$—$CH_3)$ (11).

Metal–carbon distances in fluoroalkyl compounds are significantly shorter than those in hydrocarbon analogues, evidence usually taken as implying a stronger bond for the former. Two examples of observable M—C bond shortening can be found in comparative crystallographic studies of the molybdenum complexes 1 [Mo—η^1-C_3F_7 = 2.288(9) Å] (12) and 2 [Mo—η^1-C_2H_5 = 2.397(19) Å] (13,14), and the platinum (II) complexes 3 [Pt—η^1-C_2F_5 = 2.002(8) Å] and 4 [Pt—η^1-CH_3 = 2.081(6) Å] (15). The structure of the CF_3 analogue of 3 has also been determined, but disorder between the CF_3 and Cl ligands precluded meaningful bond length comparison with 4 (15).

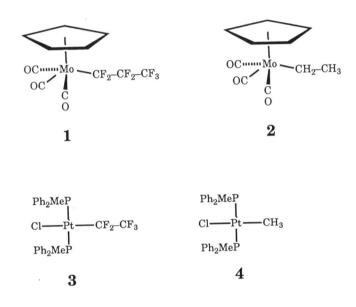

The origins of this relative bond shortening have been the subject of some controversy. Early suggestions that the CF_3 ligand is a good π acceptor due to the availability of low-lying σ^*(C—F) orbitals, and that short M—CF_3 distances, long C—F distances, and the relatively low values for v_{C-F} in the IR spectra of fluoroalkyl complexes result from back bonding between the metal and these acceptor orbitals (16–19), have now fallen into disfavor. Subsequent calculations of CO force constants in [Mn(CH_3)(CO)$_5$] and [Mn(CF_3)-(CO)$_5$] led to the conclusion that CH_3 and CF_3 are equally poor π acceptors (24), and Fenske–Hall calculations on the same pair of molecules are inconsistent with significant π-acceptor properties for the CF_3 ligand, even when the basis set is expanded to include 3p and 3d orbits on carbon (21). The latter calculations indicate that the $6a_1$ orbital used by CF_3 for σ bonding to the metal contains 13.5% C(2s) character, whereas the corresponding $3a_1$ orbital of CH_3 contains only 6.4% $C(2s)$ character. Because the electron density in the CF_3 orbital is concentrated closer to carbon, the fluoroalkyl–metal bond should be shorter (21). It has also been argued that a 7% increase in s character alone is insufficient to account for the observed (~ 0.08 Å) shortening in complex 3 compared to 4 (15,22). Because the Fenske–Hall calculations also indicate that the CF_3 carbon is more positively charged than that in the CH_3 complex, attractive electrostatic interactions could also explain the observed bond shortening (21). Notably, such an electrostatic effect should lower the energies of all bonding orbitals and could also account for the increased thermodynamic stability of fluoroalkyl complexes by stabilizing all the metal–ligand bonds (21). Photoelectron spectroscopy (PES) studies of $[Mn(CH_3)(CO)_5]$ and $[Mn(CF_3)(CO)_5]$ support this idea (23,24), and show a significant increase in the $CF_3(C_{1s})$ orbital binding energy for the latter complex (25). Finally, it should also be noted that an increase in thermal stability does not necessarily imply stronger bonds and that the steric effect of a larger fluoroalkyl ligand may also contribute to stability (21).

2. Relative Reactivities of M—R_f and M—R Bonds

Arguments based on a stronger M—C bond in fluoroalkyl complexes have also been used to explain the lack of reactivity of these compounds toward migratory insertion to coordinated carbon monoxide. For example, the reaction of $[Mn(CH_3)(CO)_5]$ with CO to give $[Mn(COCH_3)(CO)_5]$ is facile, but the corresponding reaction of $[Mn(CF_3)(CO)_5]$ does not proceed, even under high pressures of CO (26). Indeed, the $[Mn(CF_3)(CO)_5]$ complex and other transition metal fluoroalkyl complexes are synthesized by the reverse reaction, decarbonylation of the corresponding acyl precursor (9,15,16,27,28). Both thermodynamic and kinetic arguments have been advanced to explain this phenomenon. The thermodynamics of the carbonylation reaction shown in Eq. (1)

$$R\text{—}Mn(CO)_5 + CO \rightleftharpoons R\text{—}\overset{\overset{\textstyle O}{\|}}{C}\text{—}Mn(CO)_5 \qquad (1)$$

have been studied (11) and indicate that the enthalpy change (ΔH^0) is negative in sign: -54 ± 8 kJ/mol for R = CH_3 and -12 ± 7 kJ/mol for R = CF_3. However, incorporation of the exogenous CO molecule into the product results in loss of the translational component of its entropy ($\Delta S_{trans} \approx 150$ JK^{-1} mol^{-1}). Assuming that the overall entropy change for the migratory insertion reaction is probably not less than this value, ΔG^0 remains negative for CH_3 migration to CO but becomes positive for the corresponding CF_3 migration (11). An alternative explanation, based on complete neglect of differential overlap (CNDO) molecular orbital calculations, suggests a kinetic origin for the lack of reactivity of the CF_3 complex, which is not connected with any increased strength of the $Mn-CF_3$ bond (29). The calculations indicate that, although the charges on the *cis*-CO ligands remain virtually the same in the CH_3 and CF_3 complexes, the CF_3 carbon is in fact positively charged due to the electronegative fluorines (see also Ref. 21), and that migratory insertion via nucleophilic attack on an adjacent CO is suppressed for this reason (29). An extended Hückel molecular orbital (EHMO) study of the same migration reaction, which does not consider the CF_3 analogue directly, supports the idea that lowering the energy of the σ-donor orbital of the migrating group should increase the activation energy for migration (30). It is interesting that evidence for migration of a pentafluorophenyl group to CO, induced by electrochemical oxidation, has been reported recently (31), suggesting that fluoroalkyl migrations may also be facilitated by this method.

Finally, the metal–perfluoroalkyl linkage also appears to be less susceptible to facile decomposition by the α- or β-elimination pathways that dominate much of the chemistry of hydrocarbon alkyls and lead to metal hydrides. The absence of these reaction pathways, at least for the later transition metals, may reflect the relative strength of the C—F bond versus the M—F bond compared to C—H/M—H analogues (32). However, α-fluoride abstraction reactions can be accomplished with exogenous fluoride acceptors to give fluorinated carbene complexes (see Section III,B,1). One example of an apparent β-fluorine elimination reaction is shown in Eq. (2) (33) and presumably is driven by the stronger bond to fluorine formed by early transition

$$ (2) $$

metals. Similar activation reactions of δ-C—F bonds in fluoroaryl groups have been reported to give isolable examples of metal (aryl)fluoride compounds (34,35) (see also Section III,F). The isolation of the η^2-C_2F_4 complexes $[RhF(\eta^2$-$C_2F_4)(\eta^2$-$C_2H_4)]_4$ (36), $[RhF(\eta^2$-$C_2F_4)(PPh_3)_2]$ (36), and $[IrF(CO)(\eta^2$-$C_2F_4)(PPh_3)_2]$ (37), all of which contain M—F bonds, is interesting because no insertion is reported to occur to give the η^1-C_2F_5 ligand, a process that would be the reverse of a β-fluorine elimination.

B. *Difluoromethylidene and Fluoromethylidyne Complexes*

The syntheses, structures, spectroscopy, bonding, and reactivity of difluoromethylidene and other halocarbene complexes have been reviewed recently (*10*). Only some typical structures and more recent results are discussed briefly here.

1. *Terminal Difluoromethylidene Complexes*

The first terminal difluoromethylidene complex resulted from fluoride ion abstraction from $[Mo(\eta^5\text{-}C_5H_5)(CF_3)(CO)_3]$ by SbF_5, to give the spectroscopically characterized complex $[Mo(\eta^5\text{-}C_5H_5)(CF_2)(CO)_3]^+SbF_6^-$ (*38*). Similar approaches, using BF_3 as the Lewis acid, have resulted in the iron and manganese analogues $[Fe(\eta^5\text{-}C_5H_5)(CF_2)(CO)_2]^+BF_4^-$ and $[Mn(CF_2)-(CO)_5]^+BF_4^-$ (*39*). These cationic compounds contain an extremely electrophilic CF_2 ligand that is readily hydrolyzed to give a carbon monoxide. Phosphine substitution at the metal results in a more stable complex **5**, which

5   **6**

has been crystallographically characterized (*40*). In the ground-state conformation, the plane of the difluoromethylidene ligand is oriented almost perpendicular to the Fe–P axis, in order to maximize back bonding, yet the CF_2 rotates fast on the NMR time scale, interconverting the two fluorine environments and indicating that back bonding from the metal to the fluorinated carbon is poor (*40*).

In contrast, the plane of the CF_2 ligand in the osmium complex **6** lies perpendicular to the equatorial plane of the trigonal bipyramid, the conformation expected in order to maximize back bonding, and the rate of rotation of the CF_2 ligand is slow on the NMR time scale, due to more extensive back bonding from the Os(0) center (*10*).

The effect of back bonding on the reactivity of difluoromethylidene ligands is nicely illustrated by the pair of ruthenium complexes **7** and **8** (*10*). Compound **7** formally contains a Ru(II) d^6 center and the CF_2 ligand is electrophilic, whereas **8** is a complex of Ru(0), in which better back bonding to the

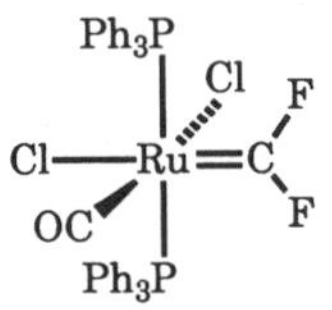

7 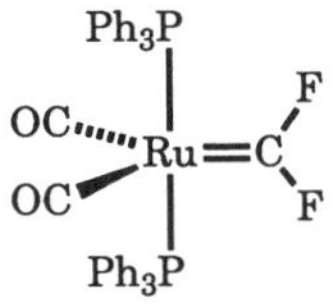   **8** 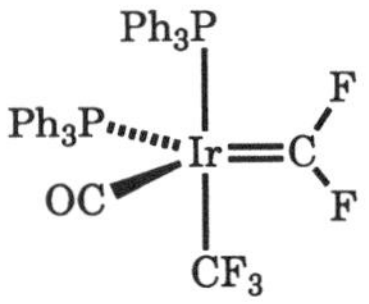   **9**

CF_2 ligand results in nucleophilic behavior. An example of a complex containing both a CF_3 and a CF_2 ligand is the Ir(I) species **9** (*10*), in which CF_2 rotation is also slow on the NMR time scale.

2. Bridging Difluoromethylidene Complexes

Various homodinuclear complexes containing one or two μ-CF_2 ligands, **10** (*41*), **11** (*41,42*), **12** (*42–44*), and **13** (*43,44*), have been characterized, and several heterodinuclear analogues are surveyed in a recent review (*10*).

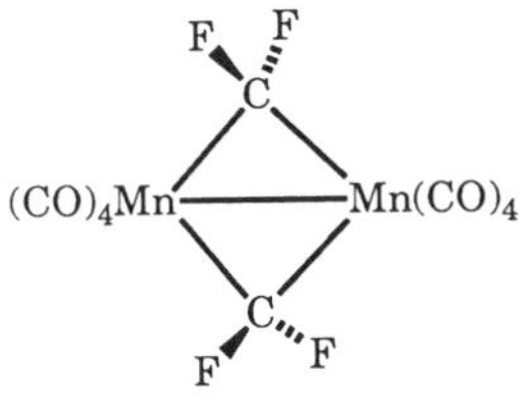

10

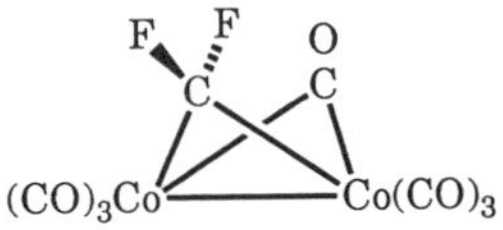

11

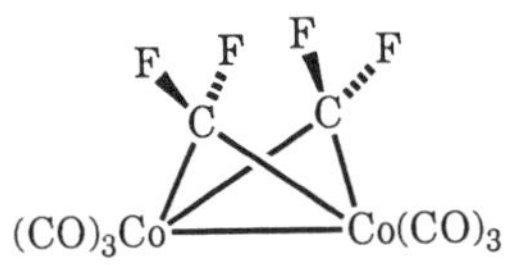

12

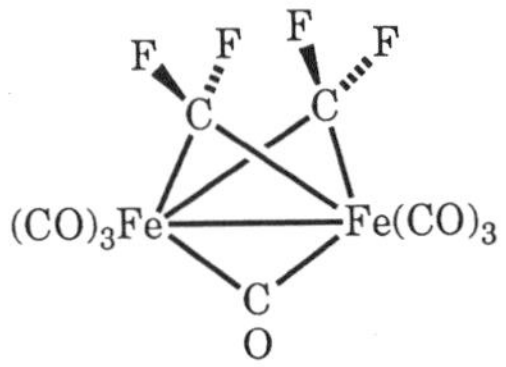

13

3. Bridging Fluoromethylidyne Complexes

No terminal fluoromethylidyne complexes have been reported, but triply bridging CF ligands are observed in the cobalt complex **14** (*45*), in various substituted analogues (*46–49*), and in the osmium complex **15** (*50*). An iron cluster **16** containing two triply bridging CF ligands has also been reported

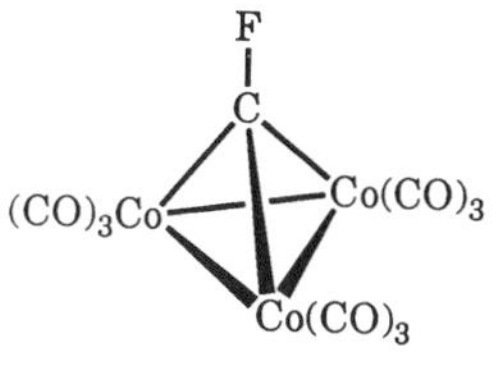

14

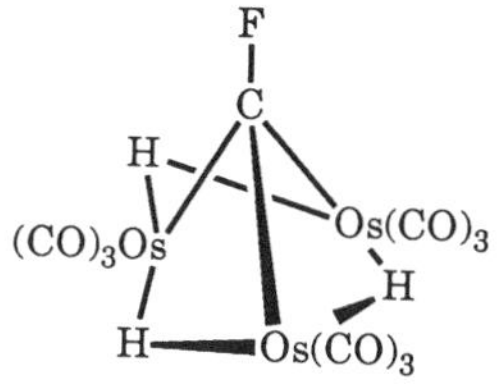

15

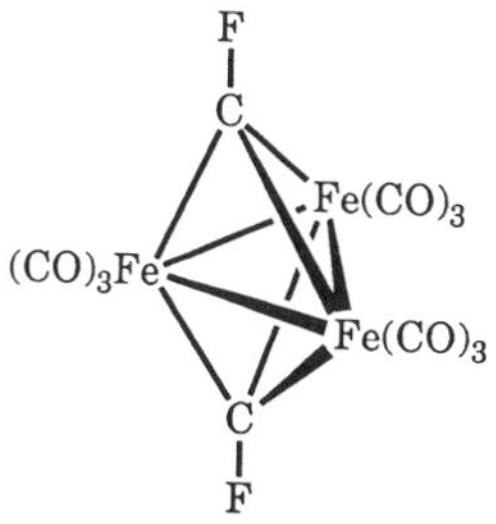

16

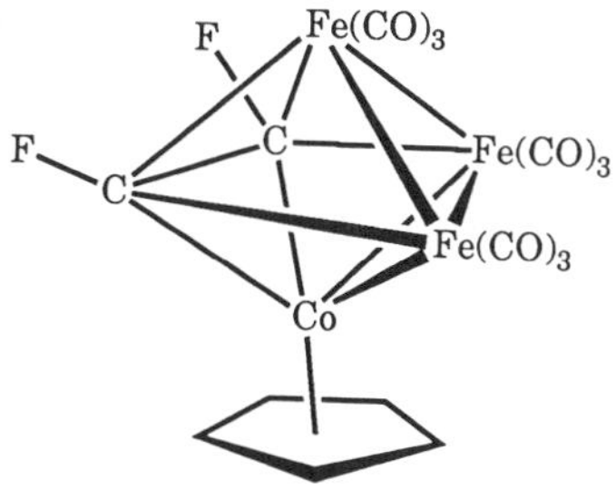

17

(51,52) to react with $[Co(\eta^5\text{-}C_5H_5)(CO)_2]$ to give the cluster **17**, in which the two CF units couple to form a difluoroacetylene subunit (52,53).

C. Perfluoroolefin Complexes

1. Structure and Bonding

Organometallic complexes in which perfluorinated ligands are bound to the metal via molecular orbitals that are formally $\pi(C=C)$ and $\pi^*(C=C)$ in character are generally observed to be more thermally stable than their hydrocarbon analogues. Several complexes containing η^2-tetrafluoroethylene are known, and the metal–olefin bond strengths in such complexes are thought to exceed those in hydrocarbon analogues, although the caveats from Section III,B concerning bond lengths and strengths are also relevant here. For example, the platinum complex $[Pt(\eta^2\text{-}C_2H_4)(PPh_3)_2]$ decomposes at 122°C to liberate ethylene and form (triphenylphosphine)platinum clusters (54), and the ethylene ligand can be displaced by exogenous ligands such as t-butylisocyanide or triphenylphosphine under mild conditions (55). In contrast, its perfluorinated analogue $[Pt(\eta^2\text{-}C_2F_4)(PPh_3)_2]$ does not decompose up to its melting point (218°C) and the tetrafluoroethylene is not displaced

by triphenylphosphine (56). However, equilibrium binding-constant studies
on [Ni(η^2-olefin){P(OR)$_3$}$_2$] complexes indicate that ethylene and perfluoro-
ethylene form complexes of almost identical stabilities and that partially
fluorinated ethylenes give less stable complexes than does ethylene (57). An
excellent review of the factors governing metal–olefin interactions in hydrocarbon and fluoroolefins, with a multitude of structural examples, has been
published (58). Pyramidalization of ligated carbon atoms bearing fluorine is
a recurring structural feature and is also predicted for fluorinated carbanions
(see Section III,D). Some compounds containing η^2-C_2F_4 ligands are also
mentioned in Section III,A,2. The coupling of tetrafluoroethylene molecules
to form metallacyclopentanes has been reviewed (3).

18   **19**

One example that exemplifies the structural differences between coordinated C_2H_4 and C_2F_4 ligands is provided by the rhodium complex **18** (59,60).
The Rh—C distances illustrate that the fluoroolefinic ligand is held closer to
the metal center [2.167(2) Å for Rh—CH (average) vs. 2.024(2) Å for Rh—CF
(average)] (60). Whereas the C=C bond distance in free C_2F_4 is actually
shorter (1.311 Å) than that in C_2H_4 (1.337 Å) (61), the reverse is true for the
ligated olefins [(F_2C=CF_2) = 1.405(7) Å; (H_2C=CH_2) = 1.358(9) Å] (60).
The fluorines are also severely bent back out of the plane containing the C—C
axis, affording a "metallacyclopropane" Rh—C_2F_4 geometry, which is more
pyramidalized at carbon (60).

2. *Dynamic Behavior*

NMR spectroscopic studies of two rhodium complexes **18** and **19** also
indicate that the binding of the C_2F_4 group to rhodium is better pictured as
a conformationally locked metallacyclopropane (59). The activation barrier
for rotation of the C_2H_4 ligands around the Rh–olefin (centroid) vector in
19 was determined to be 15.0 ± 0.2 kcal/mol. In contrast, the rotational
barrier for the C_2H_4 ligand in **18** was demonstrated to be 13.6 ± 0.6 kcal/mol

and no rotation of the coordinated C_2F_4 could be observed at 100°C on the NMR time scale (59). A large change in the fluorine–fluorine coupling constants upon coordination was also observed, indicating substantial pyramidalization of the tetrafluoroethylene carbon atoms. This, along with the apparently locked conformation of the C_2F_4 ligand, indicates the fluoroolefin–metal bonding may be better represented as occurring via two σ bonds. The lower barrier for C_2H_4 rotation in **18** as compared to **19** is consistent with the idea that the C_2F_4 ligand is an excellent π acceptor, which reduces electron density on the rhodium and diminishes back bonding to the hydrocarbon olefin (59).

Thus, solution and solid-state structural studies of such fluoroolefin compounds have helped to formulate and refine theories of metal–carbon bonding. Studies of the reactivity of metal–fluoroolefin compounds have also provided useful models and predictions for hydrocarbon systems. For example, the oxidative cyclization of fluoroolefins within the coordination sphere of a metal to give metallacyclopentane compounds was discovered many years before the importance of the corresponding reaction of hydrocarbon olefins was realized (3).

3. Ring Opening of Perfluorocyclopropene

Complexes of cyclic perfluoroolefins are also quite common. For example, the iron and platinum complexes of perfluorocyclobutene, **20** (62) and **21** (56), have been prepared. Recently the chemistry of the smallest cyclic perfluoroolefin, perfluorocyclopropene, has been explored, and initial results indicate

20   **21**

that the isolable products are not those resulting from binding of the metal to the olefinic functionality. Reaction of perfluorocyclopropene with [Pt-$(\eta^2\text{-}C_2H_4)(PR_3)_2$] results in facile displacement of the ethylene ligand at −78°C but no η^2-perfluorocyclopropene complex is observed, in contrast to hydrocarbon cyclopropenes (63,64) and perfluorocyclobutene (56). Instead,

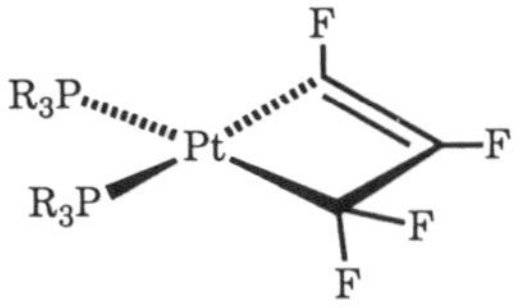

22   a. R = Ph

     b. R = Me

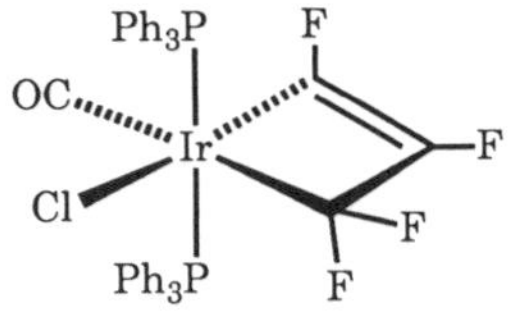

23

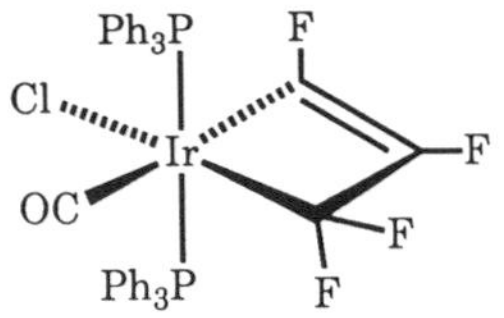

24

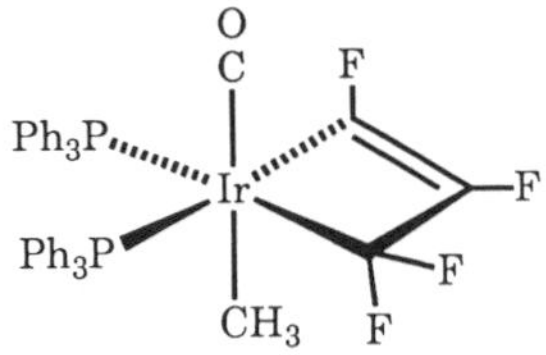

25

ring cleavage occurs to yield the platinacyclobutene complexes **22** (65). The molecular structure of **22a** is shown in Fig. 1. That this process can be thought of as a formal oxidative addition of a C—C σ bond to the metal is corroborated by the stereoselectivity observed in the corresponding reactions of perfluorocyclopropene with Ir(I) complexes. Reaction with trans-[IrCl-(CO)(PPh$_3$)$_2$] occurs stereoselectively, but not regioselectively, to give the two stereoisomers **23** and **24** in a 1:1 ratio, whereas the corresponding reaction with trans-[Ir(CH$_3$)(CO)(PPh$_3$)$_2$] affords the single stereoisomer **25** (66).The stereoselectivities are identical to those observed for unambiguously σ-bond additions of H—H and Si—H substrates to the same Ir complexes under conditions of kinetic control (67,68). Whereas H—H and Si—H additions are reversible, the C—C cleavage observed for perfluorocyclopropene appears to be irreversible, presumably due to the formation of two strong M—C bonds and release of ring strain in the cyclopropene. The differing reactivities of hydrocarbon cyclopropenes and their perfluoro analogue may result from lowering of the σ^*-orbital energies of the cyclopropene ring by the fluorine substituents, although this hypothesis requires further support of theoretical calculations at the appropriate level.

D. Perfluoroallylic Complexes

Although various η^1-pentafluoroallyl transition metal complexes such as **26** are known (69), only a single η^3-pentafluoroallyl compound, **27**, has been iso-

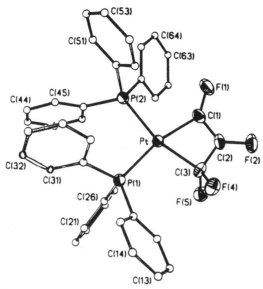

F‌IG. 1.   Molecular structure of **22a**. Carbon atoms in the phenyl rings are shown as spheres of arbitrary radius. [Reprinted with permission from Hemond *et al.* (*65*).] Selected bond lengths (Å): Pt–P(1), 2.302(2); Pt–P(2), 2.336(2); Pt–C(1), 2.039(9); Pt–C(3), 2.079(8); C(1)–C(2), 1.351(12); C(2)–C(3), 1.474(12); C(1)–F(1), 1.325(10); C(2)–F(2), 1.373(11); C(3)–F(4), 1.371(8); C(3)–F(5), 1.414(9). Selected bond angles (degrees): P(1)–Pt–P(2), 98.8(1); Pt–C(1)–C(2), 97.5(6); P(1)–Pt–C(1), 165.2(2), Pt–C(1)–F(1), 138.0(6); P(1)–Pt–C(3), 99.6(2); Pt–C(3)–C(2), 92.5(5); P(2)–Pt–C(1), 96.0(2); Pt–C(3)–F(4), 123.6(5), P(2)–Pt–C(3), 160.6(2); Pt–C(3)–F(5), 115.6(5); C(1)–Pt–C(3), 65.8(3); C(1)–C(2)–C(3), 104.7(7); F(1)–C(1)–C(2), 124.5(8); F(2)–C(2)–C(1), 130.5(8); F(2)–C(2)–C(3), 124.8(7); F(5)–C(3)–C(2), 111.0(6); F(4)–C(3)–C(2), 112.5(6); F(4)–C(3)–F(5), 102.1(6).

lated as an unstable yellow oil from the reaction of perfluoroallyl iodide with $[\text{Zn}\{\text{Co}(\text{CO})_4\}_2]$ (*70*). An intermediate η^1-perfluoroallyl complex $[\text{Co}(\eta^1\text{-}C_3F_5)(\text{CO})_4]$ was detected but not isolated. The CO stretching frequencies for **27** ($\nu_{\text{CO}} = 2115, 2064 \text{ cm}^{-1}$) are considerably higher than those observed

28   **29**

for its hydrocarbon analogue $[Co(\eta^3\text{-}C_3H_5)(CO)_3]$ (ν_{CO} = 2065, 1998 cm^{-1}) (71), indicating that the η^3-perfluoroallyl ligand is a better electron acceptor. Treatment of **27** with PPh$_3$ yields the *cis*-η^1-perfluoroallyl complex **28** and the *trans*-η^1-perfluoro-1-propenyl isomer **29**. It is suggested that both *cis*- and *trans*-η^1-perfluoroally isomers are initially formed but that the trans isomer undergoes a bimolecular fluorine shift to yield **29**. The corresponding bimolecular pathway for isomerization of the cis isomer **28** is thought to be suppressed by the bulky phosphine in the cis position (70). A similar bimolecular pathway has been suggested (72) to account for the facile isomerization of $[Mn(\eta^1\text{-}CF_2\text{---}CX{=}CF_2)(CO)_5]$ to afford $[Mn(\eta^1\text{-}CF{=}CX\text{---}CF_3)(CO)_5]$ [X = F (73,74); X = Cl (72)].

An anionic complex **30** has been synthesized from fluoride attack on Fe-$(1\text{--}4\text{-}\eta^4\text{-octafluorocyclohexa-1,3-diene})(CO)_3]$ (75), and the cobalt complex **31** is the product of the reaction of perfluorocyclopentadiene with $[Co_2\text{-}(CO)_8]$ (76). No subsequent chemistry of these compounds has been reported.

30   **31**

Ab initio MO calculations indicate that, whereas the allyl anion has the expected planar delocalized C_{2v} structure, the minimum energy conformation for its perfluoro analogue is the localized rotated structure **32** (77). The planar delocalized structure **33** lies 25.7 kcal/mol higher in energy and the perfluorocyclopropyl anion **34** lies only 2.2 kcal/mol higher in energy at the self-

32   **33**   **34**

consistent field (SCF) level; inclusion of correlation effects leads to the prediction that **34** is more stable than **32** by 2.1 kcal/mol (77). Destabilization of the planar form of the allyl anion is ascribed to the known proclivity of α-fluorocarbanions to be pyramidal (2) and is relevant to the chemistry of a number of new complexes containing fluoroallylic subunits that are surveyed later in this article.

E. Perfluoro-1,3-diene Complexes

The iron complexes **35** (78) and **36** (79) have been reported. Crystallographic studies on **35** indicate that the (η^4-diene)metal interaction is best represented by the valence bond description shown, in which the perfluorodiene acts as a

35   **36**

strong acceptor ligand and in which a formal two-electron oxidation of the metal center has occurred. These studies also confirm the "long–short–long" pattern of C—C distances in the coordinated diene and extensive rehybridization toward sp^3 for the terminal diene carbon atoms (80). The powerful acceptor properties of the perfluorodiene are also reflected in the significantly higher values for the CO stretching frequencies for **35** (ν_{CO} = 2108, 2054, and 2018 cm^{-1}) compared to its hydrocarbon analogue (78). A poorly characterized rhodium complex of perfluorocyclohexa-1,3-diene, of stoichiometry

$[Rh(C_6F_8)Cl]_2$, has also been reported (78), and an unusual product resulting from vicinal defluorination of this diene is discussed in Section III,G.

This tendency toward oxidative addition is even more conspicuous with perfluorobuta-1,3-diene, which forms the metallacyclic complex **37** rather than an η^4-diene complex (81,82). A poorly characterized cobalt complex of stoichiometry $[Co_2(C_4F_6)(CO)_6]$ has been isolated (81).

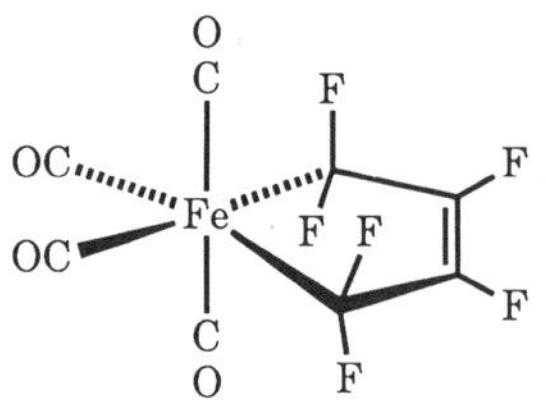

37

Curiously, no η^4-diene complexes have been obtained using perfluoro-cyclopentadiene, although a number of η^2 complexes such as **38** (32), **39** (83a), and $[Co(\eta^5\text{-}C_5H_5)(CO)(\eta^2\text{-}C_5F_6)]$ (83) have been prepared. Once again, the propensity for formal oxidative addition is revealed by the formation of complex **31** (see Section III,D) in the reaction of perfluorocyclopentadiene with

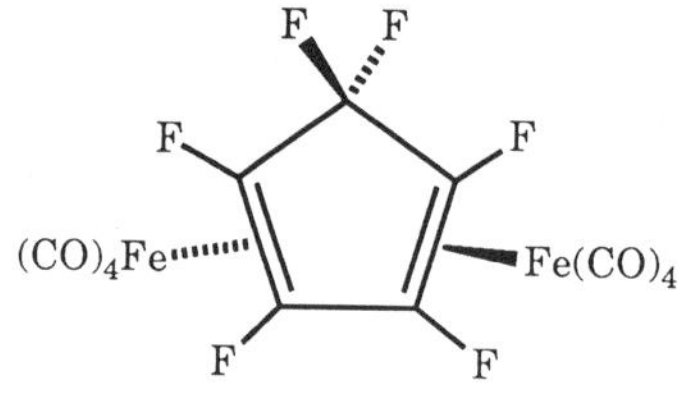

38

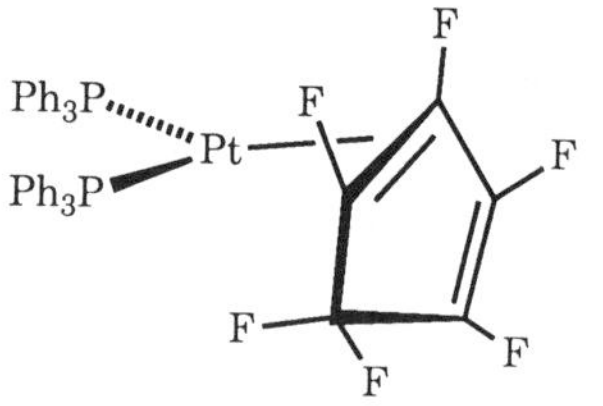

39

$[Co_2(CO)_8]$. This diene and its 5-H relative also react with transition metal hydrides to yield a mixture of insertion products and compounds derived therefrom (84,85).

In view of the ubiquity of the η^5-cyclopentadienyl ligand in organometallic chemistry, and the reports in the literature of perchlorocyclopentadienyl analogues (86), it is both curious and unfortunate that attempts (85) to prepare complexes of the η^5-pentafluorocyclopentadienyl ligand from the known $C_5F_5^-$ anion (87,88) have failed.

F. Perfluorobenzene Complexes

Isolable transition metal complexes of hexafluorobenzene are rare indeed. The only examples of η^6-perfluorobenzene complexes appear to be the chromium compounds 40 (89,90) and 41 (89), which have been prepared by metal atom vaporization techniques. Attempts to extend these methods to other metals are reported to be unsuccessful and frequently explosive (91)!

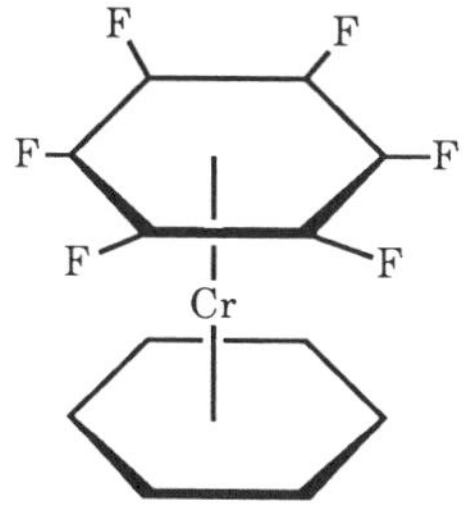

40

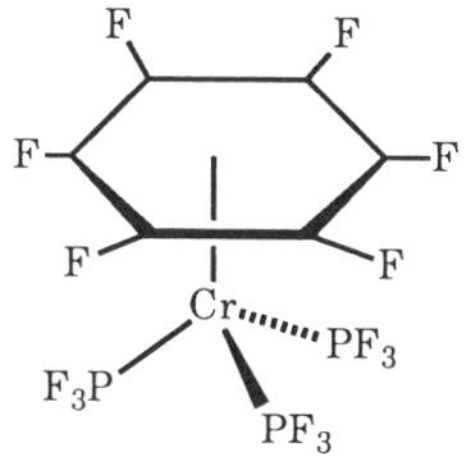

41

The rhodium complex 42, containing the first example of an η^2-hexafluorobenzene ligand, has been prepared and crystallographically characterized (92).

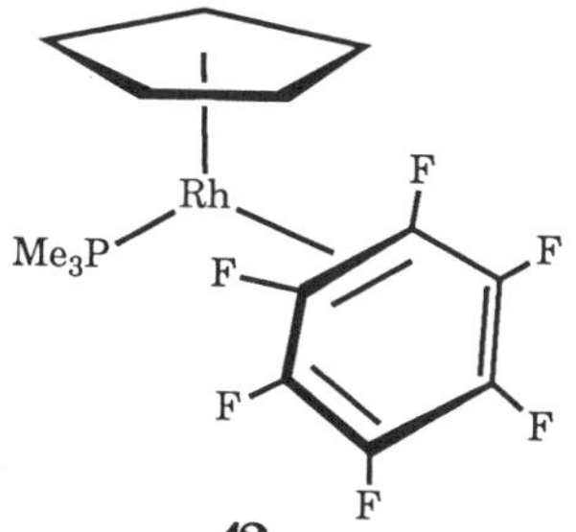

42

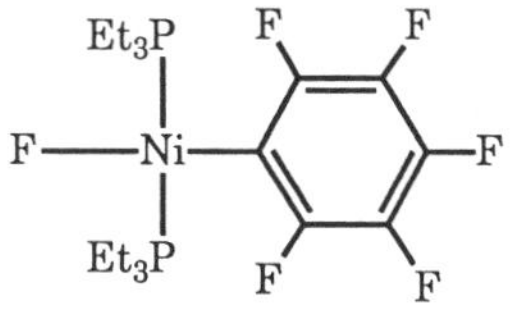

43

It is noteworthy that this latter compound is not reported to undergo C—F oxidative addition to give the corresponding (pentafluorophenyl)(fluoro)-Rh(III) complex, in contrast to a hydrocarbon analogue, which undergoes facile C—H activation to afford [Rh(η^5-C$_5$Me$_5$)(H)(η^1-C$_6$H$_5$)(PMe$_3$)] (93). However, the Ni(0) complex [Ni(1,2,5,6-η^4-COD)(PEt$_3$)$_2$] has been reported to react with perfluorobenzene to afford low yields of the thermally unstable product 43, resulting from a C—F bond activation (94). The platinum complex 44 of the Dewar valence isomer of hexafluorobenzene has also been prepared (95).

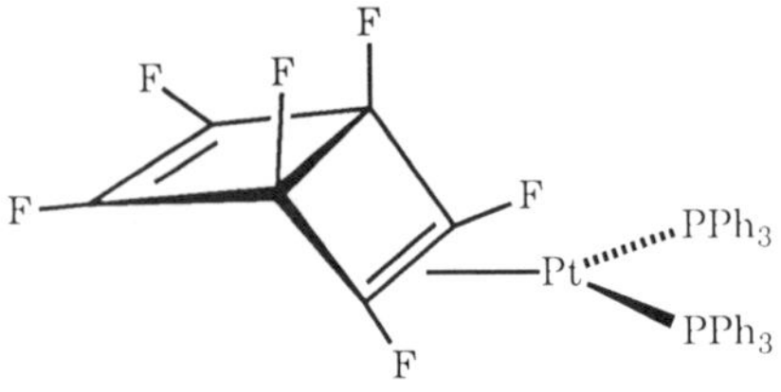

44

G. Metallation Reactions

In contrast to their hydrocarbon analogues, many unsaturated fluoro-carbons react with anionic transition metal nucleophiles to yield products arising from net displacement of fluoride ion. This synthetic approach has been successful in producing a host of fluorovinyl and fluoroaryl complexes (95–110).

In contrast to its reported reaction with perfluorocyclopentadiene to give **31** (see Section III,D), $[Co_2(CO)_8]$ has been shown to react with perfluoro-cyclohexa-1,3-diene to yield the dinuclear μ-alkyne complex **45** (111,112). The

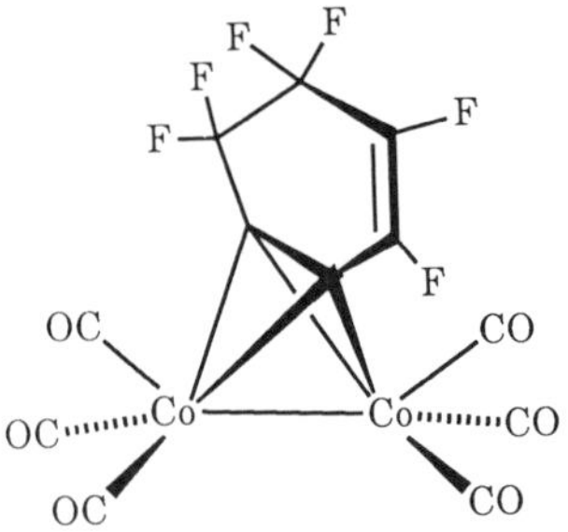

45

mechanism of this double defluorination reaction is still unknown, but gas phase studies of the reactions of a variety of transition metal ions with fluoro-carbons have been shown to result in similar vicinal defluorination pathways, with the two fluorine atoms being eliminated in the form of COF_2 (113).

H. Miscellaneous Ligands with α-Fluorines

One example of a fluoroformyl complex **46** has been prepared by the reac-tion of XeF_2 with $[Ir(CO)_3(PEt_3)_2]^+$ (114). Some routes to perfluorovinyl

46

metal complexes are discussed in Section III,G. Although no isolable mono-nuclear complexes of difluoroacetylene have been reported, gas phase species containing coordinated C_2F_2 have been observed (113). A single platinum compound **47** of monofluoroacetylene has been prepared (115). Cluster compounds containing the difluoroacetylene subunit have been synthesized (see Section III,B,3), and the tetrafluoroferrole complex **48**, which may arise by coupling of two difluoroacetylenes, has also been reported (116).

47   **48**

In contrast to the results obtained in reactions of M—CF_3 complexes with BF_3 (see Section III,B,1), the reaction of the perfluoronickelacyclopentane complex **49** appears to proceed via fluoride abstraction, followed by phosphine migration to the electron-deficient carbon to afford the α-fluorophosphonium ylid complex **50** and related derivatives (117).

49   **50**

IV

OCTAFLUOROCYCLOOCTA-1,3,5,7-TETRAENE AND ITS VALENCE ISOMERS

The hydrocarbon ligand cycloocta-1,3,5,7-tetraene (COT) occupies a place of some historical significance in organic (118) and organometallic chemistry (118,119). Its flexible hapticity, ranging from η^2 to η^8 binding to metals, and the haptotropic rearrangements observed in many of its complexes have provided synthetic and theoretical chemists with many challenges over the past three decades, and an enormous number of COT complexes has been prepared. The synthetic availability of multigram quantities of its perfluorinated analogue OFCOT **51** (120) provided an excellent opportunity to compare the transition metal chemistry of this perfluoropolyene with that of its hydrocarbon analogue.

A. Synthesis Structure of OFCOT and Its Valence Isomers

OFCOT can be synthesized by the procedure shown in Scheme 1 (120), and its solid-state structure is the expected D_{2d} puckered ring shown in Fig. 2 (121). The skeletal distances and angles are not significantly different from those

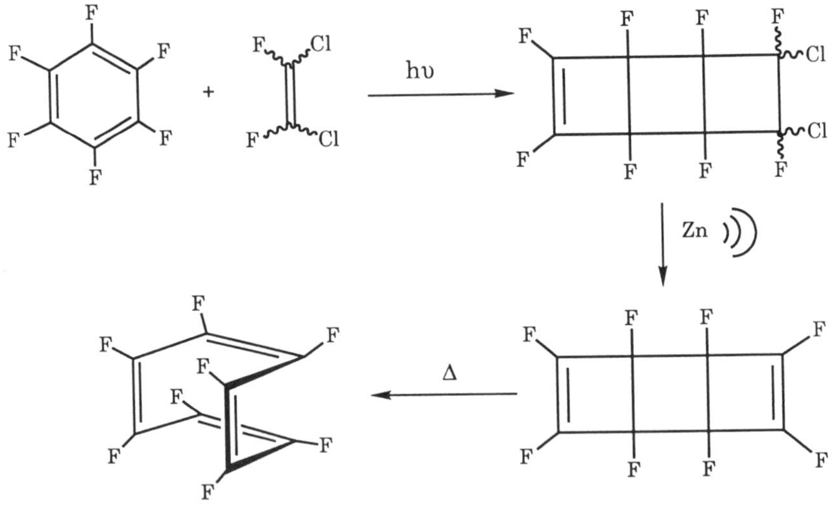

51

SCHEME 1

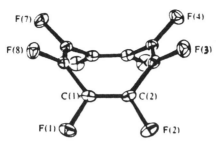

FIG. 2.   Molecular structure of OFCOT (**51**). [Reprinted with permission from Laird and Davis (*121*).] Selected average bond distances (Å) and angles (degrees) are given in Table II (Section XI,B).

found in its hydrocarbon analogue (*122*). OFCOT exhibits remarkable thermal stability, surviving unchanged after flash vacuum pyrolysis at 600°C (*123*).

$$\textbf{51} \rightleftharpoons \textbf{52} \tag{3}$$

Like its hydrocarbon analogue, it exists in thermal equilibrium with its octafluorobicyclo[4.2.0]octatriene valence isomer **52**, as shown in Eq. (3); $K_{eq} = 0.003$ at 20°C in acetone-d_6) (*123*). Valence isomer **52** can be prepared independently, but has a half-life of only 14 min at 0°C (*124*). OFCOT undergoes a photochemical ring closure to give a 20:1 mixture of the anti- and syn-tricyclic valence isomers **53** and **54**, which in turn can be converted thermally back to OFCOT at 150°C (*125*).

53   **54**

B. *Attempts to Reduce OFCOT to Its Dianion*

From a synthetic objective it is unfortunate that attempts to produce the dianion of OFCOT by various reduction procedures have not resulted in a stable dianion. Although the organic decomposition products resulting from reduction by alkali metals or sodium naphthalenide are unknown, fluoride ion is produced (*126*). However, the nine-π-electron radical anion has been produced at low temperatures by γ irradiation of OFCOT, and electron spin resonance (ESR) spectroscopy indicates that it possesses the anticipated planar delocalized D_{8h} structure **55** (*127*). The unavailability of the dianion

55

severely restricts the synthetic repertoire whereby transition metal complexes of OFCOT can be approached, and leaves only the direct reactions of OFCOT with suitable organometallic precursors as a useful preparative procedure.

V

UNSUCCESSFUL ATTEMPTS TO PREPARE TRANSITION METAL COMPLEXES OF OFCOT

Before surveying the coordination chemistry of OFCOT, some unsuccessful reactions should be noted. Photolysis of OFCOT in the presence of $[Re(\eta^5\text{-}C_5Me_5)(CO)_3]$, $[Mn(\eta^5\text{-}C_5Me_5)(CO)_2(CS)]$, $[Nb(\eta^5\text{-}C_5H_5)(CO)_4]$, $[V(\eta^5\text{-}C_5H_5)(CO)_4]$, $[V(\eta^5\text{-}C_5Me_5)(CO)_4]$ ($\lambda > 170$ nm and $\lambda > 280$ nm), $[V(CO)_6]$, $[Mo(\eta^5\text{-}C_5H_5)(CO)_3]_2$, and $[Mo(\eta^5\text{-}C_5Me_5)(CO)_3]_2$ did not afford complexes derived from OFCOT. Likewise, the thermal reaction of OFCOT with $[Cr(CO)_6]$, $[Cr(CO)_3(CH_3CN)_3]$, $[Mo(CO)_3(CH_3CN)_3]$, $[W(CO)_3(CH_3CH_2CN)_3]$, $[Mo(\eta^5\text{-}C_5H_5)(CO)_3]_2$, or $[Mo(\eta^5\text{-}C_5Me_5)\text{-}(CO)_3]_2$ did not afford complexes containing any fluorinated ligands. Thus there has been no success so far in preparing OFCOT complexes of any metals in Group 5 or 6 (*128*). Two current representatives from Group 7 are described below.

VI

MANGANESE COMPLEXES OF OFCOT AND PERFLUOROTRICYCLO[4.2.0.0^{2,5}]OCTA-3,7-DIENE

A. Synthesis and Physical Properties of η^6-OFCOT and η^6-COT Complexes of Manganese

The half-sandwich compounds $[Mn(\eta^5\text{-}C_5R_5)(CO)_3]$ undergo complete replacement of the three CO ligands by OFCOT upon photolysis in toluene solution, to afford the $(1-6)\text{-}\eta^6$-OFCOT complexes **56** (*129*). Curiously, irradiation of a toluene solution of $[Mn(\eta^5\text{-}C_5H_5)(CO)_3]$ and OFCOT at $-15°C$

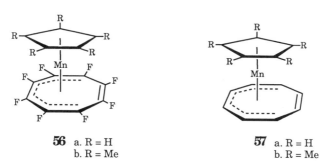

56 a. R = H
       b. R = Me

57 a. R = H
       b. R = Me

did not yield any fluorinated products arising from CO replacement. At this temperature either CO is not displaced from $[Mn(\eta^5\text{-}C_5H_5)(CO)_3]$ or recombination of photodissociated CO to reform $[Mn(\eta^5\text{-}C_5H_5)(CO)_3]$ is favored over OFCOT coordination. The observation that the photochemical reaction forms **56a** at ambient temperature and not at $-15°C$ may indicate an activation barrier for the coordination of the eight-membered ring to the coordinatively unsaturated intermediate $[Mn(\eta^5\text{-}C_5H_5)(CO)_2]$, i.e., the mechanism for the formation of **56a** may involve an initial photochemical displacement of one CO ligand followed by a thermally activated coordination of the ring with subsequent total CO replacement by the ligand to yield the observed product. No evidence for any fluorocarbon products resulting from incomplete CO displacement from $[Mn(\eta^5\text{-}C_5R_5)(CO)_3]$ has been obtained. A dicarbonyl complex containing an η^2-COT ligand, $[Mn(\eta^5\text{-}C_5H_5)(CO)_2(\eta^2\text{-}C_8H_8)]$, has been isolated from the reaction of $[Mn(\eta^5\text{-}C_5H_5)(CO)_2(THF)]$ (THF, tetrahydrofuran) with COT (*130*), but no reaction of OFCOT with $[Mn(\eta^5\text{-}C_5H_5)(CO)_2(THF)]$ has been observed (*128*).

The fluorocarbon complexes **56** are notably different in their chemical properties as compared to their hydrocarbon analogues **57**. Complex **57a** is extremely air sensitive and can only be isolated at low temperatures in

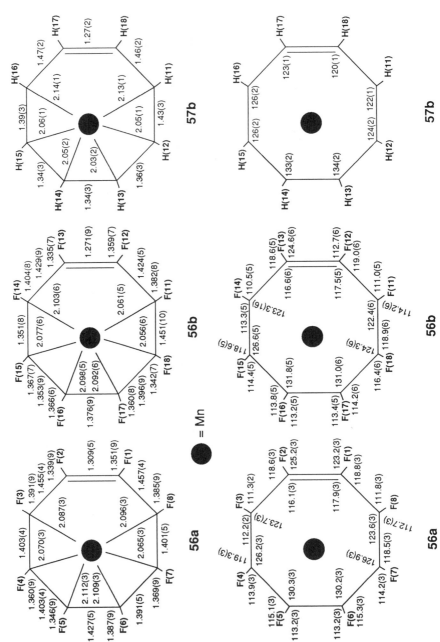

FIG. 3. Bond distances (Å, top) and angles (degrees, bottom) for **56a**, **56b**, and **57b**. [Reprinted with permission from Hemond et al.

the presence of excess COT (*131,132*). Its pentamethylcyclopentadienyl relative **57b** is more thermally stable and can be stored under nitrogen for long periods, but is exceedingly air sensitive (*129*). In contrast, the OFCOT complexes **56a** and **56b** are air stable at their melting points of 157 and 172°C, respectively, providing another rather dramatic example of the effect of perfluorination of a ligand on the physical properties of a transition metal complex.

B. *Structural Comparison between η^6-OFCOT and η^6-COT Ligands*

The two compounds **56a** and **56b** are virtually isostructural with the hydrocarbon analogue **57b** (*129*). A comparison of distances and angles involving the eight-membered rings is presented in Fig. 3 and some planes and interplanar angles are defined and described in Table I. The molecular structures of **56a**, **56b**, and **57b** (Fig. 4) clearly define the η^6 ligation of the OFCOT and COT ligands and show that the uncoordinated olefin is in the exo conformation, as observed in the related hydrocarbon complex [Mo(η^6-COT)(CO)$_3$] (*133*). The six coordinated carbons of the eight-membered rings (Plane II,

TABLE I

MOLECULAR PLANES CALCULATIONS WITH DEVIATIONS (Å) AND
INTERPLANAR ANGLES (DEGREES) FOR COMPLEXES **56a**, **56b**, AND **57b**

		56a		**56b**		**57b**
Plane I	C(9)	−0.002(6)	C(3)	0.007(8)	C(1)	−0.010(8)
	C(10)	0.001(6)	C(4)	−0.000(8)	C(2)	0.003(8)
	C(11)	−0.000(6)	C(5)	−0.007(8)	C(3)	0.006(8)
	C(12)	−0.001(6)	C(1)	0.011(8)	C(4)	−0.012(8)
	C(13)	0.002(6)	C(2)	−0.011(8)	C(5)	0.013(8)
Plane II	C(3)	0.025(8)	C(14)	−0.034(10)	C(16)	−0.034(12)
	C(4)	−0.067(8)	C(15)	0.072(10)	C(15)	0.084(12)
	C(5)	0.043(8)	C(16)	−0.035(10)	C(14)	−0.052(12)
	C(6)	0.035(8)	C(17)	−0.044(10)	C(13)	−0.045(12)
	C(7)	−0.072(8)	C(18)	0.067(10)	C(12)	0.081(12)
	C(8)	0.035(8)	C(11)	−0.026(10)	C(11)	−0.035(12)
Plane III	C(8)	−0.003(8)	C(11)	−0.005(9)	C(11)	−0.000(7)
	C(1)	0.005(8)	C(12)	0.011(9)	C(18)	0.000(7)
	C(2)	−0.005(8)	C(13)	−0.011(9)	C(17)	−0.000(7)
	C(3)	0.003(8)	C(14)	0.005(9)	C(16)	0.000(7)
I–II		9.0(2)		5.4(3)		3.6(5)
I–III		53.5(2)		57.9(3)		60.0(5)
II–III		117.5(2)		116.7(3)		116.6(5)

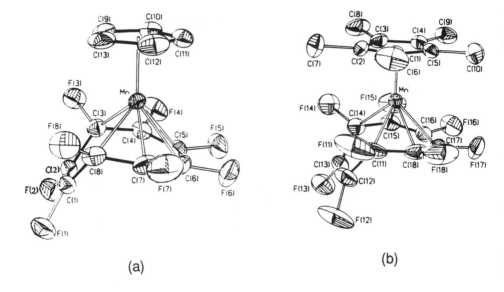

(a)   (b)

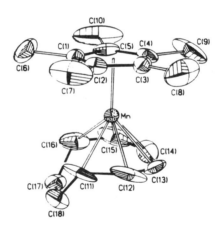

(c)

FIG. 4.   Molecular structures of (a) **56a**, (b) **56b**, and (c) **57b**. [Reprinted with permission from Hemond *et al.* (*129*).]

Table I) are virtually planar with no atom displaced by more than 0.08 Å from the least-squares plane defined by those six atoms. The dihedral angle between the plane formed by the six coordinated carbon atoms (Plane II) and that formed by the uncoordinated olefin (Plane III) are also identical (117°), within experimental error, in all three complexes (Table I). The longest C—F bond distances in **56a** and **56b** are those on the terminal ends of the coordinated triene, reflecting the considerable rehybridization toward pyramidal geometry at these carbon atoms. Surprisingly, the overall skeletal structure of **57b** is not significantly different from those of the fluorinated analogues **56**, and the observed differences in air and thermal stability are not manifested in significantly different Mn—C bond distances or in a different mode of metal–ligand bonding in **57b**.

C. Comparative NMR Studies on the Dynamic Behavior of η^6-OFCOT and η^6-COT Ligands

The temperature independence of the ^{19}F-NMR spectra of toluene-d_8 solutions of **56a** and **56b** up to 100°C prevented the use of line shape analysis to determine the nature of any haptotropic shifts of the manganese atom around the periphery of the C_8F_8 ring. Magnetization transfer experiments on these compounds using a selective ^{19}F NMR decoupler pulse were also incapable of demonstrating any site exchange on this time scale, i.e., upon irradiation of each resonance in turn, there was no observed decrease in the intensity in any of the three nondecoupled signals. The absence of coalescence in the ^{19}F NMR spectrum at 100°C did allow minimum values of $\Delta G^{\ddagger}_{398} >$ 18 kcal/mol to be calculated for any haptotropic rearrangement in either complex (*129*).

In contrast, the room-temperature ^{1}H NMR spectrum of the hydrocarbon analogue **57b** exhibited four broad resonances for the C_8H_8 ring protons, which broadened at the same rate on warming and appeared as a singlet at 79°C. A ^{13}C NMR experiment using a selective DANTE pulse sequence showed equal rates of migration from one C_8H_8 site to all other sites in the COT ring at 42°C, with a value for $\Delta G^{\ddagger} = 17.2$ kcal/mol (*134*). There are three probable interpretations of this result: (1) the metal migrates via [1,3] and [1,5] shifts, which occur at equal rates; (2) there could be random shifts via an $[Mn(\eta^3$-$C_5Me_5)(\eta^8$-$C_8H_8)]$ intermediate; and (3) a 16-electron intermediate $[Mn(\eta^5$-$C_5Me_5)(1$-4-η^4-$C_8H_8)]$ could have a long enough lifetime to permit a substantial number of symmetry-allowed [1,2] shifts, which would lead to apparently random shifts. Because it is known that the [1,3]-shift mechanism predominates for Group 6 complexes $[M(\eta^6$-$C_8H_8)(CO)_3]$ (M = Cr, W) (*135*), and that the [1,5]-shift pathway dominates for Group 8 complexes $[M(\eta^6\text{-}C_8H_8)(\eta^4\text{-diene})]$ (M = Ru, diene = norbornadiene; M = Os, diene = 1,5-cyclooctadiene) (136,137), it is not unreasonable that both dynamic pathways could be significant in the Group 7 complex **57b**.

Such an experiment is not yet possible for the OFCOT analogues, leaving us with the rather unsatisfactory qualitative conclusion that the activation free energy for haptotropic shifts appears to be higher for these complexes than for **57b**. The quantitative effect of perfluorination on the magnitude of $\Delta G^{\ddagger}$ and any effect of perfluorination on the mechanism of any such shifts remain a mystery.

D. Reaction of [Mn(η^5-C_5Me_5)(CO)$_3$] with Perfluorotricyclo[4.2.0.0^{2,5}]octa-3,7-diene

It has been shown that the *syn*-tricyclooctadiene iron carbonyl complex **58** [Eq. (4)] readily undergoes stereospecific ring opening to afford the *exo*-bicyclooctatriene complex **59**, and the *anti*-tricyclooctadiene iron carbonyl

$$\text{(4)}$$

58   **59**

$$\text{(5)}$$

60   **61**

60 likewise opens to yield the endo isomer **61** [Eq. (5)] (138). These results suggested that there might be another route to **56b** via reaction of [Mn(η^5-C_5Me_5)(CO)$_3$] with the tricyclic valence isomer **53** of OFCOT, followed by opening of the fluorinated ring. Unfortunately, when this reaction is carried out under photochemical conditions identical to those used for the production of **56**, only low yields of an air-stable, yellow solid characterized as the

62

dicarbonyl product **62** are obtained (*129*). No ring opening was observed under these reaction conditions. Similar results were obtained with an iron analogue of **62** (see Section VII,B).

VII

IRON COMPLEXES OF OFCOT AND PERFLUOROTRICYCLO[4.2.0.0²,⁵]OCTA-3,7-DIENE

A. *1,2-η²- and 1,2,3,6-η⁴-OFCOT Complexes of Iron*

1. *Synthesis*

The (1–4-η^4-COT)iron complex **63** is of some historical significance. As expected, the [Fe(CO)₃] fragment binds to four contiguous carbon atoms,

63   **64**

but the complex is fluxional in solution and represents the prototypical "ring-whizzer" (*139*). In contrast, OFCOT reacts with [Fe₂(CO)₉] to afford the nonfluxional 1,2,3,6-η^4-Fe(II) complex **64** via the isolable 1,2-η-Fe(0) complex **65** (*140*).

In view of results obtained in the rhodium chemistry of OFCOT (see Section VIII,F,1), the molecular formula and spectroscopic data for **65** do not allow a distinction to be made between the 1,2-η^2-C_8F_8 and 1,4-η^2-C_8F_8 ligation modes for the OFCOT ligand. The latter structure would also be consistent with that observed for the complex **37** formed between perfluoro-1,3-butadiene and the [Fe(CO)$_4$] fragment (see Section III,E).

No COT analogues of **64** or **65** are known in iron chemistry, but the photochemical reaction of [Os$_3$(CO)$_{12}$] with COT has been reported to yield the osmium(II) complex **66**, which rearranges to the thermodynamically stable

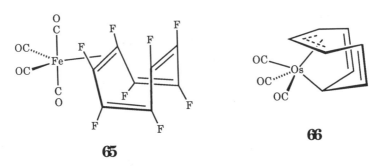

65   **66**

(1–4-η^4-COT)Os(0) isomer (*141,142*). Thus OFCOT appears to be more susceptible to formal oxidative addition reactions with iron than is COT. Notably, the 1,2,3,6-η^4-ligation of the OFCOT ligand allows one fluorinated carbon atom to achieve a formally sp^3 hybridization.

2. X-Ray Structural Studies

The solid-state structure of **65** is shown in Fig. 5 and clearly reveals it to be a simple 1,2-η^2-complex (*143*). Complex **65** exhibits the expected lengthening of the coordinated C=C bond [C(11)–C(21) = 1.435(3) Å] compared to the uncoordinated fluoroolefins [e.g., C(51)–C(61) = 1.316(4) Å], and also lengthening of the C—F bonds on the ligated carbon atoms [C(11)–F(11) = 1.389(3) Å] compared to an average value of 1.350(3) Å for the other C—F distances.

The structure of the 1,2,3,6-η^4-OFCOT complex **64** is shown in Fig. 6 (*140*). The coordination geometry around iron can be considered as octahedral, with the allylic portion of the OFCOT ligand and the Fe—C σ bond occupying three facial sites. The formal oxidation of the metal center is reflected in the higher values of the CO stretching frequencies (ν_{CO} = 2105, 2060 cm^{-1}) for complex **64** compared to those for the 1–4-η^4-COT complex **63** (ν_{CO} = 2051, 1992, 1978 cm^{-1}) (*139*).

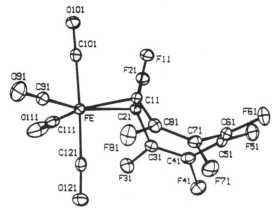

FIG. 5. Molecular structure of **65** (*143*). Selected bond distances (Å): Fe−C(11), 2.006(3); Fe−C(21), 2.016(3); Fe−C(91), 1.822(3); Fe−C(101), 1.840(3); Fe−C(111), 1.837(3); Fe−C(121), 1.817(3); C(11)−C(21), 1.435(3); C(31)−C(41). 1.313(4); C(51)−C(61), 1.316(4); C(71)−C(81), 1.328(4); C(21)−C(31), 1.477(4); C(41)−C(51), 1.449(4); C(61)−C(71), 1.454(4); C(11)−C(81), 1.472(3); C(11)−F(11), 1.389(3); C(21)−F(21), 1.389(3); C(31)−F(31), 1.344(3); C(41)−F(41), 1.352(3); C(51)−F(51), 1.349(3); C(61)−F(61), 1.350(3); C(71)−F(71), 1.350(3); C(81)−F(81), 1.346(3). Selected bond angles (degrees): C(91)−Fe−C(101), 90.02(12); C(101)−Fe−C(111), 88.72(12); C(11)−Fe−C(21), 41.82(10); C(21)−C(11)−C(81), 125.1(2); C(41)−C(31)−C(21), 126.3(3); C(31)−C(41)−C(51), 127.3(3); C(41)−C(51)−C(61), 127.7(3); C(51)−C(61)−C(71), 126.7(3); C(61)−C(71)−C(81), 127.2(2); C(11)−C(81)−C(71), 126.3(3).

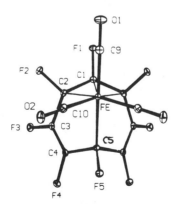

FIG. 6. Molecular structure of **64**. [Reprinted with permission from Barefoot *et al.* (*140*).] The molecule is located on a crystallographic mirror plane and has rigorous C_s symmetry, with Fe, C(1), C(5), and C(9) in the mirror plane. Selected bond distances (Å): Fe−C(1), 2.078(1); Fe−C(2), 2.071(1); Fe−C(5), 2.059(2); C(1)−C(2), 1.424(1); C(2)−C(3), 1.475(2); C(3)−C(4), 1.318(2); C(4)−C(5), 1.487(1). Selected bond angles (degrees): C(5)−Fe−C(9), 177.5(1); C(1)−Fe−C(5), 95.7(1); C(1)−Fe−C(9), 86.4(1); C(10)−Fe−C(10′), 98.4(1): C(2)−Fe−C(10). 92.1(1); C(1)−Fe−C(10), 130.8(1); C(2′)−Fe−C(10), 163.9(1).

B. *A 1,2-η²-Perfluorotricyclo[4.2.0.0²,⁵]octa-3,7-diene Complex of Iron*

Attempts to prepare **64** from the tricyclic valence isomer of OFCOT afforded only the simple olefin complex **67**, analogous to the manganese compound described above, and no subsequent ring opening could be effected (see

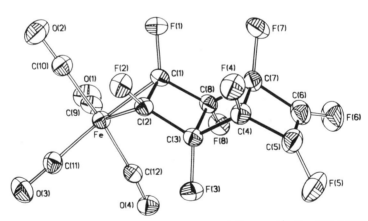

67

Section VI,D). The molecular structure of **67** is shown in Fig. 7 (*144*). As expected, the C—C distance in the coordinated fluoroolefin is significantly
longer than that in the uncoordinated double bond.

Fig. 7.   Molecular structure of **67** (*144*). Selected bond distances (Å): Fe–C(1), 2.024(5); Fe–
C(2), 2.019(4); Fe–C(9), 1.835(6); Fe–C(10), 1.837(6); Fe–C(11), 1.828(5); Fe–C(12), 1.835(5);
C(1)–C(2), 1.456(6); C(2)–C(3), 1.492(6); C(3)–C(4), 1.526(7); C(4)–C(5), 1.484(8); C(5)–C(6),
1.312(9); C(6)–C(7), 1.494(8); C(4)–C(7), 1.566(7); C(7)–C(8), 1.516(7); C(3)–C(8), 1.608(6);
C(1)–C(8), 1.488(7). Selected bond angles (degrees): C(1)–Fe–C(2), 42.2(2); Fe–C(1)–C(2),
68.7(3); Fe–C(2)–C(1), 69.1(3); C(1)–C(2)–C(3), 92.8(4); C(2)–C(1)–C(8), 93.1(4).

C. *Formation of Perfluorocyclooctadienetriyl Complexes of Iron by Nucleophilic Attack on the 1,2,3,6-η^4-OFCOT Ligand*

The previously reported susceptibility of the iron complex **35** toward fluoride ion to give the anionic compound **30** (see Section III,D) suggested that nucleophiles might react similarly with **64** to give anionic derivatives. Indeed, reaction of **64** with the soft nucleophile PMe$_3$ yields the 1:1 adduct **68** (*145*).

The regio- and stereochemistry of phosphine attack were unambiguously defined by a single-crystal X-ray diffraction study (Fig. 8). Compound **68** apparently arises by exo attack of PMe$_3$ on the internal allylic carbon of **64**, generating two new iron–carbon σ bonds, although attack at another site followed by rearrangement to the observed product cannot be excluded. Attack by anionic nucleophiles at the center carbon of cationic hydrocarbon η^3-allylic complexes is precedented, though rare (*146–148*), and analogous reactions of neutral allylic complexes are unknown. Synthesis of **68** also provided the first example of a *fac*-(trialkyl)(tricarbonyl)metal complex. This mode of phosphine reactivity appears to be unique to the Fe(II) complex **64**, because Co(III) and Rh(III) complexes containing the 1,2,3,6-η-OFCOT ligand react with donor ligands at the metal center (see Section VIII,F). Attack at the metal rather than at the fluoroallylic ligand has also been noted in the reaction of [Co(η^3-C$_3$F$_5$)(CO)$_3$] with PPh$_3$ (see Section III,D).

Harder nucleophiles undergo an analogous reaction. Treatment of **64** with [(Me$_2$N)$_3$S]$^+$[Me$_3$SiF$_2$]$^-$ [a soluble source of F$^-$ (*149*)] affords the anionic η^3-nonafluorocyclocta-2,5-diene-1,4,7-triyl complex **69** as a pale yellow oil (*145*). In contrast to **30**, which is reported to be stable in aqueous solution (*75*), **69** reacts with traces of moisture to afford the [(Me$_2$N)$_3$S]$^+$ salt of the anionic 8-oxo-heptafluorocyclocta-2,5-diene-1,4,7-triyl complex **70** (*145*). The same anion, together with HF, is produced by the reaction of **64** with H$_2$O in THF solution. The (18-crown-6)K$^+$ salt of **70** can also be synthesized directly from **64** by reaction with KOH in dimethyl sulfoxide (DMSO); crystals of this salt

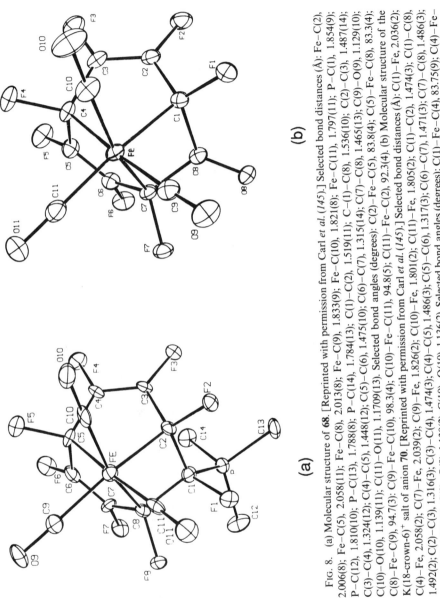

FIG. 8. (a) Molecular structure of **68**. [Reprinted with permission from Carl *et al.* (*145*).] Selected bond distances (Å): Fe–C(2), 2.006(8); Fe–C(5), 2.058(11); Fe–C(8), 2.013(8); Fe–C(9), 1.833(9); Fe–C(10), 1.821(8); Fe–C(11), 1.797(11); P–C(1), 1.854(9); P–C(12), 1.810(10); P–C(13), 1.788(8); P–C(14), 1.784(13); C(1)–C(2), 1.519(11); C–(1)–C(8), 1.536(10); C(2)–C(3), 1.487(14); C(3)–C(4), 1.324(12); C(4)–C(5), 1.448(12); C(5)–C(6), 1.475(10); C(6)–C(7), 1.315(14); C(7)–C(8), 1.465(13); C(9)–O(9), 1.129(10); C(10)–O(10), 1.139(11); C(11)–O(11), 1.1709(13). Selected bond angles (degrees): C(2)–Fe–C(5), 83.8(4); C(5)–Fe–C(8), 83.3(4); C(8)–Fe–C(9), 94.7(3); C(9)–Fe–C(10), 98.3(4); C(10)–Fe–C(11), 94.8(5); C(11)–Fe–C(2), 92.3(4). (b) Molecular structure of the K(18-crown-6)⁺ salt of anion **70**. [Reprinted with permission from Carl *et al.* (*145*).] Selected bond distances (Å): C(1)–Fe, 2.036(2); C(4)–Fe, 2.058(2); C(7)–Fe, 2.039(2); C(9)–Fe, 1.826(2); C(10)–Fe, 1.801(2); C(11)–Fe, 1.805(2); C(1)–C(2), 1.474(3); C(1)–C(8), 1.492(2); C(2)–C(3), 1.316(3); C(3)–C(4), 1.474(3); C(4)–C(5), 1.486(3); C(5)–C(6), 1.317(3); C(6)–C(7), 1.471(3); C(7)–C(8), 1.486(3); C(8)–O(8), 1.230(2); C(9)–O(9), 1.130(2); C(10)–O(10), 1.136(2). Selected bond angles (degrees): C(1)–Fe–C(4), 83.75(9); C(4)–Fe–C(7), 83.17(7); C(7)–Fe–C(9), 97.32(8); C(9)–Fe–C(10), 92.87(8); C(10)–Fe–C(11), 101.35(9); C(11)–Fe–C(1), 163.64(8).

70

proved suitable for X-ray diffraction and the resultant structure is shown in Fig. 8 (*145*).

The most unusual feature of this new family of organometallic complexes is that the polyenyl ligands bind to the iron via three σ bonds rather than through the π system of the ring. Thus, compounds **68–70** are formally iron(II) derivatives of a localized perfluorocycloocta-2,5-diene-1,4,7-triyl trianion, which is perfectly arranged to span three facial sites of an octahedron and in which four fluorinated carbon atoms achieve sp^3 hybridization. This mode of bonding is not unexpected in view of the tendency of fluorinated carbanions to adopt localized, pyramidal structures (see Section III,D). The unusually high values for the CO stretching frequencies (e.g., $v_{CO} = 2071$, 2002 cm^{-1} for the nonafluorocycloocta-2,5-diene-1,4,7-triyl complex **69**) in compounds with a formally anionic metal center are also noteworthy, implying that the fluorinated triyl ligand is a powerful electron-withdrawing ligand (*145*). The ruthenium and osmium chemistries of OFCOT remain to be explored.

VIII

COBALT AND RHODIUM COMPLEXES DERIVED FROM OFCOT AND PERFLUOROBICYCLO[4.2.0]OCTA-2,4,7-TRIENE

A. *Synthesis of Mononuclear Cyclopentadienyl and Indenyl Rhodium Complexes of OFCOT*

The reaction of OFCOT with the 18-electron Rh(I) complex [Rh(η^5-C_5H_5)(η^2-C_2H_4)$_2$] is slow at room temperature and yields a product **71** resulting from displacement of only one ethylene ligand (*150*). This is formulated as a 1,2-η^2-OFCOT complex of Rh(I), rather than a 1,4-η^2-OFCOT complex of Rh(III) (see Section VIII,F,1), based on similarities between the

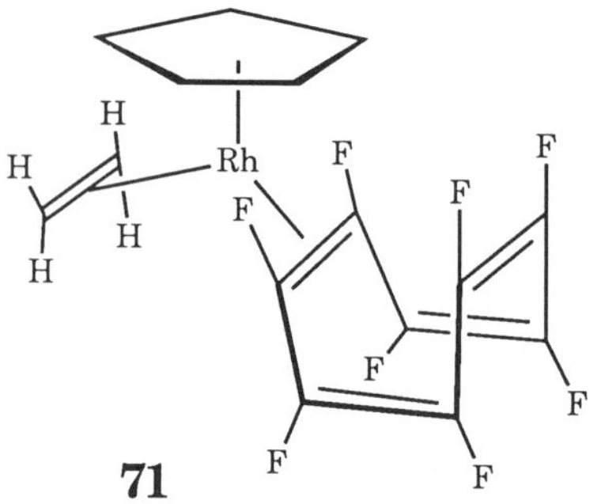

71

^{19}F NMR chemical shifts and coupling constants with the crystallographically characterized 1,2-η^2-OFCOT complex of Fe(0) (see Section VII,A,2).
Further heating of this compound or performance of the initial reaction in
refluxing hexanes yields the 1,2,5,6-η^4-OFCOT complex **72** (*150*). The η^5-
indenyl analogue **73** can be prepared similarly and has been crystallographically characterized (Fig. 9) (*151*). Discussion of the solid-state structure and
solution dynamics of this latter compound is presented in Section VIII,D,2.

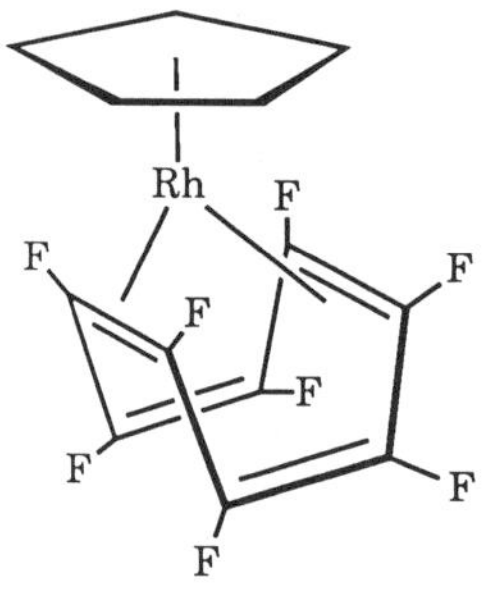

72

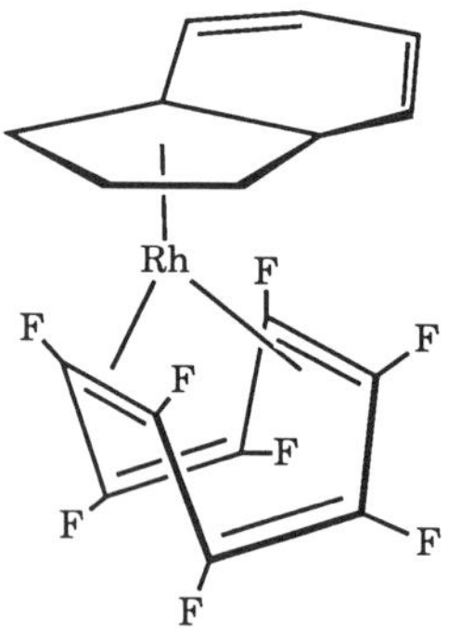

73

Either complex can be recovered unchanged from prolonged reflux in xylene,
and, in contrast to their pentamethylcyclopentadienyl analogues, neither exhibits any photochemically induced hapticity changes (see Section VIII,E,1).

B. Rhodium Complexes of Perfluorobicyclo[4.2.0]octatriene

Surprisingly, the reaction of the 16-electron dimers [RhCl(η^2-olefin)$_2$]$_2$
(olefin = ethylene, cyclooctene) with OFCOT at room temperature did not
afford a 1,2,5,6-η^4-OFCOT complex, but resulted in a sparingly soluble

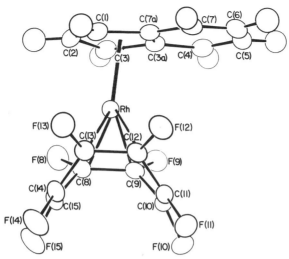

FIG. 9. Molecular structure of **73**. [Reprinted with permission from Carl *et al.* (*151*).] Selected bond distances (Å): Rh–C(1), 2.215(3); Rh–C(2), 2.227(3); Rh–C(3), 2.183(3); Rh–C(3a), 2.345(3); Rh–C(7a), 2.354(2); Rh–C(8), 2.077(2); Rh–C(9). 2.088(2); Rh–C(12), 2.099(2); Rh–C(13), 2.091(2); C(8)–C(9), 1.439(4); C(9)–C(10), 1.455(4); C(10)–C(11), 1.317(4); C(12)–C(13), 1.464(4); C(13)–C(14), 1.423(4); C(14)–C(15), 1.309(4); C(8)–F(8), 1.366(3); C(9)–F(9), 1.365(3); C(10)–F(10), 1.342(3); C(11)–F(11), 1.343(3); C(12)–F(12), 1.360(3); C(13)–F(13), 1.360(3); C(14)–F(14), 1.345(3); C(15)–F(15), 1.355(3).

brown material whose IR spectrum exhibited a band at 1769 cm^{-1} due to an uncoordinated fluoroolefin. This material could be converted to a 3:2 mixture of crystalline cyclopentadienyl derivatives **74** and **75**, the solid-state structures of which are shown in Fig. 10 (*150*). Each is a complex of the perfluorobicyclo-[4.2.0]octatriene valence isomer of OFCOT; in **74** the cyclobutene ring is

74   **75**

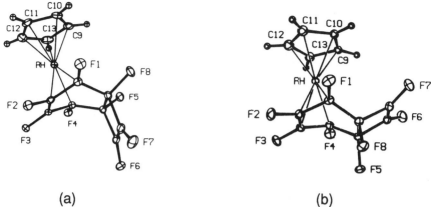

(a)   (b)

Fig. 10. (a) Molecular structure of **74**. [Reprinted with permission from Carl *et al.* (*151*).] Selected bond distances (Å): Rh–C(1), 2.086(2); Rh–C(2), 2.103(2); Rh–C(3), 2.102(2); Rh–C(4), 2.099(2); C(1)–C(2), 1.426(3); C(2)–C(3), 1.404(3); C(3)–C(4), 1.433(4); C(1)–F(1), 1.365(3); C(2)–F(2), 1.350(3); C(3)–F(3), 1.339(2); C(4)–F(4), 1.366(3). Dihedral angles (degrees): [C(1)–C(2)–C(3)–C(4)]—[C(1)–C(4)–C(5)–C(8)],   137.38(14);   [C(1)–C(4)–C(5)–C(8)]—[C(5)–C(6)–C(7)–C(8)], 124.21(14). (b) Molecular structure of **75**. [Reprinted with permission from Carl *et al.* (*151*).] Selected bond distances (Å): Rh–C(1), 2.085(2); Rh–C(2), 2.104(2); Rh–C(3), 2.103(2); Rh–C(4), 2.083(2); C(1)–C(2), 1.433(3); C(2)–C(3), 1.408(3); C(3)–C(4), 1.437(4); C(1)–F(1), 1.374(4); C(2)–F(2), 1.341(3); C(3)–F(3), 1.336(3); C(4)–F(4), 1.374(4). Dihedral angles (degrees): [C(1)–C(2)–C(3)–C(4)]—[C(1)–C(4)–C(5)–C(8)],   134.36(16);   [C(1)–C(4)–C(5)–C(8)]—[C(5)–C(6)–C(7)–C(8)], 121.03(15).

exo with respect to the Rh, whereas in **75** the cyclobutene ring is on the endo face of the coordinated diene ligand. Each structure shows considerable pyramidalization toward sp^3 at the terminal carbons of the coordinated diene moiety, with concomitant lengthening of the C—F bonds, as observed previously for an iron complex **30** of octafluorocyclohexa-1,3-diene (see Section III,E). Also consistent with the latter structure is the observation of a "long–short– long" pattern of C—C bond lengths for the internal carbon atoms of the coordinated dienes in **74** and **75**, together with significantly shorter M—C distances to the terminal diene carbons compared to the internal diene carbons. This can be rationalized if the fluorinated diene is considered to be an excellent π acceptor, with strong back bonding from the metal to the $\pi(\Psi_3)$ orbital of the diene (*150*).

Complexes **74** and **75** do not interconvert during prolonged reflux in xylene solution, nor do they isomerize to the 1,2,5,6-η^4-OFCOT complex **72**. However, when the reaction of [RhCl(η^2-olefin)$_2$]$_2$ with OFCOT was carried out in refluxing benzene, the brown residue afforded only the cyclopentadienyl complex **75** on derivatization. Both cyclopentadienyl complexes retain the IR band at ~1770 cm^{-1}, which is clearly due to the strained fluorocyclobutene functionality. By way of comparison, the corresponding IR band for the uncoordinated fluoroolefin in **72** occurs at 1729 cm^{-1} and the cyclobutene olefin stretch in the matrix isolated free ligand **52** occurs at 1773 cm^{-1} (*124*). Accordingly, it seems reasonable to formulate the initially formed brown chloride compounds as containing the same bicyclic fluorocarbon skeleton as **74** and **75**, and to conclude that ring closure has already occurred prior to transformation to the cyclopentadienyl analogues, rather than a selective co-ordination of Rh to traces of **52** present in the initial reaction mixture. Thus, the exo isomer appears to be the kinetic product of ring closure, and the endo isomer appears to be the thermodynamic product. The mechanism of ring closure and the kinetic preference for formation of the exo isomer are unclear, although the facility of exo-to-endo rearrangement in the 16-electron chloro complexes and the inability to effect this conversion in the 18-electron species **74** and **75** imply that more than two coordination sites must be available on the metal for the fluorinated ligand to undergo this reaction. It is tempting to propose that the reaction involves reversible disrotatory closure/opening, involving an intermediate η^6-OFCOT species. This would not be possible for the 18-electron complexes **74** and **75** without an energetically costly $\eta^5 \rightarrow \eta^3$ slippage of the cyclopentadienyl ring. However, it should be noted that the indenyl complex **73**, for which ring slippage to generate an η^6-OFCOT inter-mediate should be more facile, shows no proclivity for ring closure to a bicyclic isomer in refluxing octane (*150*).

C. Cyclopentadienyl Cobalt Complexes of OFCOT and the Perfluorobicyclo[3.3.0]octadienediyl Ligand

The reaction of $[Co(\eta^5\text{-}C_5H_5)(CO)_2]$ with OFCOT affords poor yields of the orange crystalline 1,2,5,6-η^4-OFCOT complex **76**, together with execrable

76   **77**

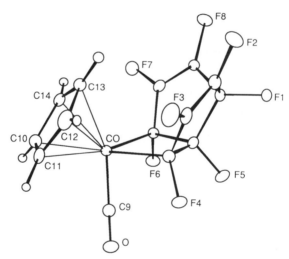

78

amounts ($<1\%$) of **77** (*152*). Most of the OFCOT is converted to perfluoro-benzocyclobutene **78** in this reaction (*153*). The molecular structure of **77** is shown in Fig. 11 (*152*), revealing that this complex results from a novel trans-annular ring closure of the OFCOT molecule to give a perfluorobicyclo-[3.3.0]octa-2,7-diene-4,6-diyl ligand and resulting in two new M—C σ bonds. Further examples of this mode of reactivity are discussed in Section VIII,F,2.

FIG. 11. Molecular structure of **77**. [Reprinted with permission from Hughes *et al.* (*152*).] Selected bond distances (Å): Co–C(4), 1.993(2); Co–C(6), 1.986(2); Co–Cp ring centroid, 1.711; range of Co–Cp carbon, 2.048(2)–2.117(2); C(4)–C(5), 1.525(3); C(5)–C(6), 1.526(3); C(3)–C(4), 1.470(2); C(6)–C(7), 1.471(2); C(2)–C(3), 1.326(2); C(7)–C(8), 1.322(2); C(1)–C(2), 1.478(3); C(8)–C(1), 1.484(3); C(1)–C(5), 1.568(2). Selected bond angles (degrees): C(4)–Co–C(6), 72.87(7); C(4)–Co–C(9), 90.68(8); C(6)–Co–C(9), 91.46(8). Selected dihedral angles (degrees): Cp ring—[Co–C(4)–C(5)–C(6)], 52.2; [C(1)–C(2)–C(3)–C(4)–C(5)]—[Co–C(4)–C(5)–C(6)], 61.5; [C(1)–C(5)–C(6)–C(7)–C(8)]—[Co–C(4)–C(5)–C(6)], 61.1; [Co–C(4)–C(6)]—[C(4)–C(5)–C(6)], 4.1.

D. Structure and Solution Dynamics of a Mononuclear Indenyl Rhodium Complex of OFCOT

1. Slip-Folding of Indenyl Ligands

It has been proposed that the enhanced reactivity of indenyl complexes of the transition metals toward ligand substitution, compared to their cyclopentadienyl analogues, is due to a slippage of the indenyl ligand from an η^5 to an η^3 bonding mode, thus making the metal center coordinatively unsaturated (154). As the indenyl ring distorts from η^5 to η^3, the arene ring gradually folds out of the plane of the allylic fragment (155). A quantitative measure of the amount of slip-folding based on X-ray crystallographic data of substituted indenyl complexes has been developed and has been related to the σ-donor/π-acceptor characteristics of the ancillary ligands (156). For example, a good σ-donor ligand such as PMe_3 should stabilize η^3 coordination by its ability to donate electron density to the metal. Conversely, a good π-acceptor ligand would remove electron density from the metal center and favor η^5 coordination of the indenyl ligand (156). The slip parameter is defined (see Fig. 9 for numbering) as $\Delta = \{$average of M–C(3a),C(7a)$\}$ − $\{$average of M–C(1),C(3)$\}$; the larger the value of Δ the more distortion to an η^3-indenyl ligand and the better the σ-donating ability of the other ligands (157). More conveniently and more relevant to solution chemistry, the slip-fold distortion of the indenyl ligand in solution can be calculated from the ^{13}C NMR spectral data. The hapticity of the indenyl ligand can be evaluated spectroscopically by comparing the ^{13}C NMR chemical shifts of the five-membered ring carbons in the metal complex with those of indene (158). For η^3-indenyl ligands C(1)–C(3) (see Fig. 9 for numbering) are relatively shielded but C(3a) and C(7a) are not (159). The chemical shift differences $\Delta\delta(C) = \delta[C(\eta\text{-indenyl})] - \delta[C(\eta\text{-indenyl})sodium]$ for a number of d^6 and d^8 metal complexes have been measured, and the values correlate well with the structurally observed hapticity. A $\Delta\delta(C3a,7a)$ of -20 to -40 ppm indicates a planar η^5-indenyl; -10 to -20 ppm a partially slipped η^5-indenyl; and $+5$ to $+30$ ppm an η^3-indenyl ligand (159).

In an attempt to assess the relative σ-donating/π-accepting ability of OFCOT toward Rh(I), these methods have been applied to the indenyl complex **73** (151).

2. Crystallographic and NMR Studies

The solid-state structure of **73** is shown in Fig. 9 (Section VIII,B) (151). As expected, there is substantial lengthening of the fluorinated C=C bond upon coordination [C=C, coordinated (average) = 1.451(8) Å vs C=C, uncoordinated (average) = 1.342(8) Å], and the C—F distances for the fluorines attached to the coordinated carbons are also significantly longer than the C—F distances for the fluorines attached to the uncoordinated carbons [C—F, coordinated (average) = 1.360(3) Å; C—F, uncoordinated (average) = 1.345(3) Å]. These observations seem to indicate that OFCOT is a fairly good π acceptor because coordination of the double bond to the metal center would populate the π* orbitals of the olefin, weakening the C=C bond and pyramidalizing the carbon atoms. In agreement, the geometry around the uncoordinated carbons is planar (sum of the angles = 360°), but that around the coordinated carbons is not (average sum of the angles = 353°). Rehybridization at carbon should also be reflected in the $^1J_{C-F}$ values observed in the $^{13}C\{^1H\}$-NMR spectrum. Curiously, the coupling constants of the coordinated carbons to the attached fluorines ($^1J_{C-F}$ = 289.9 Hz) are nearly identical to those of the uncoordinated carbons to their respective fluorines ($^1J_{C-F}$ = 290.9 Hz). In contrast, for the analogous pentamethylcyclopentadienyl cobalt complex **79a** (see Section VIII,E,1), the coupling constant of the coordinated carbons to the attached fluorines is 281 Hz and $^1J_{C-F}$ for the uncoordinated carbons is 302 Hz (*151*), perhaps reflecting the greater electron-donating ability of the pentamethylcyclopentadienyl compared to the indenyl ligand. The ability of pentamethylcyclopentadienyl to support intramolecular oxidative addition of OFCOT supports this idea (see Section VIII,E,1).

79 a. M = Co
        b. M = Rh

80 a. M = Co
        b. M = Rh

Calculation of indenyl slippage using two different methods leads to the conclusion that the indenyl slippage does not vary significantly on varying the ancillary ligands from OFCOT to other hydrocarbon olefins. Calculation of the slip parameter from the X-ray crystallographic data of **73** gives Δ = 0.150 Å (*151*), indicating that the σ-donor and π-acceptor properties of OFCOT are not significantly different from those of ethylene [Δ = 0.160 for [Rh(η^5-C$_9$H$_7$)(η^2-C$_2$H$_4$)$_2$] (*156,160*)] or 1,5-cyclooctadiene [Δ = 0.152 for

[Rh(η^5-C$_9$H$_7$)(η^4-COD)] (157)]. Calculation of the slip parameter from the ^{13}C NMR spectral data of **73** gives a $\Delta\delta$(C3a,7a) = −15 ppm (151), indicating that the hapticity of the ligand is best described as a distorted η^5-indenyl (159), in agreement with the crystallographic results.

3. Dynamic NMR Studies on Indenyl Rhodium Complexes of OFCOT and COT

Observation of only two resonances in the ^{19}F NMR spectrum of **73** at room temperature indicates rapid rotation of the indenyl ring about the Rh–indenyl axis on the NMR time scale. At −90°C, the limiting spectrum is reached and the ^{19}F NMR spectrum shows the four resonances expected from the solid-state structure. Line shape analysis of the variable temperature spectra yields values of $E_a = 8.0 \pm 0.6$ kcal/mol and $\Delta G^{\ddagger}_{190K} = 8.6 \pm 0.8$ kcal/mol for indenyl rotation (151). The value of $\Delta G^{\ddagger}_{190K}$ for indenyl rotation in [Rh(η^5-1-CH$_3$C$_9$H$_6$)(η^2-C$_2$H$_4$)$_2$] has been calculated to be 8.5 ± 0.4 kcal/mol (160). A more meaningful comparison is available from the cyclooctatetraene (COT) analogue of **73**. A variable-temperature ^{1}H NMR study from −90 to −140°C produces an E_a for indenyl rotation of 9.4 ± 0.7 kcal/mol (151), which is not significantly different from the OFCOT analogue (8.0 ± 0.6 kcal/mol) described previously.

It can be concluded that significant differences in metal–carbon bond distances to coordinated olefins are not reflected in either the slip parameter or the barrier to rotation of the indenyl ring in this series of complexes. However, the relative strengths of the interactions between the [Rh(η^5-indenyl)] fragment and OFCOT and COT are illustrated by some reactivity comparisons in Section VIII,F,2.

E. Pentamethylcyclopentadienyl Cobalt and Rhodium Complexes Derived from OFCOT

1. 1,2,5,6-η^4- and 1,2,3,6-η^4-OFCOT Complexes of Cobalt and Rhodium

Reaction of OFCOT with either [Co(η^5-C$_5$Me$_5$)(CO)$_2$] or [Rh(η^5-C$_5$Me$_5$)(η^2-C$_2$H$_4$)$_2$] in refluxing hexane afforded good yields of the 1,2,5,6-η^4-OFCOT complexes **79** (152,161). When the syntheses and workups of compounds **79** are carried out under ambient light conditions, the resultant material is invariably contaminated with a second isomeric product, assigned as the 1,2,3,6-η^4-OFCOT isomers **80** on the basis of comparison of ^{19}F NMR spectral parameters with those of the crystallographically characterized compound **64** (see Section VIII,C). This isomer results from formal oxidative addition of OFCOT to the metal. When the syntheses are carried out in the dark, only isomers **79** are produced (161). Ultraviolet irradiation of solutions of **79** results in formation of a photostationary 2.3:1.0 mixture of **79** and the oxidative addition products **80**. The ratio **79:80** does not vary on changing the metal from cobalt to rhodium or with changes in the photolysis temperature over the range -78 to $25°C$. When solutions containing mixtures of **79** and **80** are heated in the dark, clean conversion of **80** back to **79** is observed (*161*). No such isomerization behavior is observed for cyclopentadienyl or indenyl analogues (*150*). The thermal conversion of **80** to **79** clearly defines the latter as the thermodynamically preferred isomer for both cobalt and rhodium coordination. The irradiation studies indicate that a photochemically generated excited state of **79** must decay by two competitive pathways to give a photostationary mixture of **79** and **80**. The ratio of the two components of this photostationary mixture does not depend on the metal or on the temperature at which the photolysis is carried out, implying that the ratio of the rate constants for the two decay processes from the excited state is temperature independent.

2. *Comparative Electrochemical Studies on Pentamethylcyclopentadienyl Cobalt Complexes of OFCOT and COT*

The overall structures of the cobalt complex **79a** and its rhodium analogue are identical to the thermodynamically stable structures observed for the

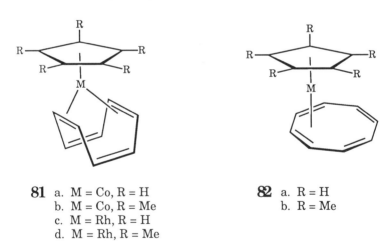

81 a. M = Co, R = H
     b. M = Co, R = Me
     c. M = Rh, R = H
     d. M = Rh, R = Me

82 a. R = H
     b. R = Me

18-electron hydrocarbon analogues **81**. In contrast to the OFCOT complexes, the cobalt complexes **81a** and **81b** are in thermal equilibrium with their fluxional $1-4-\eta^4$-COT isomers **82** (*162*).

   a. *Electrochemistry of COT Complexes.* The equilibrium mixture of cobalt complexes **81a** and **82a** undergoes reversible electrochemical reduction to give a single 19-electron radical anion, shown by ESR studies to have structure **82**⁻, in which the unpaired electron occupies an orbital located on the uncoordinated diene portion of the COT ligand (*163*). Notably, 1-electron reduction of isomer **81a** results in an extremely rapid hapticity change of the coordinated COT ligand to give **82**⁻. Extended Hückel MO calculations have provided a theoretical rationale for these observations (*164*).

    b. Electrochemistry of OFCOT Complexes. The notable absence of 1–4-η^4 coordination in any transition metal compounds of OFCOT prompted an attempt to generate such a ligation mode by electrochemical reduction of mixtures of **79a** and **80a**. Surprisingly, no evidence was found that isomer **80a** exhibits a separate wave within the range of potentials available ($\sim +2.0$ to -2.8 V vs. SCE). Furthermore, isomer **79a**, when reduced to its monoanion, showed no tendency to isomerize to an analogue of **82**⁻. Instead, **79a** reduces through an apparent ECE mechanism (two one-electron transfers interspersed by a chemical reaction) (*161*). Room-temperature cyclic voltammetry (CV) scans of **79a** negative of zero volts in dimethylformamide (DMF) or THF show two chemically reversible waves. At high sweep rates or low temperatures, the first of these waves is chemically reversible ($E^0 = -1.16$ V) and has been shown to correspond to the process **79a** $+ e \rightarrow$ **79a**⁻. The second wave is also chemically reversible ($E^0 = -1.34$ V) and is more prevalent under longer electrochemical experiment times. Bulk electrolysis confirmed this, for coulometry at -1.25 V (between the first and second waves) consumed two electrons ($n = 2.02e^-$ in DMF and $2.00e^-$ in THF) and resulted in efficient conversion to a product whose structure could not be unambiguously assigned, but whose mass spectrum showed an appreciable peak due to $[\text{Co}(C_5Me_5)(C_8F_7H)]^+$ (*161*). Notably, loss of fluoride is observed in attempts to reduce free OFCOT (see Section IV,B).

    Samples of **79a** known to contain the isomer **80a** did not show additional voltammetric waves over the range $+2.0$ to -2.8 V (*161*). Either isomer **80a** is not electroactive in the range scanned, or it has an E^0 within a few millivolts of **79a**. A more complex possibility that accounts for the absence of a separate wave for isomer **80a** requires that the E^0 of **80a** is positive of **79a**, and that the anion **80a**⁻ isomerizes very rapidly to **79a**⁻. We can only conclude, therefore, that isomer **80a** does not exhibit its own voltammetric wave under these conditions, but we cannot exclude the possibility that its electroactivity is masked either by the wave for isomer **79a** or by its electrocatalytic conversion to **79a** at the electrode surface.

    c. ESR Studies on the Structures of the Radical Anions. The radical anion **79a**⁻ has also been characterized by electron spin resonance (ESR) analysis (*161*). The frozen solution ESR spectrum is typical of d^9 cobalt π complexes and can be simulated to acceptable agreement using an axial tensor for the

g values and for the cobalt hyperfine splittings: $a_\parallel(\text{Co}) = 142$ G, $a_\perp(\text{Co}) = 33$ G; $g_\parallel = 2.170$, $g_\perp = 2.080$. Similar spectra have been observed previously for Co(0) complexes such as $[\text{Co}(\eta^5\text{-}C_5H_5)(\eta^4\text{-}C_5R_4O)]^-$ (165) and $[\text{Co}(\eta^5\text{-}C_5H_5)(1,2,5,6\text{-}\eta^4\text{-COD})]^-$ (163,164) and are strongly indicative of the half-filled orbital in **79a**$^-$ being localized predominantly on the metal. In the analogous hydrocarbon complex **82a**$^-$, the cobalt hyperfine splitting displayed very little anisotropy, consistent with a ligand–based half-filled orbital (163,164).

    d. Relative Rearrangement Rates of Radical Anions.   Even though the final product of reduction remains unidentified, the fact that it is related to **79a** by two electrons in an apparent ECE process leads to the conclusion that the two observed cathodic waves do not arise from the two different isomers **79a** and **80a** and allow the rate constant for its formation by isomerization of **79a**$^-$ to be calculated (161). The value of this rate constant (0.22 sec^{-1}) must represent the upper limit for the rate of any isomerization of radical anion **79a**$^-$ to the perfluorinated analogue of **82a**$^-$, because the reaction to form the unknown final product is much faster than any other reaction of **79a**$^-$. This leads to the conclusion that the 1,2,5,6-η^4 to 1-4-η^4 isomerization of the C_8R_8 ring bound to the $[\text{Co}(\eta^5\text{-}C_5Me_5)]$ fragment is enormously slowed when R = F compared to R = H. In the latter case, the first-order rate constant (166) for isomerization of the radical anion is 2000 sec^{-1}, at least four orders of magnitude faster than for R = F. Clearly the 1,2,5,6-η^4 form of the C_8R_8 ligand is stabilized by the substitution of fluorine for hydrogen. It is also noteworthy that the E^0 value for **79a** is very much higher than that for its hydrocarbon analogue **81a**, shifting positive by about 1200 mV and presumably reflecting a higher degree of electron acceptor ability in the fluorinated ligand. A comparison with the free ligands can also be made. The first cathodic wave (irreversible) for free OFCOT was found to be -1.50 V in DMF (Pt electrode), consistent with the value reported for CH_3CN solutions (167). This represents a positive shift of less than 200 mV from the value of the first cathodic wave of free COT, but the significance of the measured shift in potential is uncertain due to the irreversibility of the OFCOT reduction (167).

    Thus, although both COT and OFCOT form stable complexes with the $[\text{Co}(\eta^5\text{-}C_5Me_5)]$ fragment, interesting differences exist in the favored structural forms of both the neutral 18-electron complexes and in the related 19-electron monoanions.

    F. *Reactions of Pentamethylcyclopentadienyl Cobalt and Rhodium Complexes of OFCOT with Exogenous Ligands*

    It was thought that the photochemical isomerization of the cobalt and rhodium complexes **79** to **80** might proceed by a photochemical dissociation of one of the coordinated fluoroolefins of **79** to give a 16-electron intermediate in which the metal center was coordinated to the OFCOT ring in a 1,2-η fashion. Attempts to trap this intermediate as a stable 18-electron complex, by photolysing solutions of cobalt complexes **79a** and **80a** in the presence of carbon monoxide, trimethylphosphine, trimethylphosphite, and pyridine, resulted in no ligand incorporation (*168*).

1. *A Stable Perfluorocyclooctatrienediyl Complex of Rhodium*

In contrast, reaction of a THF solution containing a mixture of the rhodium isomers **79b** and **80b** with trimethylphosphine in the dark consumes only the 1,2,3,6-η^4 isomer **80b** to give the new PMe$_3$ adduct **83** (*169*). The

83

1,2,5,6-η^4 isomer **79b** is unreactive, but the overall transformation can be driven by irradiation of the mixture and converting more **79b** to **80b**. The crystal structure of **83** is shown in Fig. 12 (*169*). The OFCOT ring is bound to rhodium via C(1) and C(4) as a perfluorocycloocta-2,5,7-triene-1,4-diyl ligand, rather than by 1,2-η^2-olefin coordination as observed for an OFCOT complex of iron (see Section VII,A). This results in an extremely rigid structure for the fluorinated ligand in which the six carbon atoms [C(1) and C(4)–C(8)] are coplanar to within 0.1 Å and C(1)–C(4) are coplanar to within 0.003 Å. The dihedral angle between these two carbon atom planes is 100.8°. The structure of this rhodium complex is notably different from that of a perfluorobenzene analogue **42**, in which the arene is bound in a 1,2-η^2 rather than a 1,4-η^2 fashion (see Section III,F).

This transformation of **80b** to **83** involves an $\eta^3 \rightarrow \eta^1$ reaction of the allylic portion of the 1,2,3,6-η^4-OFCOT ligand of **80b**. It is interesting to note that the solid-state structure of **83** clearly reveals the overall stereochemistry of this transformation, indicating that the incoming PMe$_3$ ligand enters trans to the departing fluoroolefinic function rather than simply occupying the site vacated by that fluoroolefin. This could imply a concerted transformation of **80b** to **83**, but could also reflect initial $\eta^3 \rightarrow \eta^1$ conversion of **80b** followed by

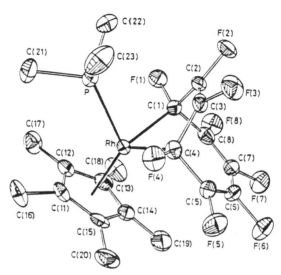

FIG. 12. Molecular structure of **83**. [Reprinted with permission from Hughes *et al.* (*169*).]
Selected bond distances (Å): Rh–P, 2.348(1); Rh–C(1), 2.088(3); Rh–C(4), 2.087(4); Rh–(centroid
C_5Me_5), 1.926(3); C(1)–C(2), 1.470(5); C(2)–C(3), 1.321(5); C(3)–C(4), 1.473(5); C(4)–C(5),
1.503(5); C(5)–C(6), 1.335(6); C(6)–C(7), 1.441(6); C(7)–C(8), 1.334(5); C(1)–C(8), 1.502(5).
Selected bond angles (degrees): C(1)–Rh–C(4), 78.3(1); (cent—C_5Me_5)—Rh–C(1), 129.8(1);
(cent—C_5Me_5)–Rh–C(4), 128.7(1); (cent—C_5Me_5)–Rh–P, 125.5(1); P–Rh–C(1), 89.4(1); P–
Rh–C(1), 89.4(1); P–Rh–C(4), 90.3(1). Dihedral angle (degrees): [C(1),C(2),C(3),C(4)]–[C(4),
C(5),C(6),C(7),C(8),C(1)], 100.8.

trapping of the 16-electron intermediate at the less sterically hindered metal
site. Interestingly, no evidence for attack by PMe_3 on the OFCOT ligand has
been obtained, in contrast with the behavior of the corresponding $(1,2,3,6-\eta^4$-
OFCOT)iron complex **64** (see Section VII,C).

2. *Perfluorobicyclo[3.3.0]octadienediyl Complexes of Cobalt and Rhodium*

Further contrasting reactivity is observed using *t*-butylisocyanide (*t*-BuNC)
as the exogenous ligand. Addition of *t*-BuNC to solutions of complexes **79**
and **80** and irradiation of the mixture cleanly afford new products **84** containing the perfluorobicyclo[3.3.0]octa-2,7-diene-4,6-diyl skeleton (*168*). The
molecular structure of an analogue is shown in Fig. 11 (Section VIII,C).
Experiments carried out in the dark confirm that the added *t*-BuNC ligand
reacts exclusively (and thermally) with the $1,2,3,6-\eta^4$ isomers **80** to form
intermediates **85**, which are isostructural with the crystallographically characterized $1,4-\eta^2$-OFCOT complex **83**. These spectroscopically identified intermediates could not be isolated but underwent a facile transannular ring
closure to give the final products **84** (*168*).

84 a. M = Co
        b. M = Rh

85 a. M = Co
        b. M = Rh

It is puzzling that only *t*-BuNC reacts in this manner. Steric crowding at the metal center might prevent trimethylphosphine from interacting with the cobalt complex **80a**, whereas the more rodlike *t*-butylisocyanide could coordinate more easily. Electronic effects may also play a part because the sterically undemanding but weak σ-donor ligand carbon monoxide could not be induced to react with **80a**. Presumably the larger rhodium center in **80b** allows coordination of PMe_3 to give **83** but it is unclear why this rhodium adduct does not undergo subsequent ring closure (*169*).

These results indicate that a mechanism that involves formation of a 1,2-η^2-OFCOT complex as the initial step in the photolysis of **79** is unlikely, because there seems to be no reason why such a species should not bind competitively with *t*-BuNC, PMe_3, or other donor ligands.

It is clear (vide supra) that intermediates **85** are precursors to ring-closed products **84**. However, it appears likely that an additional intermediate is required for this transformation because direct transannular ring closure of a cycloocta-2,5,7-triene-1,4-diyl ligand is improbable due to the rigidly constrained distance of 3.32 Å (*169*) between C(2) and C(6) (Fig. 12), the two carbon atoms that would have to constitute the new C—C bond in a ring-closed species. One speculative possibility involves transformation of the cycloocta-2,5,7-triene-1,4-diyl structure to a zwitterionic species as shown in Scheme 2. This structure would allow considerably more flexibility than the rigidly constrained structure of **85**, and attack of the negatively charged carbon (C6) on the central carbon (C2) of the coordinated allylic portion of the fluorocarbon ligand would lead to the observed ring-closed product. Notably, attack of C(6) at the metal is blocked by the added ligand. As described in Section VII,C, the central carbon of the fluoroallylic ligand is indeed susceptible to nucleophilic attack.

No photochemical rearrangements or transannular ring closure chemistry were observed with the corresponding cyclopentadienyl analogues (see Section VIII,A), illustrating the key role of the pentamethylcyclopentadienyl

SCHEME 2.   (M = Co, Rh; X = C_5Me_5; L = t-BuNC).

ligand. This is perhaps not surprising because it is exclusively the $1,2,3,6$-η^4-
OFCOT isomer **80** that undergoes thermal reaction with exogenous ligands,
and this isomer is not accessible from **72** or **76**, presumably because cyclopentadienyl does not stabilize the M(III) oxidation state as well as its penta-
methyl analogue. However, the indenyl analogue **73** does react with t-BuNC

to afford the final product **86** via the spectroscopically observed intermediate **87** (*168*). Evidently slippage of the indenyl ligand allows the *t*-BuNC to coordinate to the metal center and subsequently generate the ring-closed product. This mode of reactivity can be contrasted with the reaction of the hydrocarbon analogue [Rh(η^5-C$_9$H$_7$)(1,2,5,6-η^4-COT)] with *t*-BuNC to displace the COT ligand and form [Rh(η^5-C$_9$H$_7$)(*t*-BuNC)$_2$], indicating that the metal–ligand bonding in the OFCOT complex may indeed be stronger than in the COT analogue (*168*).

Finally it should be noted that formation of the perfluorobicyclo[3.3.0]-octa-2,7-diene-4,6-diyl ligand allows pyramidalization of four fluorinated carbons and may reflect a thermodynamic preference for *sp*^{2}-hybridized carbon atoms in coordinated OFCOT to undergo rehybridization to sp^3, provided that the ancillary ligands present on the metal can support an increase in the formal oxidation state and that the constraints of the 18-electron rule are obeyed. The origins of this thermodynamic effect for uncoordinated fluorinated alkenes have been discussed in detail (*2*). Extensions to nickel, palladium, and platinum systems are described in Section IX.

G. Dinuclear Indenyl Rhodium Complexes with Bridging OFCOT Ligands

1. Synthesis and Structure

Reaction of two equivalents of [Rh(η^5-C$_9$H$_7$)(η^2-C$_2$H$_4$)$_2$] with OFCOT yields, in addition to the yellow mononuclear complex **73**, a red dinuclear species **88** whose molecular structure is shown in Fig. 13 (*151*). The compound contains a dinuclear core with a Rh(1)–Rh(2) distance of 2.719 Å. The OFCOT ligand spans the metal–metal bond with Rh(1) bound to C(5), C(6), and C(7) in an η^3-allyl fashion, and with Rh(2) bound to C(4) in an η^1 fashion and to C(1) and C(8) in an η^2 fashion. As expected, the coordinated C=C bond is substantially lengthened upon coordination [C(1)–C(8) = 1.392 Å vs C(2)–C(3) = 1.290 Å].

88

89 a. R = H
    b. R = Me

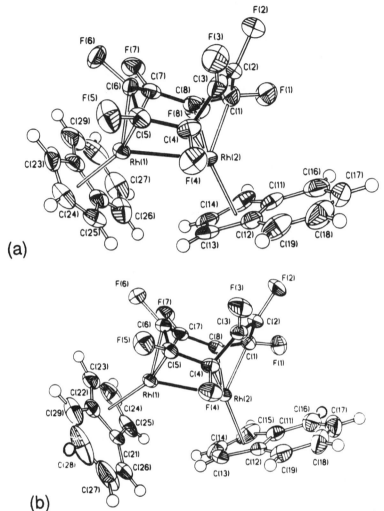

(a)

(b)

FIG. 13.   (a) Molecular structure of **88**. [Reprinted with permission from Carl *et al.* (*151*).] Selected bond distances (Å): Rh(1)–Rh(2), 2.719(1); Rh(1)–C(5), 2.099(8); Rh(1)–C(6), 2.083(8); Rh(1)–C(7), 2.122(7); Rh(2)–C(4), 2.103(9); Rh(2)–C(1), 2.077(9); Rh(2)–C(8), 2.031(8); C(1)–C(2), 1.42(1); C(2)–C(3), 1.29(1); C(3)–C(4), 1.47(1); C(4)–C(5), 1.50(1); C(5)–C(6), 1.41(1); C(6)–C(7), 1.42(1); C(7)–C(8), 1.44(1); C(8)–C(1), 1.39(1). Selected bond angles (degrees): C(1)–C(2)–C(3), 121.6(8); C(2)–C(3)–C(4), 117.9(8); C(3)–C(4)–C(5), 112.2(7); C(4)–C(5)–C(6), 119.7(7); C(5)–C(6)–C(7), 120.8(7); C(6)–C(7)–C(8), 126.6(7); C(7)–C(8)–C(1), 128.2(8); C(8)–C(1)–C(2), 124.0(8). (b) Molecular structure of **95**. [Reprinted with permission from Carl *et al.* (*151*).] Selected bond distances (Å): Rh(1)–Rh(2), 2.713(1); Rh(1)–C(5), 2.092(4); Rh(1)–C(6), 2.074(4); Rh(1)–C(7), 2.117(4); Rh(2)–C(4), 2.077(4); Rh(2)–C(1), 2.114(4); Rh(2)–C(8), 2.108(4); C(1)–C(2), 1.466(6); C(2)–C(3), 1.319(6); C(3)–C(4), 1.480(6); C(4)–C(5), 1.487(6); C(5)–C(6), 1.421(6); C(6)–C(7), 1.427(6); C(7)–C(8), 1.471(6); C(8)–C(1), 1.423(6). Selected bond angles (degrees): C(1)–C(2)–C(3), 119.7(4); C(2)–C(3)–C(4), 117.6(4); C(3)–C(4)–C(5), 111.7(3); C(4)–C(5)–C(6), 120.7(4); C(5)–C(6)–C(7), 120.2(3); C(6)–C(7)–C(8), 129.2(4); C(7)–C(8)–C(1), 125.2(4); C(8)–C(1)–C(2), 129.6(4).

Hydrocarbon analogues that exhibit this type of bonding are plentiful, including the dinuclear rhodium complexes of COT **89** (*170*). The structurally characterized analogue **89a** (*170*) provides a useful structural comparison to **88**. One notable difference is the shortened metal–olefin bond distance in **88** [**89a**, $Rh-CH_{olefin}$ (average) = 2.128 Å; **88**, $Rh-CF_{olefin}$ (average) = 2.054 Å]. This can be attributed to increased back bonding from the metal center to the electron-deficient olefin as discussed previously. Depletion of electron density at the metal centers by the fluorinated ligand also weakens the metal–metal interaction, as evidenced by the longer Rh–Rh distance in the fluorinated complex [**89a**, Rh(1)–Rh(2) = 2.719(1) Å; **88**, Rh–Rh = 2.689(1) Å].

2. Stereochemical Nonrigidity of the Bridging OFCOT Ligand

The observant reader will undoubtedly have noted that no evidence of any stereochemical nonrigidity (on the NMR time scale) has been obtained for any of the mononuclear complexes derived from OFCOT. However, the dinuclear complex **88** does exhibit fluxional behavior (*151*). If the solid-state structure of **88** were rigidly maintained in solution, the ^{19}F-NMR spectrum should show eight resonances due to symmetry inequivalent fluorines. However, only four broad resonances are present in the ^{19}F NMR spectrum at room temperature, indicative of a dynamic process that interconverts the two enantiomorphs of **88**. Two mechanisms have been proposed for such interconversions in COT analogues: a "glide" and a "twitch" (Scheme 3). The

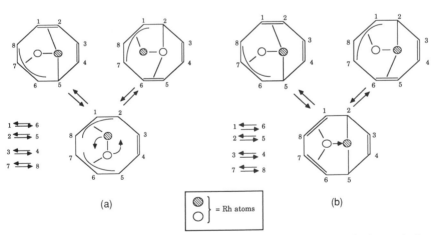

SCHEME 3.   Glide and twitch mechanisms for site exchange. (a), Glide mechanism: pairwise exchange of F sites and of metal sites; (b), twitch mechanism: pairwise exchange of F sites but not of metal sites.

"twitch" process is found in all dinuclear COT complexes of this type that have been examined (170–172).

At −40°C, eight sharp fluorine resonances and two sets of indenyl resonances were observed as expected for the solid-state structure of **88** (151). At −93°C, two sets of eight resonances (20:3 relative ratio) are seen in the ^{19}F-NMR spectrum. As discussed in Section VIII,D,3, it is possible to slow indenyl rotation on the NMR time scale at low temperatures. At −40°C, it seems that indenyl rotation is still fast but the fluxional process of the OFCOT ring has slowed, explaining the single set of eight resonances. However, at −93°C, the indenyl rotation may have slowed sufficiently to allow observation of a second isomer in which the indenyl ligands adopt a different conformation with respect to each other. At +90°C, observation of four sets of fluorine resonances and two sets of indenyl resonances indicates that the glide mechanism is not operative and the twitch mechanism is probably responsible for the interconversion of the enantiomorphs of **88**. Line shape analysis yields an E_a of 16.3 ± 0.4 kcal/mol and a $\Delta G^{\ddagger}_{247K}$ of 13.7 ± 0.4 kcal/mol (151). No activation parameters were obtained for the analogous fluxional process in the COT complex **89** (170), so the most direct comparison is that for the cyclooctatriene analogue **90** ($\Delta G^{\ddagger}_{247K}$ = 12.9 kcal/mol) (173). Although no uncertainty

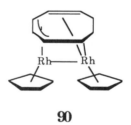

90

was reported for the latter value, it appears that, unlike olefin rotation (Section III,C,2) and haptotropic rearrangements in η^6-C_6X_6 complexes (Section VI,C), perfluorination does not appreciably affect the activation barrier for stereochemical nonrigidity in dinuclear rhodium complexes of this general type.

3. Structure of an Intermediate in the Formation of the Dinuclear Rhodium OFCOT Complex from the Mononuclear Precursor

Reaction of one equivalent of the isolated mononuclear species **73** with one equivalent of [Rh(η^5-C_9H_7)(η^2-C_2H_4)$_2$] leads to clean formation of the dinuclear species **88** (151), presumably via an intermediate in which both rhodiums are coordinated to the same side of the perfluorinated ring. This cis

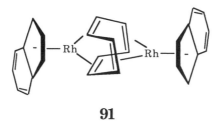

91

coordination in the intermediate is necessary because no trans product analogous to the COT complex **91** (*170*) was observed in the reaction. This intermediate was not observable under the reaction conditions, but addition of carbon monoxide to a solution of **88** in THF yielded the two isomeric adducts **92** and **93**, with a color change from red to yellow. The molecular structure of the major isomer **92** is shown in Fig. 14 (*151*). Both rhodiums are still coordinated in a cis fashion to the OFCOT ring, but there is no longer a Rh—Rh

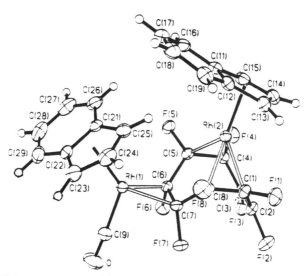

FIG. 14. Molecular structure of **92**. [Reprinted with permission from Carl *et al.* (*151*).] Selected bond distances (Å): Rh(1)–C(6), 2.079(2); Rh(1)–C(7), 2.050(3); Rh(2)–C(1), 2.076(3); Rh(2)–C(4), 2.071(3); Rh(2)–C(5), 2.117(3); Rh(2)–C(8), 2.112(3); C(1)–C(2), 1.473(4); C(2)–C(3), 1.308(4); C(3)–C(4), 1.463(4); C(4)–C(5), 1.429(4); C(5)–C(6), 1.475(3); C(6)–C(7), 1.434(3); C(7)–C(8), 1.472(4); C(8)–C(1), 1.422(4). Selected bond angles (degrees): C(1)–C(2)–C(3), 119.7(3); C(2)–C(3)–C(4), 119.4(3); C(3)–C(4)–C(5), 122.2(2); C(4)–C(5)–C(6), 119.9(2); C(5)–C(6)–C(7), 115.6(2); C(6)–C(7)–C(8), 117.3(2); C(7)–C(8)–C(1), 119.9(2); C(8)–C(1)–C(2), 121.7(3).

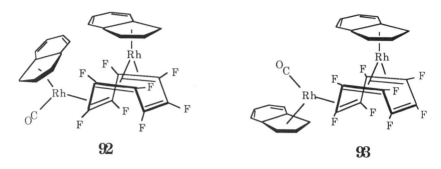

92   **93**

bond. Rh(2) is bound in a 1,2,5,6-η fashion to the perfluorinated ring, and Rh(1) is bound in an η^2 fashion to an olefinic bond of the ring and to the added CO ligand. The assumed structure of the minor isomer **93** is related to that of **92** by rotation about the Rh(1)–olefin axis.

The structure of **92** nicely models what an intermediate in the formation of the dinuclear species **88** might look like structurally, i.e., resulting from attack of a coordinatively unsaturated [Rh(C$_9$H$_7$)L] fragment on the less hindered exo face of one of the uncoordinated fluoroolefins of the mononuclear species **73**. Loss of L and formation of a metal–metal bond followed by or in conjunction with rearrangement of the bonding to the OFCOT ring would lead to **88**. In agreement with this hypothesis, heating a solution of **92** does indeed lead to dissociation of the CO ligand and reformation of **88**. Notably formation of the trans dinuclear complex **91** should be less favorable, because it would involve attack by a second metal fragment on the more hindered endo face of an uncoordinated olefin of [Rh(η^5-C$_9$H$_7$)(1,2,5,6-η^4-COT)]; the fact that **91** is indeed formed (*170*) is surprising, therefore.

Attempts to add ethylene to **88** in hopes of obtaining the actual intermediate were only partially successful. Spectroscopically, it appeared that the product

94

94 of this reaction was structurally akin to **92**, but ethylene rapidly dissociated in solution even at −20°C to give **88** (*151*).

4. Selective C−F Bond Activation

Attempts to purify **88** via column chromatography on silica gel or alumina led to a reaction on the column support to produce a new complex **95**, whose structure is shown in Fig. 13 (*151*). Clearly, formation of **95** from **88** involves

95

a selective fluorine–hydrogen exchange. This is remarkable because C—F bonds are considerably stronger than their C—H analogues (*2*). Presumably, a strong fluorine bond to another element must be formed to make up for the bond energy difference between the C—X bonds, although the fate of the fluorine is unknown, as is the source of the hydrogen. Contact time on the column support also influences production of **95**. Rapid chromatography over a span of 5 min allows recovery of **88** unchanged, whereas chromatography over a 3-hr span leads to formation of **95** (*151*).

Additional chromatography allowed for isolation of minor amounts of a second isomeric complex that could not be totally separated from the major isomer **95**. The location of the lone hydrogen remains unknown but it is clearly a positional isomer of the major product. This isomer along with what appear to be other heptafluoro isomers can be produced by heating solutions of **95** up to 80°C. Scrambling of the H site occurs around the ring to give a mixture of heptafluoro products, confirming that the initial production of the major isomer **95** on the column was kinetically selective. The mechanism of this remarkable reaction is still unclear.

Complex **95** and its isomers are not fluxional on the NMR time scale as evidenced by seven sharp resonances in the ^{19}F NMR spectrum (*151*). The replacement of a fluorine by a hydrogen presumably causes a thermodynamic preference for a single isomer.

IX

NICKEL, PALLADIUM, AND PLATINUM COMPLEXES
DERIVED FROM OFCOT

A. *Perfluorobicyclo[3.3.0]octadienediyl Complexes of Platinum*

Although reaction of [Pt(COD)$_2$] alone with OFCOT did not produce any metal–fluorocarbon products, addition of two molar equivalents of PPh$_3$, AsPh$_3$, or t-BuNC to [Pt(COD)$_2$] followed by addition of OFCOT yielded

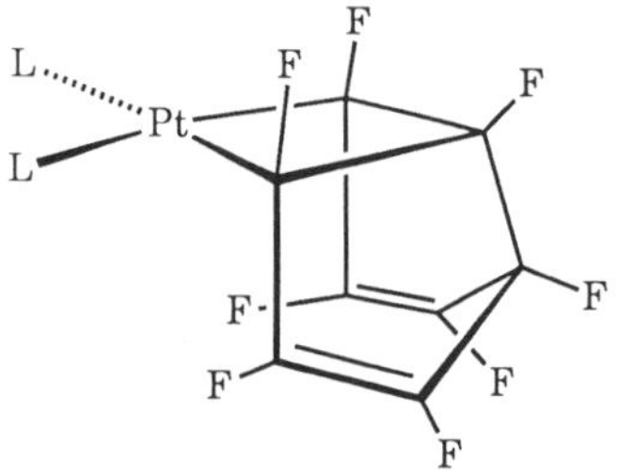

96 a. L = PPh$_3$
    b. L = AsPh$_3$
    c. L = t-BuNC
    d. L = PPh$_2$Me
    e. L = PPhMe$_2$

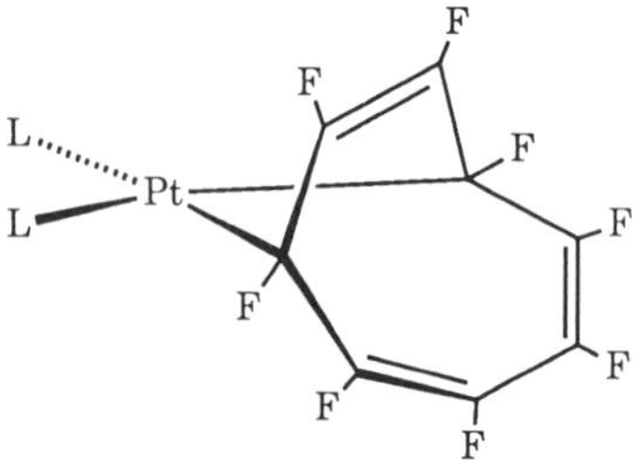

97 a. L = PPh$_3$
    b. L = PPh$_2$Me
    c. L = PPhMe$_2$

the 16-electron Pt(II) compounds **96a–c** (*174*), which contain the same perfluorobicyclo[3.3.0]octa-2,7-diene-4,6-diyl ligand found in several 18-electron Co(III) and Rh(III) complexes (see Section VIII,F,2). Reactions of [Pt(η^2-C$_2$H$_4$)(PPh$_3$)$_2$] or [PtL$_4$] (L = PPh$_3$, PPh$_2$Me, or PMe$_2$Ph) with OFCOT also lead eventually to 16-electron ring-closed products **96a,d,e**, via the spectroscopically observed intermediates **97** (*174*).

It was hoped that reductive elimination of the fluorocarbon ligand from compounds **96** might afford the elusive valence isomer of OFCOT, perfluorosemibullvalene **98**. All attempts to induce formation of **98** from these Pt(II)

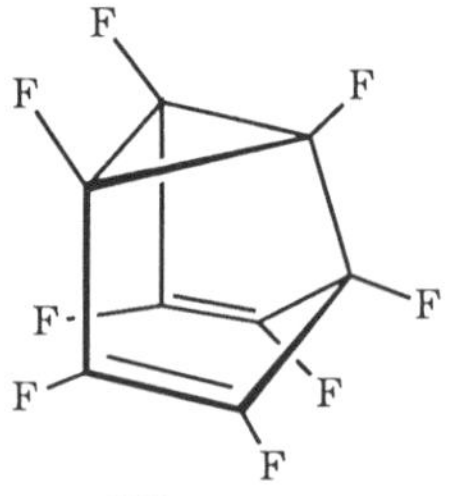

98

complexes yield only free OFCOT. Other attempts to form perfluorosemibullvalene under mild conditions from organic precursors have also failed, suggesting that it is a fragile molecule with respect to isomerization to OFCOT (*123*). Thus, we cannot distinguish whether reductive elimination occurs to give perfluorosemibullvalene, which then rapidly undergoes ring opening, or whether OFCOT is formed directly by a retrotransannular reaction during displacement from the metal.

B. *Perfluorobicyclo[3.3.0]octadienediyl Complexes of Palladium*

Similarly, reaction of tris(dibenzylideneacetone)dipalladium with two equivalents of *t*-butylisocyanide per Pd atom, followed by addition of OFCOT, leads to an analogous ring-closed product **99a**. Reaction of [Pd(PPh$_3$)$_4$] with OFCOT afforded the phosphine analogue **99b** (*174*). No attempts have been made to probe any further chemistry of these molecules.

99   a. L = t-BuNC
        b. L = PPh$_3$

C. *1,2,5,6-η⁴-OFCOT Complexes of Nickel*

Reaction of [Ni(COD)$_2$] or [Ni$_2$(COT)$_2$] with OFCOT leads to decomposition of the nickel starting materials, with formation of a nickel mirror; no metal–fluorocarbon products were formed. However, reaction of [Ni(COD)$_2$] with two equivalents of PMe$_3$ or PMe$_2$Ph, followed by addition of OFCOT, leads to formation of the (1,2,5,6-η⁴-OFCOT)Ni(0) complexes **100** (*174,175*). In contrast to behavior observed for some cobalt and rhodium analogues (see Section VIII,E,1), attempts to isomerize the OFCOT ring in compounds **100** via heating or ultraviolet irradiation lead only to recovery of starting materials. Attempts to make the corresponding PMePh$_2$, PPh$_3$, PCy$_3$ (Cy = cyclohexyl), P(OPh)$_3$, or P(OMe)$_3$ derivatives via [Ni(COD)$_2$] were unsuccessful, as were attempts to produce fluorocarbon complexes from the reactions of [Ni(η²-C$_2$H$_4$)(PPh$_3$)$_2$] or [Ni(η²-C$_2$H$_4$)(PPh$_2$Me)$_2$] with

100 a. L = PMe$_3$
        b. L = PPhMe$_2$

OFCOT. No evidence was obtained for oxidative addition of a C—F bond
to the nickel center, in contrast to the reported reaction of perfluorobenzene
with [Ni(COD)(PEt$_3$)$_2$] (see Section III,F).

D. *Perfluorobicyclo[3.3.0]octadienediyl Complexes of Nickel*

In further contrast to the results obtained using phosphine ligands, treatment of [Ni(COD)$_2$] with 2 molar equivalents of *t*-butylisocyanide, followed
by OFCOT, yields exclusively the Ni(II) complex **101** in which OFCOT has
undergone ring closure to give the now familiar perfluorobicyclo[3.3.0]octa-
2,7-diene-4,6-diyl ligand (*174,175*).

101

It was intriguing to see whether replacement of the *t*-butylisocyanide ligands in **101** with PMe$_3$ would lead to simple ligand substitution, with retention of the ring-closed structure, or whether substitution by phosphines would
cause ring opening to give the 1,2,5,6-η^4-OFCOT complex **100a**. Surprisingly,
neither result is obtained, and even an excess of PMe$_3$ produces only the
five-coordinate 1:1 adduct **102a** (*174,175*). The molecular structure of **102a** is
shown in Fig. 15 (*175*). The coordination geometry around nickel is approximately square–pyramidal with the PMe$_3$ in the apical position.

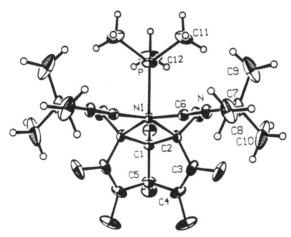

102 a. L = PMe₃
         b. L = PPh₃
         c. L = t-BuNC

103

Addition of triphenylphosphine or *t*-butylisocyanide to **101** affords the analogous five-coordinate complexes **102b** and **102c**. These species represent rare examples of stable five-coordinate *cis*-dialkyl complexes of Ni(II), although stable trigonal–bipyramidal *trans*-dialkyls of nickel are well known (*176–178*), and they provide a structural model for a putative square–pyramidal intermediate in reductive elimination reactions of *cis*-dialkyl

FIG. 15. Molecular structure of **102a**. [Reprinted with permission from Hughes *et al.* (*175*).] Selected bond distances (Å): Ni–P, 2.3628(14); Ni–C(2), 1.955(3); Ni–C(6), 1.863(3); C(6)–N(6), 1.155(4); C(7)–N(6), 1.454(4); C(1)–C(2), 1.527(4); C(1)–C(5), 1.578(6); C(2)–C(3), 1.462(5); C(3)–C(4), 1.312(5); C(4)–C(5), 1.471(4). Selected bond angles (degrees): P–Ni–C(2), 97.56(10); P–Ni–C(6), 98.30(9); C(2)–Ni–C(6), 93.01(12); C(2)–Ni–C(2'), 72.98(11); C(2)–C(1)–C(2'), 99.2(3); C(6)–Ni–C(6'), 96.56(13); C(6)–N(6)–C(7), 176.7(3).

nickel(II) complexes (175). It is also notable that complexes 102 are considerably more stable than is the hydrocarbon nickelacyclopentane complex 103, which decomposes above 9°C in solution (179).

The apical phosphine ligands in 102a and 102b dissociate reversibly in solution at room temperature, but the basal t-butylisocyanide ligands do not undergo ligand substitution at room temperature, even in the presence of excess phosphine. Assuming that ligand dissociation occurs from the apical site, it can be concluded that there is no energetically accessible pathway for interconversion of basal and apical sites that would allow dissociation of the t-BuNC ligands, and their subsequent replacement by phosphine. A Berry pseudorotation is perhaps unlikely because the metallacyclobutane ring would have to span basal/apical sites, and a turnstile rotation of the three monodentate ligands is also precluded by these observations. However, only a single proton resonance for the t-butylisocyanide ligands is observed for the tris(t-BuNC) complex 102c, even at $-80°C$, and the possibility of rapid turnstile or Berry pseudorotation in this particular complex cannot be excluded (174,175).

A pathway by which OFCOT might be transformed to the final perfluorobicyclo[3.3.0]octa-2,7-diene-4,6-diyl ligand has been discussed in Section VIII,F,2. For the Co and Rh systems, access to the observed diyl intermediates analogous to 97 occurs via exogenous ligand-induced $\eta^3 \rightarrow \eta^1$ rearrangement of 1,2,3,6-η^4-OFCOT complexes. No spectroscopic evidence for analogous 1,2,3,6-η^4-OFCOT coordination has been obtained in any of the Ni, Pd, or Pt systems reported here, although their presence among other unobserved intermediates en route to the spectroscopically observed diyl intermediates 97 cannot be discounted.

In the platinum chemistry of OFCOT, ancillary ligands with varying steric and electronic effects, such as phosphines, triphenylarsine, and t-BuNC, all give rise to a transannular ring-closure reaction. Notably, $[Pt(COD)_2]$ is unreactive, perhaps because the metal center is less basic. We conclude that in all cases the Pt(0) fragment "PtL_2" is electron rich enough to be capable of an oxidative addition reaction with OFCOT to afford a 16-electron square–planar Pt(II)–diyl intermediate (e.g., 97) rather than a 1,2,3,6-η^4-OFCOT complex. Thus in contrast to analogous Co(III) and Rh(III) systems, in which 1,2,3,6-η^4-OFCOT ligation is required to maintain the 18-electron pseudooctahedral coordination geometry around the metal, Pt(II) prefers the 16-electron square–planar structure that can accommodate the 1,4-η-diyl ligand and allows subsequent ring closure to occur spontaneously without requiring an exogenous ligand. A similar argument can be made for the two palladium systems examined (174).

The nickel chemistry of OFCOT is much more dependent on the nature of the ancillary ligand. Less basic phosphines (PPh_3, PPh_2Me) and phosphites [$P(OPh)_3$, $P(OMe)_3$] and the sterically demanding phosphine (PCy_3) result in no observable binding of OFCOT, whereas more basic phosphines ($PPhMe_2$, PMe_3) with smaller cone angles result in 1,2,5,6-η^4 binding of OFCOT without any oxidative addition. Thus the ability of "NiL_2" to bind OFCOT appears to be controlled by both the steric requirements of the phosphine and the basicity of the metal center. It is also noteworthy that, in contrast to 1,2,5,6-η^4-OFCOT ligands bound to the [$M(\eta^5$-$C_5Me_5)$] fragment (M = Co, Rh), their Ni(0) analogues **100** do not undergo photochemical (or thermal) rearrangement to 1,2,3,6-η^4-OFCOT isomers, and consequently no subsequent ring closure has been observed. Of all the ligands examined, only t-BuNC affords a ring-closed product containing square–planar Ni(II) (*174*).

X

METALLATION OF OFCOT AND ITS VALENCE ISOMERS

A. *Hydrocarbon Complexes Containing the*
   η^1-Cyclooctatetraenyl Ligand

Cyclooctatetraenyl complexes in which the metal is attached to the organic ring framework via a σ bond to carbon are rare. The racemic iron compound **104** has been prepared, and its diastereoisomeric phosphine derivatives **105** (only one diastereoisomer shown) have been synthesized by photochemical replacement of a CO ligand in **104** by PR_3 (*180*). Compound **104** exists as two

104   **105**

enantiomers by virtue of the conformational asymmetry of the monosubstituted cyclooctatetraene ring, and **105** exists as a diastereoisomeric mixture due to the additional asymmetric center at iron.

B. *Isodynamical Processes in Substituted Cyclooctatetraenes*

In such monosubstituted systems two isodynamical phenomena are possible due to the nonplanarity of the ring and the alternating single and double bonds (see Scheme 4) (181), and if they occur at suitable rates they can be monitored by NMR techniques. The first process, ring inversion (RI), interconverts enantiomers **104a** and **104b** through a planar intermediate **106**, but does not cause site exchange of ring carbon atoms (or their substituents). Its operation is manifested by site exchange of diastereotopic carbonyl sites (CO_A and CO_B) on the metal atom. Second, during the lifetime of **106** a bond shift (BS) isomerization process may also occur via transition state **107**. This does give rise to site exchange within the cyclooctatetraene ring, interconverting positions 2 with 8, 3 with 7, and 4 with 6. For compounds such as **104**, RI with or without BS interconverts enantiomers **104a** and **104b**: for diastereoisomeric pairs such as **105**, RI epimerizes only one chiral center and would interconvert, for example, the *RS* and *RR* diastereoisomers, which have different NMR resonances.

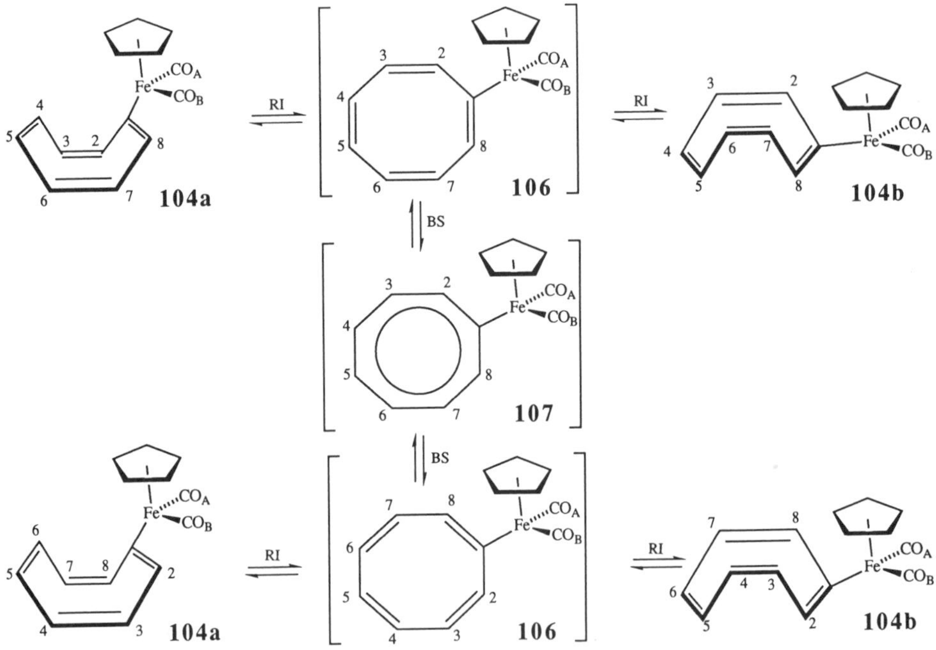

SCHEME 4.   Isodynamical processes in substituted cyclooctatetraenes.

For both **104** and **105**, values of $\Delta G^{\ddagger}(\text{RI}) \sim 17$ kcal/mol and $\Delta G^{\ddagger}(\text{BS}) \sim 18$ kcal/mol have been determined from NMR measurements (*180*). Analogous processes have been studied extensively in cyclooctatetraenes bearing other substituents (*181*). By way of comparison, ethoxycyclooctatetraene has $\Delta G^{\ddagger}(\text{RI}) = 12.5$ kcal/mol and $\Delta G^{\ddagger}(\text{BS}) = 16.0$ kcal/mol (*182*). Values for other monosubstituted derivatives (*118*) differ by $\sim \pm 2$ kcal/mol and more heavily substituted compounds have substantially higher barriers for both processes, due to steric buttressing effects in the planar intermediate (*181*). Thus, while the transition metal substituent in **104** or **105** does increase the barriers for RI and BS relative to smaller substituents, this effect is not large enough to prevent observation of both these processes on the NMR time scale (*180*). Prior to our work, the effect of polyfluorination on the energetics of these isodynamic processes had not been evaluated.

C. Monometallation of OFCOT to give η^{1}-Heptafluorocyclooctatetraenyl Complexes and Their Heptafluorobicyclo[4.2.0]octatrienyl Valence Isomers

The known reactions of fluorinated olefins, arenes, and heterocycles with metal carbonyl anions, to afford fluorovinyl or fluoroaryl complexes resulting from net displacement of fluoride ion (see Section III,G), prompted us to attempt such substitution reactions with OFCOT, in order to generate the required metal-substituted heptafluorocyclooctatetraenes.

1. Synthesis

Reactions of OFCOT with the metal carbonyl anions $[\text{Mn(CO)}_5]^-$ and $[\text{M}(\eta^5\text{-C}_5\text{R}_5)(\text{CO})_2]^-$ (M = Fe, R = H, Me; M = Ru, R = H) generate the monocyclic η^1-heptafluorocyclooctatetraenyl complexes **108a–c** (*183*), which are in equilibrium in solution with small amounts ($\sim 10\%$) of their bicyclic valence isomers **109a–c** (*184*). For example, at 20°C in CDCl$_3$, $K_{\text{eq}} = [\textbf{109a}]/[\textbf{108a}] \sim 0.1$ (*184*), a value much larger than that observed for the equilibrium involving free OFCOT, for which $K_{\text{eq}} = [\textbf{52}]/[\textbf{51}] = 0.003$ at 20° in acetone-d_6 (*124*). The dynamic nature of this equilibrium is confirmed by the temperature dependence of the ratio of the iron complexes **108b** and **109b**, for which $K_{\text{eq}}(\text{CDCl}_3) = [\textbf{109b}]/[\textbf{108b}] = 0.13$ (21°C), 0.15 (1°C), and 0.18 (-55°C) (184). Curiously, the room-temperature reaction of Na$^+[\text{Re(CO)}_5]^-$ with OFCOT affords only the monocyclic complex **108d**, and no evidence was obtained for the presence of a bicyclic valence isomer **109d**. The reason for this is unclear. The cobalt analogues **108e–h** and **109e–h** are also obtained by reaction of the appropriate metal carbonyl anion with

108 a. M = Mn(CO)$_5$
      b. M = Fe(C$_5$H$_5$)(CO)$_2$
      c. M = Fe(C$_5$Me$_5$)(CO)$_2$
      d. M = Re(CO)$_5$
      e. M = Co(CO)$_3$(PPh$_3$)
      f. M = Co(CO)$_3$(PPh$_2$Me)
      g. M = Co(CO)$_3$(PPhMe$_2$)
      h. M = Co(CO)$_3$(PMe$_3$)

109 a. M = Mn(CO)$_5$
      b. M = Fe(C$_5$H$_5$)(CO)$_2$
      c. M = Fe(C$_5$Me$_5$)(CO)$_2$
      d. M = Re(CO)$_5$
      e. M = Co(CO)$_3$(PPh$_3$)
      f. M = Co(CO)$_3$(PPh$_2$Me)
      g. M = Co(CO)$_3$(PPhMe$_2$)
      h. M = Co(CO)$_3$(PMe$_3$)

OFCOT. Also formed in these latter reactions are dinuclear compounds containing the μ-hexafluorocyclooctatrienyne ligand; these are discussed in Section XI,B.

The bicyclic isomers **109** are clearly formed by electrocyclic ring closure of their initially formed monocyclic valence isomers and not from the nucleophilic attack of the transition metal anion on traces of **52** in equilibrium with OFCOT, as shown by the low-temperature reaction of [Fe(η^5-C$_5$H$_5$)-(CO)$_2$]$^-$. At $-78°$C, the ^{19}F NMR spectrum of the crude reaction mixture indicates exclusive formation of the monocyclic complex **108b**, but on warming the solution to room temperature the resonances of the bicyclic valence isomer **109b** grow in (*184*).

2. Crystallographic Studies of η^1-Heptafluorocyclooctatetraenyl Complexes of Iron

The expected tub conformation for the fluorinated ring is confirmed by the crystal structures of two complexes, **108b** and **108c**, which are shown in Fig. 16 (*184*). The C$_8$F$_7$ ligand in **108b** adopts the expected tub conformation with a dihedral angle (α) of 42.4(6)° for the planes defined by [C(1),C(2),C(5),C(6)] and [C(6),C(7),C(8),C(1)], and 39.6(6)° for [C(1),C(2),C(5),C(6)] and [C(2),-C(3),C(4),C(5)], as compared to the corresponding angles in OFCOT of 41.8(2)° and 41.4(2)° (*121*). The Fe—C σ bond (sp^2) in **108b**, 1.977(10) Å, is shorter than the corresponding Fe—C σ-bond (sp^2) distance of 2.11(1) Å to the phenyl ring in [Fe(η^5-C$_5$H$_5$)(η^1-C$_6$H$_5$)(CO)(PPh$_3$)] (*185*), but longer than

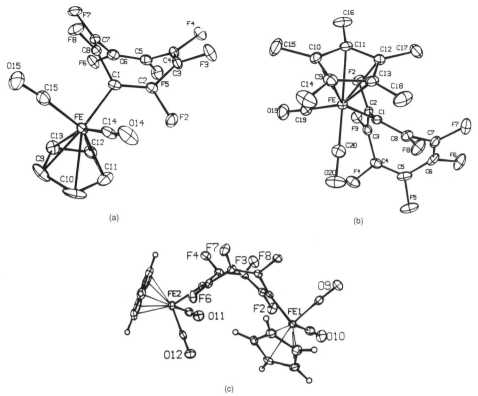

(a)

(b)

(c)

FIG. 16. (a) Molecular structure of **108b** (*184*). Selected bond distances (Å): Fe–C(1), 1.977(1); C(1)–C(2), 1.335(10); C(2)–C(3), 1.443(11); C(3)–C(4), 1.325(11); C(4)–C(5), 1.451(11); C(5)–C(6), 1.319(11); C(6)–C(7), 1.458(2); C(7)–C(8), 1.313(10); C(8)–C(1), 1.462(8). Selected bond angles (degrees): Fe–C(1)–C(2), 122.9(7); Fe–C(1)–C(8). 118.2(7); C(8)–C(1)–C(2), 118.7(6); C(1)–C(2)–C(3), 131.1(6); C(2)–C(3)–C(4), 127.2(7); C(3)–C(4)–C(5), 127.5(7); C(4)–C(5)–C(6), 126.9(8); C(5)–C(6)–C(7), 126.4(7); C(6)–C(7)–C(8), 126.7(6); C(7)–C(8)–C(1), 129.3(7). (b) Molecular structure of **108c** (*184*). Selected bond distances (Å): Fe–C(1), 2.001(2); C(1)–C(2), 1.328(3); C(2)–C(3), 1.465(3); C(3)–C(4), 1.311(4); C(4)–C(5), 1.450(3); C(5)–C(6), 1.315(4); C(6)–C(7), 1.444(3); C(7)–C(8), 1.326(3); C(8)–C(1), 1.454(3). Selected bond angles (degrees): Fe–C(1)–C(2), 125.0(2); Fe–C(1)–C(8), 117.8(2); C(8)–C(1)–C(2), 117.1(2); C(1)–C(2)–C(3), 131.0(2); C(2)–C(3)–C(4), 127.9(2); C(3)–C(4)–C(5), 125.6(2); C(4)–C(5)–C(6), 125.8(6); C(5)–C(6)–C(7), 126.3(2); C(6)–C(7)–C(8), 128.2(2); C(7)–C(8)–C(1), 130.8(2). (c) Molecular structure of **112** (*184*). Selected bond distances (Å): Fe(1)–C(1), 1.994(4); Fe(2)–C(5), 1.991(4); C(1)–C(2), 1.330(6); C(2)–C(3), 1.449(6); C(3)–C(4), 1.324(6); C(4)–C(5), 1.459(7); C(5)–C(6), 1.330(6); C(6)–C(7), 1.452(6); C(7)–C(8), 1.315(6); C(8)–C(1), 1.464(6). Selected bond angles (degrees): Fe–C(1)–C(2), 128.0(3); Fe–C(1)–C(8), 116.0(3); Fe(2)–C(5)–C(4), 117.8(3); Fe(2)–C(5)–C(6), 125.5(4); C(8)–C(1)–C(2), 115.8(4); C(1)–C(2)–C(3), 130.4(3); C(2)–C(3)–C(4), 126.7(4); C(3)–C(4)–C(5), 128.9(4); C(4)–C(5)–C(6), 116.6(4), C(5)–C(6)–C(7), 130.3(4); C(6)–C(7)–C(8), 126.3(4); C(7)–C(8)–C(1), 129.2(4).

that to the pentafluorocyclobutenyl ring [1.936(9) Å] in [Fe(η^5-C$_5$H$_5$)(η^1-C$_4$F$_5$)(CO)$_2$] (*110*). The individual C—C and C—F bond distances in **108b** are not significantly (>3 ESD) different than those in OFCOT (*121*). The bond length of the C=C double bond containing the iron fragment in **108b**, 1.335(10) Å, is the longest C=C bond distance and is slightly longer than the C=C bond distances in OFCOT. The C(2)–C(1)–C(8) bond angle of 118.7(6)° is significantly smaller than the corresponding angles in OFCOT, which range between 126.3(2)° and 127.4(2)°, whereas the adjacent C(1)–C(2)–C(3) and C(1)–C(8)–C(7) bond angles of 131.1(6)° and 129.3(7)°, respectively, are larger than the other C—C—C bond angles in **108b** or OFCOT. This general phenomenon of contraction of the bond angle at the ligated carbon atom, with expansion of the adjacent angles, has been observed in a variety of other systems, including the pentafluorocyclobutenyl complex [Fe(η^5-C$_5$H$_5$)(η^1-C$_4$F$_5$)(CO)$_2$] (*110*) and several pentafluorophenyl complexes (*186*).

The iron atom in **108c** is also in an approximately octahedral environment with the C$_8$F$_7$ ligand in the tub conformation. The values for α of 41.0(2)° for [C(1),C(2),C(3),C(8)] and [C(3),C(4),C(7),C(8)] and 40.7(2)° for [C(4),C(5),C(6),C(7)] and [C(3),C(4),C(7),C(8)] are similar to the corresponding values of 42.4(6)° and 39.6(6)° in **108b**. The individual Fe—C, C—C, C=C, and C—F bond distances and angles in **108c** are not significantly different from those in **108b**. The bulkier C$_5$Me$_5$ ligand, which is on the side of the metal away from the fluorinated ring, has negligible steric interaction with the C$_8$F$_7$ ring.

3. Ligand Substitution Reactions of η^1-Heptafluorocyclooctatetraenyl Complexes of Iron

Irradiation of hexane solutions containing **108b** and triphenylphosphine affords **110a** as a 1:1 mixture of diastereoisomers. An analogous reaction with

110 a. L = PPh$_3$
      b. L = AsPh$_3$

triphenylarsine affords **110b** as a 2:1 diastereoisomeric mixture (*184*). A single recrystallization of each mixture affords pure samples of a single diastereoisomer. Clearly the stability of these purified diastereoisomers at room temperature indicates that RI is not just slow on the NMR time scale, but is a chemically slow event. Small peaks due to the corresponding bicyclic valence isomers are observed in the NMR spectra of both pure and mixed diastereoisomers.

4. Reaction of η¹-Heptafluorocyclooctatetraenyl Ligands with Chromatography Supports

Chromatography of the iron complexes **108b** and **108c** and the cobalt analogues **108e–h** on Florisil effects partial conversion to the ketone derivatives **111** (*184*). The mechanism of this reaction is unclear.

111 a. M = $Fe(C_5H_5)(CO)_2$
   b. M = $Fe(C_5Me_5)(CO)_2$
   c. M = $Co(CO)_3(PPh_3)$
   d. M = $Co(CO)_3(PPh_2Me)$
   e. M = $Co(CO)_3(PPhMe_2)$
   f. M = $Co(CO)_3(PMe_3)$

D. Selective Dimetallation of OFCOT

The reaction of $[Fe(\eta^5\text{-}C_5H_5)(CO)_2]^-$ with OFCOT also yields small amounts of the disubstituted complex **112**, which can also be prepared by treating **108b** with a second equivalent of nucleophile (*184,187*). Therefore, **112** is formed by consecutive displacement of fluorides from OFCOT, the second displacement being selective in producing only the 1,5-diyl isomer, whose molecular structure is shown in Fig. 16 (*184*). The overall structural features of the fluorinated ring are not significantly different from those in **108b** and **108c**. The weaker nucleophiles $[M(CO)_5]^-$ (M = Mn, Re) do not react with their respective monosubstituted complexes **108a** and **108d** to give any identifiable analogues of **112**. Another type of dimetallation product is described in Section XI,B.

112

E. *Mono- and Dimetallation of Perfluorotricyclo[4.2.0.0^{2,5}]octa-3,7-diene*

A second valence isomer of **108a** can be synthesized by reaction of $Na^+[Mn(CO)_5]^-$ with *anti*-octafluorotricyclo[4.2.0.0^{2,5}]octa-3,7-diene **53**, to give the tricyclic product **113a** (*184*). The molecular structure of **113a** is

113 a. M = Mn(CO)$_5$
         b. M = Fe(C$_5$H$_5$)(CO)$_2$

shown in Fig. 17. The coordination geometry around the manganese atom is octahedral and the overall anti configuration of the tricyclic ring is unchanged from that of the unsubstituted fluorocarbon **53**. The overall geometry of the [Mn(CO)$_5$] fragment is not significantly different from that observed in [Mn$_2$(CO)$_{10}$] (*188*), and the Mn—C(1) σ-bond distance of 2.031(3) Å is not significantly different from the corresponding bond length [1.95(3) Å] observed in the related complex [Mn(η^1-*cis*-CF=CFH)(CO)$_5$] (*189*).

In contrast, the reaction of [Fe(η^5-C$_5$H$_5$)(CO)$_2$]$^-$ with the tricyclic fluorocarbon **53** affords not only the monosubstituted complex **113b** but also a

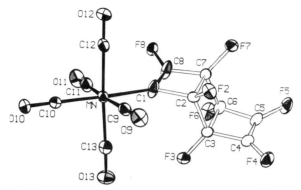

FIG. 17. Molecular structure of **113a** (*184*). Selected bond distances (Å): Mn–C(1), 2.031(4); C(1)–C(8), 1.333(4); C(1)–C(2), 1.578(5); C(2)–C(3), 1.528(6); C(3)–C(4), 1.502(8); C(4)–C(5), 1.318(8); C(5)–C(6), 1.488(8); C(6)–C(7), 1.528(7); C(7)–C(8), 1.518(6). Selected bond angles (degrees): Mn–C(1)–C(8), 137.1(3); Mn–C(1)–C(2), 133.1(2); C(2)–C(1)–C(8), 88.6(3); C(1)–C(8)–C(7), 100.1(3).

1:1 mixture of the disubstituted complexes **114** and **115** (*184*). Thus, displacement of the second fluoride from **113b** is not stereospecific, in contrast to the corresponding displacement of fluoride from **108b**.

114

115

F. *Isodynamical Processes in η^1-Heptafluorocyclooctatetraenyl Ligands*

Because the ^{19}F NMR spectra of all η^1-heptafluorocyclooctatetraenyl complexes exhibit seven resonances, with no line broadening or magnetization transfer observed at 80°C, a rapid BS process for these complexes on the NMR time scale is clearly ruled out. These observations provide no information about the isodynamical RI process.

However, at 80°C the $^{13}C\{^{19}F\}$ NMR spectrum of **108b** displays eight cyclooctatetraene ring carbon resonances as well as two resonances due to diastereotopic CO ligands. Notably, the carbonyl signals could not be observed clearly in the $^{13}C\{^1H\}$ NMR spectrum due to coupling to fluorines. The observation of two CO signals indicates that the RI process must also be slow on the NMR time scale, and this absence of coalescence in the $^{13}C\{^{19}F\}$ NMR spectrum at 80°C allows a minimum value of $\Delta G_{RI}^{\ddagger} \geq 18.9$ kcal/mol to be calculated. However, the actual value for $\Delta G_{RI}^{\ddagger}$ may be significantly greater, because the equation used to obtain $\Delta G_{RI}^{\ddagger}$ is dependent upon the line separation between the exchangeable nuclei, which in this case is small (7 Hz). The observed conformational stability of the pure diastereoisomers **110** (see Section X,C,3) at room temperature also indicates that RI is a chemically slow event (*184*).

Studies on the epimerization of a purified diastereomer of **110b** allow the following kinetic parameters to be calculated for the process by which it is transformed to its epimer: $E_a = 23.0$ (± 0.9) kcal/mol, $\ln A = 26.1$ (± 1.4), $\Delta G^{\ddagger} = 24.4$ (± 0.3) kcal/mol, $\Delta H^{\ddagger} = 22.4$ (± 0.9) kcal/mol, and $\Delta S^{\ddagger} = -6.6$ (± 1.8) e.u. At first sight it would seem that these data are valid for the process of ring inversion. However, a control experiment in which **110b** is heated under an atmosphere of ^{13}CO shows that labeled CO is incorporated into both **110b** and its epimer, indicating that diastereomer interconversion may not result from inversion of the conformationally asymmetric fluorinated ring, but rather from epimerization at the metal center via a CO dissociation/recombination reaction. Clearly, if epimerization at the metal center is the sole mechanism responsible for diastereomer interconversion, these kinetic parameters represent *minimum* values for RI of the eight-membered ring (*184*).

Thus we can conclude that heptafluorination of the eight-membered ring in these mononuclear iron complexes raises the barrier to the RI process as compared to the hydrocarbon analogues (see Section X,B). This difference ($\Delta\Delta G_{RI}^{\ddagger} \geq 7.1$ kcal/mol) must be attributed solely to the presence of fluorines on the ring. Two effects may contribute to this phenomenon: steric buttressing between slightly larger (than hydrogen) fluorine substituents in going from the ground-state tub conformation to the planar intermediate, and increased dipole–dipole repulsion between the more polar C—F bonds in the latter intermediate.

The corresponding parameters for the BS process in these systems cannot be evaluated using ^{19}F NMR experiments because the rate is too slow. However, kinetic parameters for the BS process in heptafluorocyclooccta-1,3,5,7-tetraene **116** have been shown by magnetization transfer techniques to be

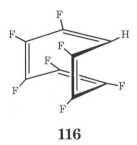

116

$E_{aBS} = 20\ (\pm 0.9)$ and $\Delta H^{\ddagger}_{BS} = 21.1\ (\pm 0.9)$ kcal/mol (190). Because BS activation energies are slightly larger than those for RI (see Section X,B), the *difference* in BS activation energy for heptafluorocyclooctatetraene relative to the parent hydrocarbon molecule, cyclooctatetraene ($\Delta E_a = 10.6$ kcal/mol), is consistent with the RI results obtained for these organometallic analogues.

XI

VICINAL DEFLUORINATION OF OFCOT
TO GIVE μ-HEXAFLUOROCYCLOOCTATRIENEYNE
COMPLEXES OF COBALT

A. *Structures of Hydrocarbon Cyclooctatrieneynes*

Nonplanar ground-state structures for cyclooctatetraene and most of its derivatives are well established (*118*), and crystallographic studies of OFCOT have shown that fluorination does not result in significant alteration of the skeletal structure (see Section IV,A). The question of whether the corresponding dehydro[8]annulenes (cycloocta-3,5,7-triene-1-ynes) should be planar or puckered has aroused considerable experimental and theoretical interest. Experimental evidence has been presented for the existence of dehydrocyclooctatetraene **117** as a reaction intermediate (*191,192*) and several derivatives with fused benzene rings have been isolated (*193*). Theoretical calculations of the structure of the parent compound **117** and its valence isomers have been carried out at the modified neglect of differential overlap (MNDO) level, resulting in the prediction of a highly strained, planar structure (*194*). In

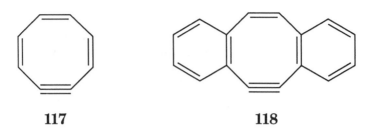

117   **118**

agreement, the crystallographically determined structure of 5,6-didehydrodi-benzo[a,e]cyclooctene **118** shows that it possesses a planar eight-membered ring (*195–197*).

B. (μ-Hexafluorocyclooctatrieneyne)dicobalt Complexes

1. Synthesis

Strained cycloalkynes can be stabilized by coordination to one or more transition metal centers (*198*). The unusual vicinal defluorination reaction of perfluoro-1,3-cyclohexadiene with $[Co_2(CO)_8]$ to give the μ-alkyne complex **45** (see Section III,E) prompted a study of the reactions of OFCOT with cobalt carbonyl precursors.

The μ-hexafluorocyclooctatrieneyne complexes **119** are obtained in variable yields from the reactions of the appropriate neutral dimers $[Co_2(CO)_6L_2]$ or mononuclear anions $[Co(CO)_3L]^-$ with OFCOT (*187,199*). In the metal

119 a. L = CO
         b. L = PPh₃
         c. L = PPhMe₂
         d. L = PMe₃
         e. L = PPh₂Me

anion reactions, with the notable exception of $[Co(CO)_4]^-$, equilibrium mixtures of η^1-heptafluorocyclooctatetraenyl complexes and their bicyclic valence isomers are also obtained (199) (see Section X,C,1). Treatment of the η^1-heptafluorocyclooctatetraenyl complex **108e** with a second equivalent of $[(Co(PPh_3)(CO)_3)]^-$ results in recovery of starting materials, demonstrating that the 1,2-disubstituted ring in **119** is not produced by a consecutive displacement reaction (187,199). This result contrasts with formation of a 1,5-disubstituted ring in the reaction of two equivalents of $[Fe(\eta^5\text{-}C_5H_5)(CO)_2]^-$ with OFCOT (see Section X,D).

2. Crystallographic Studies

The molecular structures of four dinuclear compounds containing the μ-hexafluorocyclooctatrieneyne ligand are shown in Fig. 18 (199). Selected average geometrical features of the fluorinated rings in these four structures,

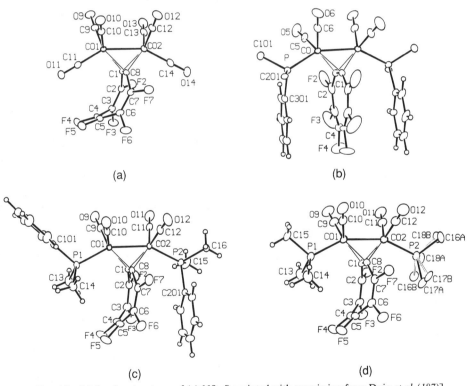

(a)   (b)

(c)   (d)

FIG. 18. Molecular structures of (a) **119a** [reprinted with permission from Doig et al. (187)], (b) **119b** [reprinted with permission from Doig et al. (187)], (c) **119c** (199), and (d) **119d** (199). Selected average structural parameters for all complexes are presented in Table II.

TABLE II

Comparison of Average Structural Parameters for
OFCOT (51), 119a, 119d, 119c, and 119b

	OFCOT	119a	119d	119c	119b
—C≡C—C[a]	—	131.9	135.7	136.4	139.8
—C=C—C[a]	126.7	127.5	129.5	130.2	133.3
—C=C—F[a]	119.1	118.2	117.1	116.9	114.6
—C—C—F[a]	114.0	114.0	113.1	113.5	112.0
(—C=C—F)—(—C—C—F)[a]	5.1	4.2	4.0	3.4	2.6
C≡C[b]	—	1.371	1.351	1.362	1.348
C=C[b]	1.322	1.332	1.326	1.328	1.348
C—C[b]	1.447	1.446	1.441	1.439	1.419
C—F[b]	1.346	1.350	1.359	1.364	1.359
F···F across C=C[b]	2.63	2.56	2.57	2.56	2.45
F···F across C—C[b]	2.76	2.73	2.65	2.60	2.39
α_1[a,c]	41.4	33.0	20.5	20.4	0.9
α_2[a,c]	41.4	42.4	38.4	33.6	1.8

[a] Degrees.

[b] Angstroms.

[c] See text for definition of α_1 and α_2.

compared with those of the uncomplexed fluorocarbon octafluorocyclooctatetraene, appear in Table II.

The bonding of the hexafluorocyclooctatrienyne ligand to the dicobalt framework is identical in all four compounds, with the eight-membered ring bound via its formal triple bond. The perpendicular geometry between the C≡C triple bond and the Co—Co bond is consistent with alkyne–dicobalt geometry as described in previous literature and as predicted by theory (200). The Co—Co bond lengths are significantly shorter than the Co—Co bond distance of 2.52 Å in [Co$_2$(CO)$_8$] (201), but are similar to Co—Co distances found in related μ-alkyne complexes, e.g., 2.463(1) Å in [(μ_2-1η,2η-di-t-butyl-acetylene)hexacarbonyldicobalt] (202) and 2.471 Å in the fluorocarbon complex 45 (112).

The fluorinated ring adopts a tub-shaped conformation in structures 119a, 119c, and 119d and is planar in 119b. This nonplanarity is quantified in Table II by the dihedral angles α_1 and α_2, where α_1 is the angle between planes [C1,C2,C7,C8] and [C2,C3,C6,C7] and α_2 is the angle between planes [C2,C3,C6,C7] and [C3,C4,C5,C6]. The ring becomes systematically flattened with increasing steric bulk of the ligand trans to Co—Co (i.e., the ligand closest to the C$_8$ ring). Thus in the series 119a → 119d → 119c → 119b, α_1 ranges from 33.0(3)° → 20.5(3)° → 20.4(2)° → 0.9(3)°, while α_2 ranges from

$42.4(3)° \rightarrow 38.4(3)° \rightarrow 33.6(2)° \rightarrow 1.8(3)°$. The corresponding dihedral angle in OFCOT is $41.4(2)°$ (see Section IV,A). Clearly, steric interactions with the ancillary ligands bound to cobalt play a major role in determining the conformation of the hexafluorocyclooctatrieneyne ring (*199*).

In complex **119b** the distance between C(1) on the C_8F_6 ligand and C(301) on the phenyl ring is 3.34 Å, which is less than the sum of their van der Waals radii (3.40 Å). This type of structure could be indicative of a favorable orbital overlap between the orbitals on the phenyl rings and those of the C_8F_6 ligand, or this short distance may simply be imposed on the rings by the steric requirements of the phosphine ligands and the fluorinated ring. However, for complex **119c** the observation that the C_8F_6 ring and the phenyl ring on only one phosphine ligand are in the "eclipsed" arrangement implies that both steric and electronic factors may play a role in determining the optimum conformation of the fluorinated ligand. If steric factors alone were dominant, then the preferred structure of **119c** should be that in which both phenyl groups were directed away from the eight-membered ring. In such a structure the values of α_1 and α_2 should be approximately equal to those found in the PMe_3 complex **119d**. The values of α_1 for **119c** and **119d** are indeed equal, but the value of α_2 is significantly smaller in **119c** (33.6°) than in **119d** (38.4°). This is surprising because the steric effect of the eclipsing phenyl ring in **119c** should be larger than that of the corresponding methyl group in **119d**, and may indicate the presence of a weak attractive interaction between the fluorinated ring and this phenyl group in the former compound. However, any such attractive interactions are insufficient to prevent rapid inversion of the fluorinated ring or rapid rotation about the Co—P bonds in solution (see Section XI,B,3).

3. Isodynamical Processes in μ-Hexafluorocyclooctatrieneyne Complexes

Although the fluorinated rings in complexes **119a,c,d** are puckered in the solid state, observation of only a single "virtual triplet" ^1H NMR resonance for the methyl protons on the phosphine ligands is consistent either with a planar cyclooctatrieneyne ring in solution, or with a low-energy barrier for inversion of the fluorinated ring (*199*). The ^{31}P NMR resonances for these compounds are very broad, presumably due to the quadrupolar cobalt nuclei, and are not useful in detecting the presence or absence of dynamic ring inversion. Although these data do not allow a distinction to be made between these two possibilities, it is clear that the activation energy for deformation of the puckered hexafluorocyclooctatrieneyne must be considerably lower than that observed for η^1-heptafluorocyclooctatetraenyl analogues (see Section X,F).

120

The magnitude of any attractive interaction between aryl rings on phosphorus and the fluorocarbon ring must also be small, because the low-temperature ^{1}H-NMR spectrum of complex **120** exhibits only a sharp singlet resonance for the methyl groups on the p-tolyl rings (*199*). Assuming that the ground-state structure of **120** is the same as that of its triphenylphosphine analogue **119b**, two different methyl signals in a ratio of 1:2 would be observed, corresponding to the p-tolyl groups that form part of the sandwich and those that do not. Clearly, any electronic π-system interaction or steric impediment between aryl and C_8F_6 rings is very weak in solution.

It is clear that the most favorable conformation for the μ-hexafluorocyclooctatrieneyne ligand in complexes **119** is a severely puckered one, in the absence of steric and possibly electronic effects associated with ancillary ligands on the metals. Both solution NMR and crystallographic studies indicate that the energy required to invert or otherwise distort the fluorinated ring toward a more planar structure is significantly lower than that for a η^1-heptafluorocyclooctatetraenyl ligand (see Section X,F). Although the fluorinated ligand in complexes **119** would be considered formally to be a cyclooctatrieneyne were it not coordinated to the dicobalt framework, use of the acetylenic π electrons to bind to the bimetallic subunit may leave the coordinated molecule resembling a cyclooctatriene rather than a cyclooctatrieneyne. In contrast to cyclooctatrieneynes, the structure of cyclocta-1,3,5-triene is known to be puckered (*203*), with a barrier to ring inversion, $\Delta G^{\ddagger}$ ($-145°C$) = 6.2 (± 0.5) kcal/mol (*204*). This inversion barrier is much lower than that observed for derivatives of cyclooctatetraene (see Section X,B), and recognition that the fluorinated ring in complexes **119** may more closely resemble a cyclooctatriene than a cyclooctatetraene provides one rationale for the difference in its dynamic behavior compared to its η^1-heptafluorocyclooctatetraenyl relatives.

XII

CONCLUDING REMARKS

Studies on the organo-transition metal chemistry of OFCOT and other fluorocarbons indicate that organometallic molecules with perfluorinated carbon skeletons possess structural and chemical properties equally as rich as their hydrocarbon analogues. Their structures and chemistry have helped our understanding of the effects of fluorine substitution on the physical and chemical behavior of organometallic compounds.

Organometallic chemistry using hydrocarbon ligands is still a burgeoning area, with novel discoveries reported regularly. There is no reason to suppose that perfluorocarbon organometallic chemistry will not continue to do likewise, even though it remains an area that is neglected by the majority of the world's organometallic chemists. The results surveyed here suggest that further research efforts will not result in mundane chemistry that simply parallels that of hydrocarbon analogues. Potentially exciting complexes, perhaps most notably those containing the η^5-perfluorocyclopentadienyl ligand, remain elusive challenges for the synthetic chemist.

ACKNOWLEDGMENTS

I wish to thank my talented and resourseful postdoctoral, graduate, and undergraduate students, whose names appear as coauthors in many of the references to this review, for their efforts in unraveling most of the chemistry discussed here. The crystallographic assistance of Professors Ray Davis (University of Texas at Austin), Arnie Rheingold (University of Delaware), and Todd Marder and Nick Taylor (University of Waterloo) has been invaluable, as has the electrochemical expertise of Professor Bill Geiger (University of Vermont) and the NMR talents of Dr. Brian Mann (University of Sheffield). Generous financial assistance by the National Science Foundation, the United States Air Force Office of Scientific Research, the Petroleum Research Fund administered by the American Chemical Society, the Alfred P. Sloan Foundation, the Research Corporation, and Dartmouth College, as well as handsome loans of precious metal salts from Matthey–Bishop, are also acknowledged with gratitude.

Thanks are also due to Professor Robert N. Haszeldine and Dr. Eric Banks, who originally stimulated my interest in fluorinated molecules during my undergraduate years at UMIST, and Dr. Michael Green, who reignited my curiosity as a postdoctoral student. Finally, I am indebted to my colleague, Professor Dave Lemal, who showed me how to make OFCOT, for his enthusiasm for the fundamental principles of chemistry, and his continued support and encouragement.

REFERENCES

1. G. W. Parshall, "Homogeneous Catalysis." Wiley, New York, 1970.
2. B. E. Smart, in "The Chemistry of Functional Groups, Supplement D" (S. Patai and Z. Rappoport, eds.), Chapter 14. Wiley, New York, 1983.
3. F. G. A. Stone, Pure Appl. Chem. 30, 551 (1972), and references cited therein.
4. D. Lentz and R. Marschall, Chem. Ber. 122, 1223 (1989), and references cited therein.
5. R. Usón, J. Forniés, and M. Tomás, J. Organomet. Chem., 358, 525 (1988).

6. M. I. Bruce and F. G. A. Stone, *Prep. Inorg. React.* **4**, 177 (1968).
7. P. M. Treichel and F. G. A. Stone, *Adv. Organomet. Chem.* **1**, 143 (1964).
8. G. Wilkinson, F. G. A. Stone, and E. W. Abel, eds., "Comprehensive Organometallic Chemistry." Pergamon, Oxford, 1982.
9. J. A. Morrison, *Adv. Inorg. Chem. Radiochem.* **27**, 293 (1983).
10. P. J. Brothers and W. R. Roper, *Chem. Rev.* **88**, 1293 (1988).
11. J. A. Connor, M. T. Zafarani-Moattar, J. Bickerton, N. I. El Saied, S. Suradi, R. Carson, G. Al Takhin, and H. A. Skinner, *Organometallics* **1**, 1166 (1982).
12. M. R. Churchill and J. P. Fennessey, *Inorg. Chem.* **6**, 1213 (1967).
13. M. J. Bennett and R. Mason, *Proc. Chem. Soc., London*, 273 (1963).
14. M. J. Bennett, Ph.D. Thesis, University of Sheffield, England, 1965.
15. M. A. Bennett, H.-K. Chee, and G. B. Robertson, *Inorg. Chem.* **18**, 1061 (1979).
16. R. B. King and M. B. Bisnette, *J. Organomet. Chem.* **2**, 15 (1964).
17. F. A. Cotton and J. A. McCleverty, *J. Organomet. Chem.* **4**, 490 (1965).
18. F. A. Cotton and R. M. Wing, *J. Organomet. Chem.* **9**, 511 (1967).
19. H. C. Clark and J. H. Tsai, *J. Organomet. Chem.* **7**, 515 (1967).
20. W. A. G. Graham, *Inorg. Chem.* **7**, 315 (1968).
21. M. B. Hall and R. F. Fenske, *Inorg. Chem.* **11**, 768 (1972).
22. M. A. Bennett, H.-K. Chee, J. C. Jeffery, and G. B. Robertson, *Inorg. Chem.* **18**, 1071 (1979).
23. D. L. Lichtenberger and R. F. Fenske, *Inorg. Chem.* **13**, 486 (1974).
24. S. Evans, J. C. Green, M. L. H. Green, A. F. Orchard, and D. W. Turner, *Discuss. Faraday Soc.* **47**, 112 (1969).
25. J. A. Connor, M. B. Hall, I. H. Hillier, W. N. E. Meredith, M. Barber, and Q. Herd, *J. Chem. Soc., Faraday Trans. 2* **70**, 1677 (1973).
26. F. Calderazzo, *Angew. Chem.* **89**, 305 (1977).
27. R. B. King, *Acc. Chem. Res.* **3**, 417 (1970).
28. C. M. Lukehart, G. P. Torrence, and J. V. Zeile, *Inorg. Synth.* **18**, 57 (1978).
29. D. Saddei, H.-J. Freund, and G. Hohlneicher, *J. Organomet. Chem.* **186**, 63 (1980).
30. H. Berke and R. Hoffmann, *J. Am. Chem. Soc.* **100**, 7224 (1978).
31. S. L. Gipson and K. Kneten, *Inorg. Chim. Acta* **157**, 143 (1989).
32. N. M. Doherty and N. W. Hoffman, *Chem. Rev.* (submitted for publication).
33. S. L. Buchwald, unpublished observations.
34. T. G. Richmond, C. E. Osterberg, and A. M. Arif, *J. Am. Chem. Soc.* **109**, 8091 (1987).
35. C. M. Anderson, R. J. Puddephatt, G. Ferguson, and A. B. Lough, *J. Chem. Soc., Chem. Commun.*, 1297 (1989).
36. R. R. Burch, R. L. Harlow, and S. D. Ittel, *Organometallics* **6**, 982 (1987).
37. C. T. Mortimer, J. L. McNaughton, J. Burgess, M. J. Hacker, and R. D. W. Kemmitt, *J. Organomet. Chem.* **47**, 439 (1973).
38. D. L. Reger and M. D. Dukes, *J. Organomet. Chem.* **153**, 67 (1978).
39. T. G. Richmond, A. M. Crespi, and D. F. Shriver, *Organometallics* **3**, 314 (1984).
40. A. M. Crespi and D. F. Shriver, *Organometallics* **4**, 1830 (1985).
41. W. Schulze, H. Hartl, and K. Seppelt, *Angew. Chem., Int. Ed. Engl.* **25**, 185 (1986).
42. F. Seel and R. D. Flaccus, *J. Fluorine Chem.* **12**, 81 (1978).
43. F. Seel and G.-V. Röschenthaler, *Z. Anorg. Allg. Chem.* **386**, 297 (1971).
44. F. Seel and G.-V. Röschenthaler, *Angew. Chem., Int. Ed. Engl.* **9**, 166 (1970).
45. W. T. Dent, L. A. Duncanson, R. G. Guy, H. W. B. Reed, and B. L. Shaw, *Proc. Chem. Soc., London*, 169 (1961).
46. B. R. Penfold and B. H. Robinson, *Acc. Chem. Res.* **6**, 73 (1973).
47. P. A. Dawson, B. H. Robinson, and J. Simpson, *J. Chem. Soc., Dalton Trans.*, 1762 (1979).

48. H. Beurich, R. Blumhofer, and H. Vahrenkamp, *Chem. Ber.* **115**, 2409 (1982).
49. H. Beurich and H. Vahrenkamp, *Chem. Ber.* **115**, 2385 (1982).
50. H.-J. Kneuper, D. S. Strickland, and J. R. Shapley, *Inorg. Chem.* **27**, 1110 (1988).
51. D. Lenz, I. Brüdgam, and H. Hartl, *Angew. Chem., Int. Ed. Engl.* **24**, 119 (1985).
52. D. Lenz and M. Michael, *J. Organomet. Chem.* **372**, 109 (1989).
53. D. Lenz and M. Michael, *Angew. Chem., Int. Ed. Engl.* **27**, 845 (1988).
54. C. D. Cook and G. S. Jauhal, *J. Am. Chem. Soc.* **90**, 1464 (1968).
55. G. A. Larkin, R. Mason, and M. G. H. Wallbridge, *J. Chem. Soc., Dalton Trans.*, 2305 (1975).
56. M. Green, R. B. L. Osborn, A. J. Rest, and F. G. A. Stone, *J. Chem. Soc. A*, 2525 (1968).
57. C. A. Tolman, *J. Am. Chem. Soc.* **96**, 2780 (1974).
58. S. D. Ittel and J. A. Ibers, *Adv. Organomet. Chem.* **14**, 33 (1976).
59. R. Cramer, J. B. Kline, and J. D. Roberts, *J. Am. Chem. Soc.* **91**, 2519 (1969).
60. L. J. Guggenberger and R. Cramer, *J. Am. Chem. Soc.* **94**, 3779 (1972).
61. J. L. Carlos, Jr., R. R. Karl, Jr., and S. H. Bauer, *J. Chem. Soc., Faraday Trans. 2* **70**, 177 (1974).
62. R. Fields, M. M. Germain, R. N. Haszeldine, and P. W. Wiggans, *J. Chem. Soc. A*, 1969 (1970).
63. J. P. Visser, A. J. Schipperijn, J. Lukas, D. Bright, and J. J. De Boer, *J. Chem. Soc., Chem. Commun.*, 1266 (1971).
64. J. J. De Boer and D. Bright, *J. Chem. Soc., Dalton Trans.*, 662 (1975).
65. R. C. Hemond, R. P. Hughes, D. J. Robinson, and A. L. Rheingold, *Organometallics* **7**, 2239 (1988).
66. R. P. Hughes, M. E. King, D. J. Robinson, and J. M. Spotts, *J. Am. Chem. Soc.* **111**, 8919 (1989).
67. M. J. Burk, M. P. McGrath, R. Wheeler, and R. H. Crabtree, *J. Am. Chem. Soc.* **110**, 5034 (1988).
68. P. P. Deutsch and R. Eisenberg, *Chem. Rev.* **88**, 1147 (1988).
69. D. W. McBride, E. Dudek, and F. G. A. Stone, *J. Chem. Soc.*, 1752 (1964).
70. K. Stanley and D. W. McBride, *Can. J. Chem.* **53**, 2537 (1975).
71. D. C. Andrews and G. Davidson, *J. Chem. Soc., Dalton Trans.*, 1381 (1972).
72. H. Goldwhite, D. G. Rowsell, and C. Valdez, *J. Organomet. Chem.* **12**, 133 (1968).
73. H. D. Kaesz, R. B. King, and F. G. A. Stone, *Z. Naturforsch. B: Anorg. Chem., Org. Chem., Biochem., Biophys., Biol.* **15B**, 682 (1960).
74. W. R. McClellan, *J. Am. Chem. Soc.* **83**, 1598 (1961).
75. G. W. Parshall and G. Wilkinson, *J. Chem. Soc.*, 1132 (1962).
76. P. B. Hitchcock and R. Mason, *J. Chem. Soc., Chem. Commun.*, 503 (1966).
77. D. A. Dixon, T. Fukunaga, and B. E. Smart, *J. Phys. Org. Chem.* **1**, 153 (1988).
78. H. H. Hoehn, L. Pratt, K. F. Watterson, and G. Wilkinson, *J. Chem. Soc.*, 2738 (1961).
79. P. Dodman and J. C. Tatlow, *J. Organomet. Chem.* **67**, 87 (1974).
80. M. R. Churchill and R. Mason, *Proc. Ry. Soc. London, Ser. A* **301**, 433 (1967).
81. R. L. Hunt, D. M. Roundhill, and G. Wilkinson, *J. Chem. Soc. A*, 982 (1967).
82. P. B. Hitchcock and R. Mason, *J. Chem. Soc., Chem. Commun.*, 242 (1967).
83. R. E. Banks, T. Harrison, R. N. Haszeldine, A. B. P. Lever, T. F. Smith, and J. B. Walton, *J. Chem. Soc., Chem. Commun.*, 30 (1965).
83a. R. P. Hughes and D. J. Robinson, unpublished observations.
84. R. Fields, M. Green, T. Harrison, R. N. Haszeldine, A. Jones, and A. B. P. Lever, *J. Chem. Soc. A*, 49 (1970).
85. G. Paprott, S. Lehmann, and K. Seppelt, *Chem. Ber.* **121**, 727 (1988).

86. W. Priebsch, M. Hoch, and D. Rehder, *Chem. Ber.* **121**, 1971 (1988), and references cited therein.
87. G. Paprott and K. Seppelt, *J. Am. Chem. Soc.* **106**, 4060 (1984).
88. G. Paprott, D. Lenz, and K. Seppelt, *Chem. Ber.* **117**, 1153 (1984).
89. R. Middleton, J. R. Hull, S. R. Simpson, C. H. Tomlinson, and P. L. Timms, *J. Chem. Soc., Dalton Trans.*, 120 (1973).
90. M. J. McGlinchey and T.-S. Tan, *J. Am. Chem. Soc.* **98**, 2271 (1976).
91. K. J. Klabunde and H. F. Efner, *J. Fluorine Chem.* **4**, 114 (1974).
92. S. T. Belt, S. B. Duckett, M. Helliwell, and R. N. Perutz, *J. Chem. Soc., Chem. Commun.*, 928 (1989).
93. W. D. Jones and F. J. Feher, *J. Am. Chem. Soc.* **106**, 1650 (1984).
94. D. R. Fahey and J. E. Mahan, *J. Am. Chem. Soc.* **99**, 2501 (1977).
95. B. L. Booth, R. N. Haszeldine, and N. I. Tucker, *J. Organomet. Chem.* **11**, P5 (1968).
96. P. W. Jolly, M. I. Bruce, and F. G. A. Stone, *J. Chem. Soc.*, 5830 (1965).
97. M. I. Bruce, P. W. Jolly, and F. G. A. Stone, *J. Chem. Soc. A*, 1602 (1966).
98. M. I. Bruce, and F. G. A. Stone, *Angew. Chem., Int. Ed. Engl.* **7**, 747 (1968).
99. D. J. Cooke, M. Green, N. Mayne, and F. G. A. Stone, *J. Chem. Soc. A* 1771 (1968).
100. W. R. Cullen, *Fluorine Chem. Rev.* **3**, 73 (1969).
101. B. L. Booth, R. N. Haszeldine, and I. Perkins, *J. Chem. Soc. A*, 937 (1971).
102. M. I. Bruce, B. L. Goodall, D. N. Sharrocks, and F. G. A. Stone, *J. Organomet. Chem.* **39**, 139 (1972).
103. R. L. Bennett, M. I. Bruce, and R. C. F. Gardner, *J. Chem. Soc., Dalton Trans.*, 2653 (1973).
104. B. L. Booth, R. N. Haszeldine, and M. B. Taylor, *J. Chem. Soc. A*, 1974 (1970).
105. B. L. Booth, R. N. Haszeldine, R. N. Perkins, *J. Chem. Soc., Dalton Trans.*, 1843 (1975).
106. B. L. Booth, R. N. Haszeldine, and I. Perkins, *J. Chem. Soc., Dalton Trans.*, 1847 (1975).
107. B. L. Booth, S. Casey, and R. N. Haszeldine, *J. Organomet. Chem.* **226**, 289 (1982).
108. B. L. Booth, S. Casey, R. P. Critchley, and R. N. Haszeldine, *J. Organomet. Chem.* **226**, 301 (1982).
109. G. A. Artamkina, A. Yu. Mil'chenko, I. P. Beletskaya, and O. A. Reutov, *J. Organomet. Chem.* **311**, 199 (1986).
110. M. I. Bruce, M. J. Liddell, M. R. Snow, and E. R. T. Tiekink, *J. Organomet. Chem.* **354**, 103 (1988).
111. R. L. Hunt and G. Wilkinson, *Inorg. Chem.* **4**, 1270 (1965).
112. N. A. Bailey, M. R. Churchill, R. Hunt, R. Mason, and G. Wilkinson, *Proc. Chem. Soc., London*, 401 (1964).
113. M. T. Jones and R. N. McDonald, *Organometallics* **7**, 1221 (1988).
114. A. J. Blake, R. W. Cockman, E. A. V. Ebsworth, and J. H. Holloway, *J. Chem. Soc., Chem. Commun.*, 529 (1988).
115. K. Sünkel, *J. Organomet. Chem.* **348**, C12 (1988).
116. D. Lenz and H. Michael, *J. Organomet. Chem.* **346**, C37 (1988).
117. R. R. Burch, J. C. Calabrese, and S. D. Ittel, *Organometallics* **7**, 1642 (1988).
118. G. I. Fray and R. G. Saxton, "The Chemistry of Cyclooctatetraene and Its Derivatives." Cambridge Univ. Press, London and New York, 1978.
119. G. Deganello, "Transition Metal Complexes of Cyclic Polyolefins." Academic Press, New York, 1979.
120. D. M. Lemal, J. M. Buzby, A. C. Barefoot, III, M. W. Grayston, and E. D. Laganis, *J. Org. Chem.* **45**, 3118 (1980).
121. B. B. Laird and R. E. Davis, *Acta Crystallogr., Sect. B* **B38**, 678 (1982).
122. M. Traetteberg, *Acta Chem. Scand.* **20**, 1724 (1966).

123. D. M. Lemal, unpublished observations.
124. R. F. Waldron, A. C. Barefoot, III, and D. M. Lemal, *J. Am. Chem. Soc.* **106,** 8301 (1984).
125. A. C. Barefoot, III, W. D. Saunders, J. M. Buzby, M. W. Grayston, and D. M. Lemal, *J. Org. Chem.* **45,** 4292 (1980).
126. A. C. Barefoot, III, M. A. Thesis, Dartmouth College, Hanover, New Hampshire, 1978.
127. B. W. Walther, F. Williams, and D. M. Lemal, *J. Am. Chem. Soc.* **106,** 548 (1984).
128. R. C. Hemond and R. P. Hughes, unpublished observations.
129. R. C. Hemond, R. P. Hughes, and A. L. Rheingold, *Organometallics* **8,** 1261 (1989).
130. I. A. Benson, S. A. R. Knox, R. F. D. Stansfield, and P. Woodward, *J. Chem. Soc., Chem. Commun.,* 404 (1977).
131. P. L. Pauson and J. A. Segal, *J. Organomet. Chem.* **63,** C13 (1973).
132. P. L. Pauson and J. A. Segal, *J. Chem. Soc., Dalton Trans.,* 2387 (1975).
133. J. S. McKechnie and I. C. Paul, *J. Am. Chem. Soc.* **90,** 903 (1968).
134. B. E. Mann, unpublished observations.
135. J. A. Gibson and B. E. Mann, *J. Chem. Soc., Dalton Trans.,* 1021 (1979).
136. M. Grassi, B. E. Mann, and C. M. Spencer, *J. Chem. Soc., Chem. Commun.,* 1169 (1985).
137. M. Grassi, B. E. Mann, B. T. Pickup, and C. M. Spencer, *J. Chem. Soc., Dalton Trans.,* 2649 (1987).
138. R. Slegeir, R. Case, J. S. McKennis, and R. Pettit, *J. Am. Chem. Soc.* **96,** 287 (1974).
139. T. A. Manuel and F. G. A. Stone, *J. Am. Chem. Soc.* **82,** 366 (1960).
140. A. C. Barefoot, III, E. W. Corcoran, Jr., R. P. Hughes, D. M. Lemal, B. B. Laird, and R. E. Davis, *J. Am. Chem. Soc.* **103,** 970 (1981).
141. M. I. Bruce, M. Cooke, M. Green, and D. L. Westlake, *J. Chem. Soc. A,* 987 (1969).
142. M. I. Bruce, M. Cooke, and M. Green, *Angew. Chem., Int. Ed. Engl.* **7,** 639 (1968).
143. R. T. Carl, R. P. Hughes, and R. E. Davis, unpublished observations.
144. R. T. Carl, R. P. Hughes, and A. L. Rheingold, unpublished observations.
145. R. T. Carl, R. P. Hughes, J. A. Johnson, R. E. Davis, and R. P. Kashyap, *J. Am. Chem. Soc.* **109,** 6875 (1987).
146. M. Ephritikhine, B. R. Francis, M. L. H. Green, R. E. Mackenzie, and M. J. Smith, *J. Chem. Soc., Dalton Trans.,* 1131 (1977).
147. R. A. Periana and R. G. Bergman, *J. Am. Chem. Soc.* **108,** 7346 (1986).
148. M. D. Curtis and O. Eisenstein, *Organometallics* **3,** 887 (1984).
149. W. J. Middleton, *Org. Synth.* **64,** 221 (1984).
150. N. M. Doherty, B. E. Ewels, R. P. Hughes, D. E. Samkoff, W. D. Saunders, R. E. Davis, and B. B. Laird, *Organometallics* **4,** 1606 (1985).
151. R. T. Carl, R. P. Hughes, A. L. Rheingold, T. B. Marder, and N. J. Taylor, *Organometallics* **7,** 1613 (1988).
152. R. P. Hughes, D. E. Samkoff, R. E. Davis, and B. B. Laird, *Organometallics* **2,** 195 (1983).
153. R. T. Carl and R. P. Hughes, unpublished observations.
154. J. W. Faller, R. H. Crabtree, and A. Habib, *Organometallics* **4,** 929 (1985), and references cited therein.
155. J. Merola, R. T. Kacmarcik, and D. Van-Engan, *J. Am. Chem. Soc.* **108,** 329 (1986), and references cited therein.
156. T. B. Marder, J. C. Calabrese, D. C. Roe, and T. H. Tulip, *Organometallics* **7,** 2012 (1987).
157. A. K. Kakkar, S. F. Jones, N. J. Taylor, S. Collins, and T. B. Marder, *J. Chem. Soc., Chem. Commun.* 1454 (1989).
158. F. Kohler, *Chem. Ber.* **107,** 570 (1974).
159. R. T. Baker and T. H. Tulip, *Organometallics* **5,** 839 (1986).

160. M. Mlekuz, P. Bougeard, B. G. Sayer, M. J. McGlinchey, C. A. Rodger, M. R. Churchill, J. W. Ziller, S.-K. Kang, and T. A. Albright, *Organometallics* **5**, 1656 (1986).
161. R. T. Carl, S. J. Doig, W. E. Geiger, R. C. Hemond, R. P. Hughes, R. S. Kelly, and D. E. Samkoff, *Organometallics* **6**, 611 (1987).
162. J. E. Moraczewski and W. E. Geiger, *J. Am. Chem. Soc.* **101**, 3407 (1979).
163. J. E. Moraczewski and W. E. Geiger, *J. Am. Chem. Soc.* **103**, 4779 (1981).
164. T. A. Albright, J. E. Moraczewski, W. E. Geiger, and B. Tulyathan, *J. Am. Chem. Soc.* **103**, 4787 (1981).
165. H. van Willigen, W. E. Geiger, and M. D. Rausch, *Inorg. Chem.* **16**, 581 (1977).
166. M. Grezeszcuk, D. E. Smith, and W. E. Geiger, *J. Am. Chem. Soc.* **105**, 1772 (1983).
167. C. McDade, Senior Fellowship Thesis, Dartmouth College, Hanover, New Hampshire, 1980.
168. R. T. Carl, R. P. Hughes, and D. E. Samkoff, *Organometallics* **7**, 1625 (1988).
169. R. P. Hughes, R. T. Carl, R. C. Hemond, D. E. Samkoff, and A. L. Rheingold, *J. Chem. Soc., Chem. Commun.*, 306 (1986).
170. J. H. Bieri, T. Egolf, W. von Phillipsborn, U. Piantini, R. Prewo, U. Ruppoli, and A. Salzer, *Organometallics* **5**, 2413 (1986).
171. F. A. Cotton, A. Davison, T. J. Marks, and A. Musco, *J. Am. Chem. Soc.* **91**, 9598 (1969).
172. F. A. Cotton and W. T. Edwards, *J. Am. Chem. Soc.* **90**, 5412 (1968).
173. J. Evans, B. F. G. Johnson, J. Lewis, and R. Watt, *J. Chem. Soc., Dalton Trans.*, 2368 (1974).
174. R. T. Carl, E. W. Corcoran, Jr., R. P. Hughes, and D. E. Samkoff, *Organometallics* **9**, 838 (1990).
175. R. P. Hughes, R. T. Carl, D. E. Samkoff, R. E. Davis, and K. D. Holland, *Organometallics* **5**, 1053 (1986).
176. E. A. Jeffery, *Aust. J. Chem.* **26**, 219 (1973).
177. H. F. Klein and H. H. Karsch, *Chem. Ber.* **106**, 1433 (1973).
178. H. H. Karsch, H. F. Klein, and H. Schmidbaur, *Chem. Ber.* **107**, 93 (1974).
179. R. H. Grubbs, A. Miyashita, M. M. Liv, and P. L. Burk, *J. Am. Chem. Soc.* **99**, 3863 (1977).
180. M. D. Radcliffe and W. M. Jones, *Organometallics* **2**, 1053 (1983).
181. L. A. Paquette, *Pure Appl. Chem.* **54**, 987 (1982), and references cited therein.
182. J. F. M. Oth, *Pure Appl. Chem.* **25**, 573 (1971).
183. S. J. Doig, R. P. Hughes, S. L. Patt, D. E. Samkoff, and W. L. Smith, *J. Organomet. Chem.* **250**, C1 (1983).
184. R. P. Hughes, R. T. Carl, S. J. Doig, R. C. Hemond, D. E. Samkoff, W. L. Smith, L. C. Stewart, R. E. Davis, K. D. Holland, P. Dickens, and R. P. Kashyap, *Organometallics* (in press).
185. R. L. Avoyan, Yu. A. Chapovsii, and Yu. T. Struchkov, *J. Struct. Chem. (Engl. Transl.)* **7**, 538 (1966).
186. P. G. Jones, *J. Organomet. Chem.* **345**, 405 (1988).
187. S. J. Doig, R. P. Hughes, R. E. Davis, S. M. Gadol, and K. D. Holland, *Organometallics* **3**, 1921 (1984).
188. M. R. Churchill, K. N. Amoh, and H. J. Wasserman, *Inorg. Chem.* **20**, 405 (1981).
189. F. W. B. Einstein, H. Luth, and J. Trotter, *J. Chem. Soc. A*, 89 (1967).
190. T. Spector, Ph.D. Thesis, Dartmouth College, Hanover, New Hampshire, 1987.
191. A. Krebs, *Angew. Chem., Int. Ed. Engl.* **4**, 954 (1965).
192. A. Krebs and D. Byrd, *Justus Liebig's Ann. Chem.* **707**, 66 (1967).
193. N. Z. Huang and F. Sondheimer, *Acc. Chem. Res.* **15**, 96 (1982), and references cited therein.
194. M. J. S. Dewar and K. M. Merz, Jr., *J. Am. Chem. Soc.* **107**, 6175 (1985).
195. R. A. G. de Graaf, S. Gorter, C. Romers, H. N. C. Wong, and F. Sondheimer, *J. Chem. Soc., Perkin Trans. 2*, 478 (1981).
196. R. Destro, T. Pilati, and M. Simonetta, *J. Am. Chem. Soc.* **97**, 658 (1975).
197. R. Destro, T. Pilati, and M. Simonetta, *Acta Crystallogr., Sect. B* **B33**, 447 (1977).

198. M. A. Bennett and H. P. Schwemlein, *Angew. Chem., Int. Ed. Engl.* **29,** 1296 (1989) and references cited therein.
199. R. P. Hughes, S. J. Doig, R. C. Hemond, W. L. Smith, R. E. Davis, S. M. Gadol, and K. D. Holland, *Organometallics* (in press).
200. D. M. Hoffmann, R. Hoffmann, and C. R. Fisel, *J. Am. Chem. Soc.* **104,** 3858 (1982), and references cited therein.
201. G. G. Summer, H. P. Klug, and L. E. Alexander, *Acta Crystallogr.* **17,** 732 (1964).
202. F. A. Cotton, J. D. Jamerson, and B. R. Stults, *J. Am. Chem. Soc.* **98,** 1774 (1976).
203. A. C. Cope and F. A. Hochstein, *J. Am. Chem. Soc.* **72,** 2515 (1950).
204. M. St. Jacques and R. Prud'homme, *Tetrahedron Lett.,* 4833 (1970).

ADVANCES IN ORGANOMETALLIC CHEMISTRY, VOL. 31

Alkyl- and Aryl-Substituted Main-Group Metal Amides

M. VEITH

Anorganische Chemie
Universität des Saarlandes
D-6600 Saarbrücken, Federal Republic of Germany

I

INTRODUCTION

Main-group metal alkyls and aryls are valuable tools in preparative organic and inorganic chemistry, and many industrial applications are known for these reactive substances. In addition, they have attracted considerable attention from a more fundamental and theoretical point of view, because the high Lewis acidity at the metal center is the source of singular structures and bonding (1–3).

The formal substitution of one of the alkyl or aryl groups at the metal by nonmetallic and electronegative groups leads to compounds that have both electrophilic and nucleophilic centers in the molecule. The Grignard reagents are the best known and most important class of such compounds (4,5). Because of their high polarity and their acid/base properties, they must be handled in coordinating solvents. Moreover, they undergo exchange reactions that are known as Schlenck equilibria [Eq. (1)] (6–8). In Eq. (1), D is the donor solvent, X is the halogen, and R is the organic group.

$$
\begin{array}{c}
\overset{\displaystyle D}{\underset{\displaystyle D}{R-Mg-X}} + \overset{\displaystyle D}{\underset{\displaystyle D}{R-Mg-X}} \rightleftharpoons \underset{R\;\;\;R\;\;\;D}{\overset{D\;\;\;X\;\;\;X}{Mg\;\;\;Mg}} \rightleftharpoons \underset{R\;\;\;R\;\;\;D}{\overset{D\;\;\;D\;\;\;X\;\;\;X}{Mg}} + \underset{D\;\;\;D}{Mg} \quad (1)
\end{array}
$$

A consequence of these complex intermolecular equilibria [Eq. (1) is only one of several occurring in solution] is that their use in preparative chemistry is somewhat limited with respect to high-yield procedures. Another limiting effect of these reagents is their dependence on the presence of coordinating solvents.

Substitution of the halogen X by another electronegative group affords compounds that are stable with respect to intermolecular ligand transfer; these compounds do not require donor solvents for stabilization. In particular, ligands with a nitrogen atom attached to the metal fulfill at least one of the

269

two required properties. The amide function in these metal amides (9) may be derived from alkyl or aryl amines or from trialkylsilyl amines (silazanes). Silyl amides appear to stabilize the electron deficiency at the metal center most efficiently, an effect that is ascribed to the electron-donating properties of the silyl ligands (9).

In this article we describe the syntheses, structures, and physical properties of metal alkyls or aryls that have a further "chelating" amino ligand attached to the metallic center. These substances have been mainly developed in our laboratories in order to accomplish the following objectives:

1. Suppress coordination by solvent molecules at the metal.
2. Improve their solubility in nonpolar solvents.
3. Minimize the coordination number at the metal.

The substituents at the nitrogen atoms are invariably *tert*-butyl groups (see objective 3) in combination with silicon, which simultaneously functions as an electron donor as well as a bridging atom to other intramolecular basic centers X (nitrogen or oxygen atoms). Formula **1** shows the general structural framework of the compounds.

1

In Section II, the structures of alkyl and aryl metal amides of the metallic elements Mg, Ca, Sr, Ba, Al, Ga, In, and Tl are briefly discussed from a general point of view. Some analogous compounds of other elements are included for comparison.

II

STRUCTURAL ASPECTS OF ALKYL AND ARYL METAL AMIDES

A. Compounds of the Alkaline Earth Elements

Whereas a considerable number of pure amides of Mg, Ca, and Ba are known, the alkyl or aryl amides of these elements have been little investigated (9). Compounds of the general formula $RMgNR'_2$ are usually coordination polymers, unless ether molecules are added to the compound or bulky R groups are used. For example, $MeMg(NMe_2)$ seems to be polymeric, whereas $[iPrMg(NiPr_2)]_2$ is dimeric (10), and the addition of tetrahydrofuran (THF) to $EtMgNPh_2$ leads to the formation of monomeric $EtMgNPh_2 \cdot (THF)_2$ (10). If chelating ligands such as $MeNCH_2NMe_2$ are used, dimers such as $[MeMgNMe(CH_2)_2NMe_2]_2$ (2) are obtained. The crystal structure of this compound shows that it contains a four-membered Mg_2N_2 ring as the central unit (11).

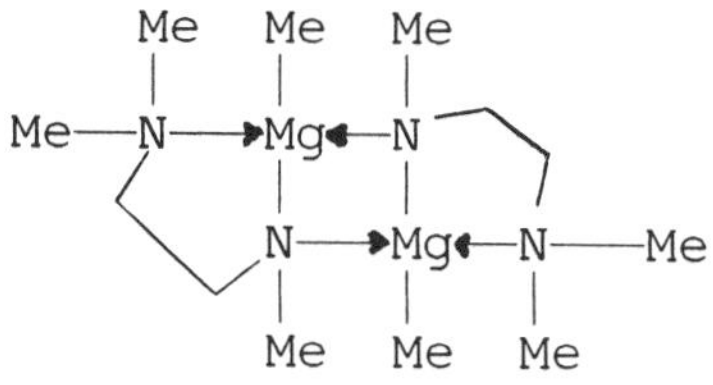

2

As can be seen from the structure, the dimerization proceeds via nitrogen–magnesium interactions. To our knowledge, dimerization via Mg—R—Mg two-electron three-center bonds in alkylmagnesium amides has neither previously been found in the solid state nor proved by X-ray techniques, although the existence of such bonds has been claimed (10).

The existence of alkyl- or aryl calcium amides has not as yet been unambiguously demonstrated. One reason for this seems to be because calcium—carbon bonds are much less stable than calcium—nitrogen bonds, as is illustrated by the following reaction [Eq. (2)] (12).

$$Ph_3CCaCl \cdot 2THF + R_2NH \rightarrow R_2NCaCl + Ph_3C—H + 2THF \qquad (2)$$

B. Compounds of Aluminum, Gallium, Indium, and Thallium

1. Simple Organometal Amides

The chemistry of alkyl- and arylaluminum, gallium, indium, and thallium amides has been investigated much more than the comparable chemistry of Group II elements (9,13–18). Alkylaluminum amides especially have a very rich and diverse structural chemistry (9,13–15). When a trialkylalane reacts with primary or secondary amines, the amides 3, 3′, and 4 are formed in addition to a simple adduct [Eqs. (3) and (4)].

$$R_3Al + H_2N—R \xrightarrow[-RH]{} 1/n(R_2Al—NHR)_n \xrightarrow[-RH]{} 1/n(RAl—NR)_n \qquad (3)$$
$$\qquad\qquad\qquad\qquad\qquad 3 \qquad\qquad\qquad\qquad 4$$

$$R_3Al + HNR_2 \xrightarrow[-RH]{} 1/n(R_2Al—NR_2)_n \qquad (4)$$
$$\qquad\qquad\qquad\qquad 3'$$

Compounds of type **3** or **3′** are cyclic and those of type **4** are polycyclic, as shown by spectroscopic data or X-ray structure determinations. $(Me_2-AlNMe_2)_2$ forms a planar four-membered cycle (*19*) but $(Me_2AlNHMe)_3$ has a six-membered Al_3N_3 ring, one isomeric form of which adopts a chair (*20*) and the other a skew-boat conformation (*21*). The two isomers interconvert, as may be seen in temperature-dependent NMR spectra (*21*). The six-membered Al_3N_3 ring of trimeric ethylenimino(dimethyl)alane also has a chair $\rightleftharpoons$ boat conformation (*22*). In the four-membered rings, the angles at aluminum and nitrogen approach 90°, whereas in the trimers the angles are different, with the smaller value found at the heavier atom (Al, 100–108°; N, 115–123°) (*14*). The Al—N bond lengths (of the order 1.93–1.90 Å) are quite constant and show no dependence on the ring size.

The polycyclic oligomers $(RAlNR)_n$ (type **4** compounds) may adopt an Al_nN_n skeleton that resembles either a cube ($n = 4$), a hexagonal prism ($n = 6$), or two more complex cages with $n = 7$ or $n = 8$ (*14,23*). No monomers have ever been isolated, but borazole-like $[MeAl(2,6-iPr_2C_6H_3)]_3$ has been obtained, which has a planar $Al_3N_3C_6$ core (X-ray structure) (*82*). As in the compounds of type **3** or **3′**, both the aluminum and nitrogen atoms always have the coordination number four. When the substituents on the secondary amine in Eq. (4) are bulky or when trimethylsilyl groups are employed, compounds are isolated that have only one amide function in the product [Eq. (5), R = Ph, $SiMe_3$] (*24,25*).

$$[Me_3Al]_2 + HNR_2 \xrightarrow[-CH_4]{} Me_3Al\cdot Me_2AlNR_2 \qquad (5)$$
$$\mathbf{5}$$

A structure determination of compound **5** (R = Ph) reveals a four-membered Al_2NC ring as the central unit in which the two aluminum atoms are connected by two different bridging atoms. Whereas the nitrogen atom functions as a base (four-electron three-center bond), the methyl group forms a two-electron three-center bond (Al–N = 2.00 Å; Al–C bridge = 2.14 Å) (*24*). The corresponding gallium derivative has also been isolated (*26*).

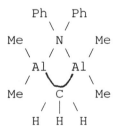

In addition to compounds of type **3** and **3′**, some of which have been claimed to be monomeric [e.g., $Me_2AlN(SiEt_3)_2$ (*27*), Me_2Al—$NPhSiMe_3$ (*28*), and $iBu_2AlN(SiMe)_2$ (*29*)], compounds of type **6** and **7** have also been synthesized but have been poorly characterized structurally (*9,15*).

$$R—Al \begin{array}{c} \diagup \; NR_2 \,' \\ \diagdown \; NR_2 \,' \end{array}$$

$$\begin{array}{c} R_2\,Al \\ \diagdown \\ \diagup \\ R_2\,Al \end{array} N—R'$$

6   **7**

Many of the structural principles found for organoaluminum amides also hold for the corresponding compounds of gallium, indium, and thallium (*15–18*). As in the case of the aluminum compounds, the amides tend to oligomerize through intermolecular Lewis acid/base interactions. There are nevertheless remarkable differences. Whereas $Me_2AlNHEt$ in benzene is a cyclic trimer (*30*), the corresponding thallium compound $Me_2TlNHEt$ appears to be dimeric, as assessed by spectroscopic methods (*31*). In order to obtain a compound that is monomeric, all the alkyl groups in an alkyl metal amide need to be substituted by very bulky groups, for example, $Al[N(SiMe_3)_2]_3$ (*32*) and $Tl[N(SiMe_3)_2]_3$ (*33*). In both these species the metal has the low coordination number three.

2. Cyclic and "Chelated" Organometal Amides

Aluminum, gallium, indium, and thallium have been successfully incorporated in cyclic systems, although little structural information has been obtained (*15*). Hoberg (*34*) has prepared many such ring compounds, including, for example, **8a** and **8b**, and has demonstrated the former to be dimers from molecular mass determinations and from mass spectra.

8a   **8b**

There are numerous examples of chelating ligands that have been used to obtain monometal-centered rings (*15*). However, a problem that arises from such attempts is that chelation may not necessarily take place at the same metal center and may lead to an intermolecular chelate, as is shown in formula **10**, rather than to the form **9**.

9 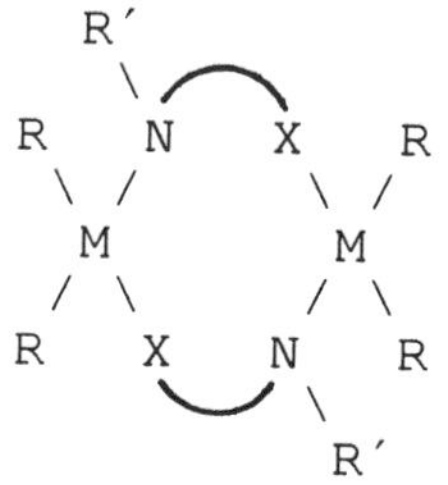   10

The two forms **9** and **10** may, however, be in equilibrium, as has been demonstrated in some cases. Some representative examples of the "intermolecular chelate" **10** are depicted in formulas **11–16**. Many of these structures have

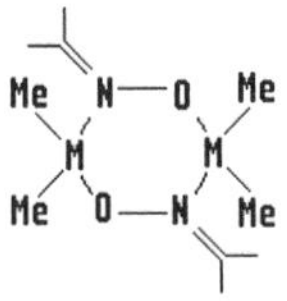

M = Al, Tl

11 [35]

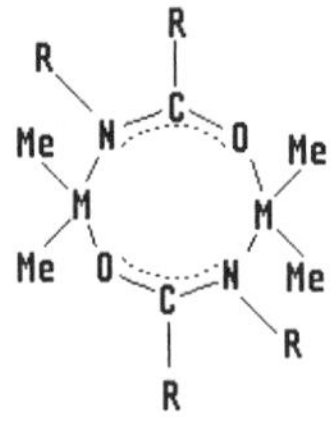

M = Ga, In
R = Me [36], R = Ph [37,38]

12

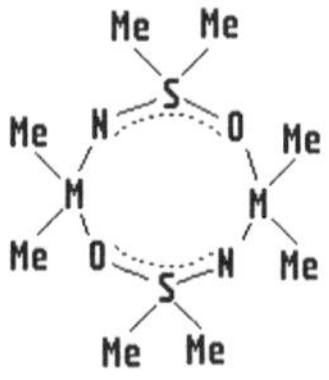

M = Ga, In

13 [39]

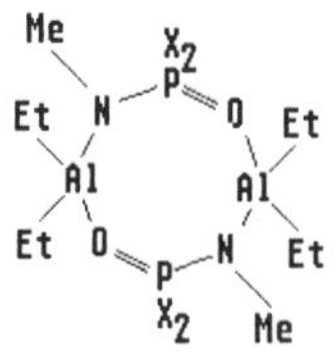

X = OEt, OCH$_2$CHMe$_2$, NMe$_2$

14 [40]

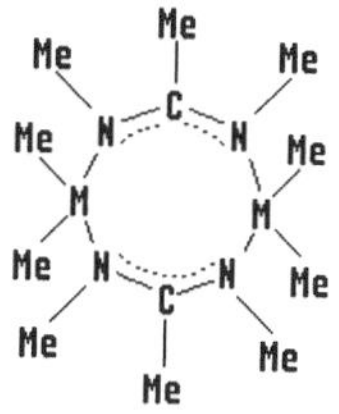

M = Al, Ga

15 [41,42]

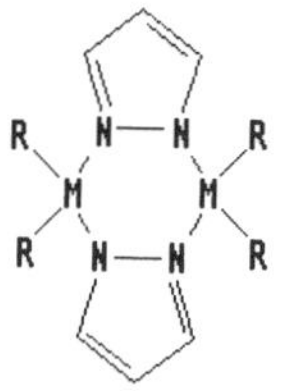

M = Al, Ga, Tl
R = Me, Cl

16 [43-47]

been confirmed by X-ray diffraction determinations. Although the ring systems of the six-membered cycles are planar, the corresponding eight-membered rings may be significantly distorted (see references).

In addition to these dimeric molecules some mononuclear metal compounds of type **9** have been found either by molecular weight determinations in noncoordinating solvents or by X-ray structure analyses (**17–21**). Some of

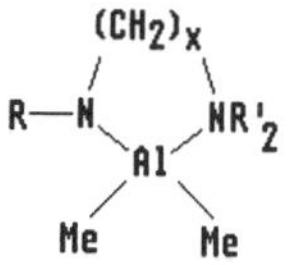

R = Me, R' = Me, x = 3
R = Et, R' = Me, Et, x = 2

17 [48]

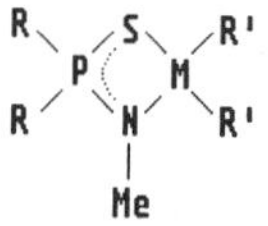

M = Al, R = CHMe$_2$, R' = Et
M = Al, R = tBu, R' = Et
M = Al, Ga, In, R = R' = Me

18 [40,49,50]

R = Me$_2$CHCH$_2$O
R = CHMe$_2$

19 [40]

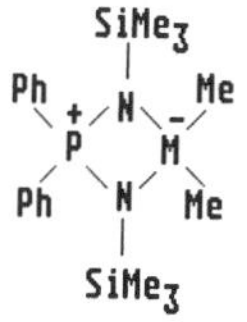

M = Al, Ga

20 [50]

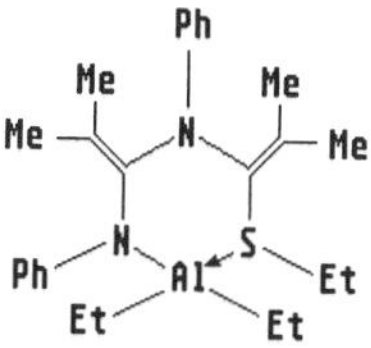

21 [51]

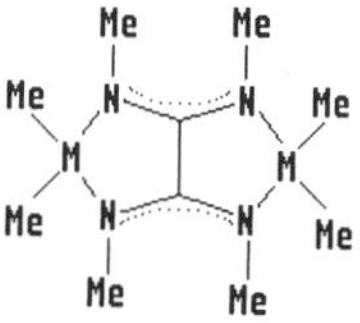

M = Al, Ga, In

22 [52]

M = Al, R = Me,Et
M = Ga, R = Me

23 [53,54]   **24 [54]**   **24' [55]**

the ligands used may chelate two metals in the same molecule, as found in **22–24**.

That cyclic compounds **17** and **19** are in equilibrium with their corresponding dimers has been deduced from NMR spectra (*40,48*). If sulfur is the donor atom (**18** and **21**), the stabilization at the metal seems to be more efficient than in the case of an oxygen donor (**19**). Two remarkable isomeric forms, **23** and **24**, have been shown to exist even in the solid state for the gallium derivative of N, N'-dimethyloxamide [X-ray structure investigations (*54*); see also compounds **22** and **24'** (*52,55*)].

The arrows and dots used in formulas **11–24'** are arbitrary and do not represent any chemical information. Often in these chelates a distinction between two-electron two-center bonds and two-electron donor bonds cannot be made by comparing bond lengths. A more general view of acid/base interactions and the representation of formulas has been given elsewhere (*56*).

In addition, macrocycles have been used to chelate alkyl metal compounds efficiently. Methyltetraphenylporphinatoindium contains an indium atom that is coordinated by the four nitrogen atoms of the porphyrin and has a fifth methyl ligand (*57*).

III

ORGANOMETALLIC DERIVATIVES OF SPECIAL SILAZANES

A. *Synthesis of the Compounds*

In order to incorporate the metal alkyl or aryl in a ring system of nitrogen and silicon atoms, we have used, over the past 10 years, the well-known amines **25** (*58*) and **26** (*59*), or have developed synthetic routes to oxoamines **27** (*60*)

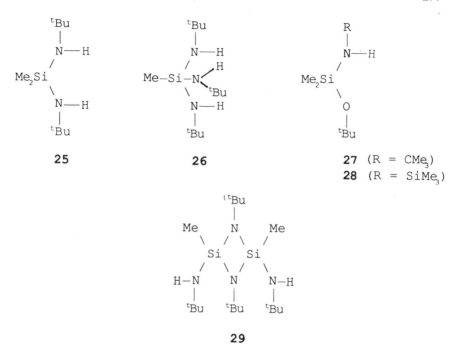

25

26

27 (R = CMe₃)
28 (R = SiMe₃)

29

and **28** (*61*) and to the cyclic "cis" amine **29** (*62*). These amines have been used to synthesize metallacycles by the routes depicted in Eqs. (6)–(14) (*63–69*).

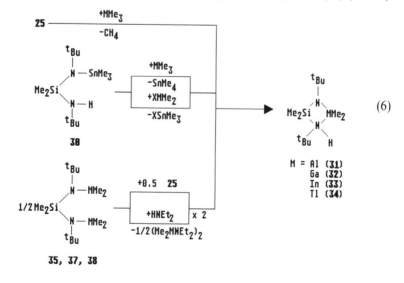

(6)

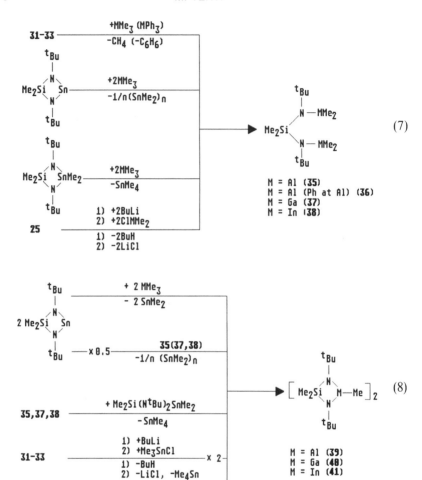

The products **31–54** of these reactions can be handled under dry nitrogen and they analyze correctly, are soluble in nonpolar solvents such as pentane, cyclohexane, benzene, and toluene, and are often sublimable without decomposition under reduced pressure. They may be prepared in high yields and large quantities. The synthetic procedures starting with a stannazane [**30**, $Me_2Si(NtBu)_2Sn$ or $Me_2Si(NtBu)_2SnMe_2$] afford high yields of product. The reaction temperature in Eq. (12) must be strictly controlled, because compound **49**, and to a lesser extent compound **50**, tend to disproportionate into the symmetrical products [Eq. (15)] (67).

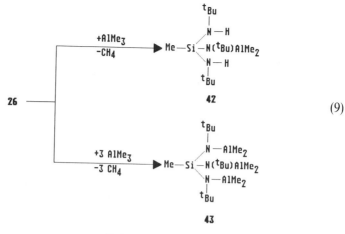

$$(9)$$

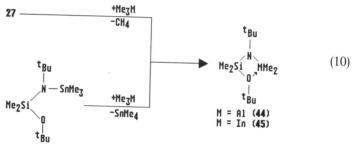

$$(10)$$

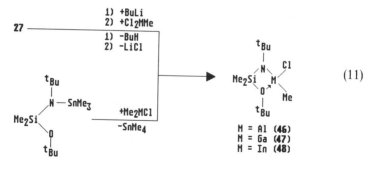

$$(11)$$

$$(12)$$

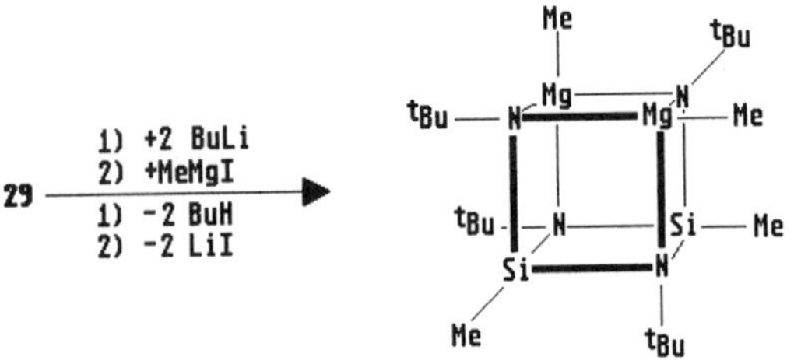

$$29 \xrightarrow[\substack{1)\ -2\ BuH \\ 2)\ -2\ LiI}]{\substack{1)\ +2\ BuLi \\ 2)\ +MeMgI}}$$

(13)

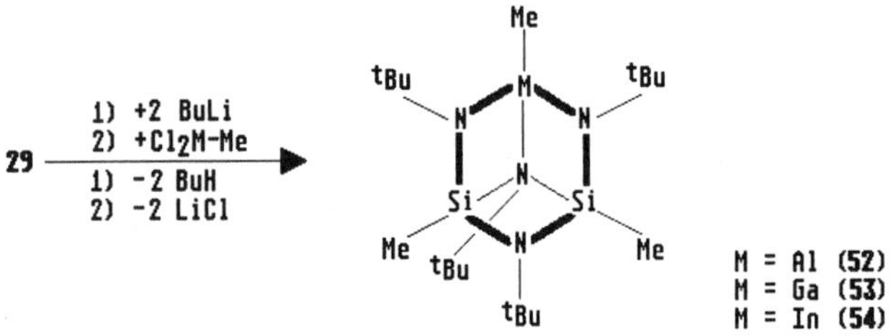

$$29 \xrightarrow[\substack{1)\ -2\ BuH \\ 2)\ -2\ LiCl}]{\substack{1)\ +2\ BuLi \\ 2)\ +Cl_2M-Me}}$$

M = Al (52)
M = Ga (53)
M = In (54)

(14)

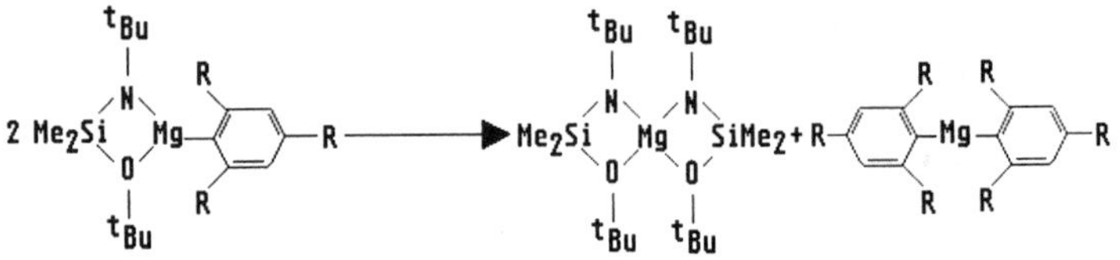

$$2\ Me_2Si \cdots \longrightarrow Me_2Si \cdots SiMe_2 + R - \cdots - Mg - \cdots - R$$

(15)

49,58

B. Structures of the Compounds

The structures of compounds **31–54** may be deduced indirectly by spectroscopic methods (NMR, IR, or MS) or may be determined by X-ray crystallography. Frequently the structures found in solution are different from those existing in the crystalline state due to intramolecular bond fluctuations, as explained in Section III,D.

The molecules **31–34** are all monomeric and cyclic, as shown by infrared and ^{1}H and multinuclear NMR spectra *(63,70)*. Unfortunately, the quality of crystals of these compounds was not sufficient for X-ray diffraction studies and therefore we used a derivative of **31**, $Me_2Si[tBu(H)N](NtBu)AlCl_2$, to establish unambiguously the structural details *(64)*. As depicted in Fig. 1, a four-membered SiN_2Al ring is the central unit. The two nitrogen atoms N1 and N2 differ in their coordination numbers (three and four), the trigonal–planar coordinated N1 effecting shorter bonds to Al and Si than those effected

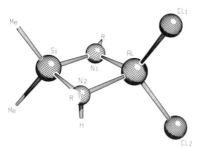

FIG. 1. The four-membered ring in Me$_2$Si(NtBu)[NtBu(H)]AlCl$_2$. As in Figs. 2–14, the *tert*-butyl groups are denoted by R and the methyl groups are denoted by Me.

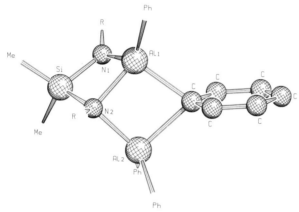

FIG. 2. The polycycle in Me$_2$Si(NtBu)$_2$(AlPh$_2$)$_2$ (36). The hydrogen atoms at the bridging phenyl group are not shown.

by the tetrahedrally distorted N2, as expected [N1–Al = 1.830(2) Å, N1–Si = 1.712(2) Å, N2–Al = 1.994(2) Å, and N2–Si = 1.861(2) Å] (64).

The dimetallasilazanes 35, 37, and 38 are not suitable for X-ray structure determinations because they form plastic crystalline phases, as shown for 35 (cubic crystal system, point symmetry for 35 m3m or 43m) (63). In contrast, the structure of the phenyl derivative 36 has been established, and the central skeleton is shown in Fig. 2 (64). Two four-membered rings (Si, N1, N2, Al1 and Al1, N2, Al2, and C29) share an edge, and C29 connects a four-membered with a six-membered ring as a spirocenter (64). The bonding in this compound can be derived from Al$_2$Ph$_6$ (71), where one bridging and one terminal phenyl group have been replaced by nitrogen atoms (formulas 55 and 56).

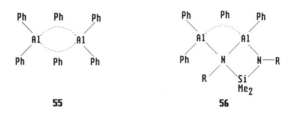

55   **56**

The Al1–C29–Al2 and Al1–N2–Al2 bridges are almost symmetrical with an Al–C distance of 2.18(3) Å and an Al–N distance of 1.99(1) Å. As expected, the bond distances to the terminal atoms are remarkably shorter (64). Although C29 may be visualized as the center of a two-electron three-center bond, N2 uses two electron pairs to coordinate to the aluminum atoms.

Figure 3 represents the only dimer formed in this series of metallacycles. The coordination number of the indium atom is increased by interaction with the nitrogen atom of a second molecule. The dimer **41** has a center of inversion and three edge-fused four-membered rings. This type of structure is very common in metallacycles that are derived from the ligand **25** (56). The central N_2In_2 ring is almost square–planar [In–N = 2.27(1) Å] and the In–C dis-

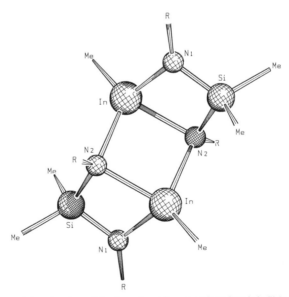

FIG. 3.   The escalator structure of the three fused four-membered cycle in [Me$_2$Si(NtBu)$_2$In—Me]$_2$ (**41**).

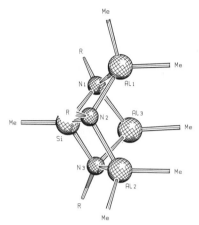

FIG. 4. The open cage of MeSi(N(tBu)AlMe₂)₃ (**43**). The second methyl group of Al3 is hidden behind this atom.

tance is 2.153(5) Å. The homologous compounds **39** and **40** are also dimeric, **39** forming crystals isomorphous with **41** (*63,64*).

Compound **43** is a polycycle that can be derived from a cube having one corner missing (Fig. 4). The distortion from the point symmetry 3m (C_{3v}) is small, the pseudo-threefold axis passing through the Si—C bond. All atoms of the ring have the coordination number four in a tetrahedrally distorted environment (*65*). The Al–N distances are almost the same [2.02(1) Å], whereas the Al–C distances are slightly shorter in the direction of the three-fold axis [1.95(1) Å] than those that are almost perpendicular to it [1.98(1) Å]. The chelates **44**–**48** have been shown by cryoscopy in benzene to be mono-meric. Their ring structure can be deduced from NMR spectra (*66*) (see also Section III,D).

No suitable crystals have been obtained of compound **51**. Nevertheless, we have been able to synthesize the corresponding bromine derivative (the methyl groups at the aluminum being replaced by bromine atoms) (*69*) and have de-termined its crystal structure by X-ray diffraction. Figure 5 shows the central $Si_2N_4Mg_2$ cage, which may be derived from a cube, the corners being occu-pied alternatively by metals and nonmetals. The Mg–N distances are different [Mg–N2 = 2.210(6) Å and Mg–N1(N1′) = 2.147(6) Å] and are considerably longer than the Si–N distances [1.72–1.78 Å], which accounts for the distor-tion from ideal cubic geometry.

Of compounds **52**–**54**, the aluminum and gallium derivatives have been fully characterized in the solid state by X-ray diffraction (*62,69*). In Fig. 6, the

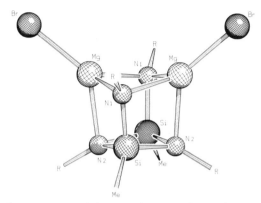

FIG. 5.   The cube arrangement of the noncarbon atoms in $(MeSi)_2(BrMg)_2(NtBu)_4$ **(59)**.

polycyclic Si_2N_4Ga skeleton of **53** is depicted; it resembles the skeleton of the aluminum compound **52**. The gallium atom is in fourfold coordination by three nitrogen atoms and one carbon atom of the methyl group. As expected, the N3—Ga donor bond is significantly longer than are the N(2,4)—Ga bond distances [Ga–N3 = 2.119(3) Å and Ga–N(2,4) = 1.927(2) Å]. In the aluminum derivative **52**, the corresponding distances are Al–N3 = 1.795(7) Å and Al–N(2,4) = 1.733(6) Å (62,69). The Ga—C bond in **53** [2.009(4) Å] is again longer than the Al—C bond in **52** [1.85(1) Å].

C. Synthetic Uses; Polymetal Silazanes

Reactive centers in a cyclic or chelated alkyl or aryl metal silazane, such as compounds **31–54**, are (1) the polar metal–carbon bond, (2) the polar metal–nitrogen bond, and (3) other bonds in the silazane, especially nitrogen–hydrogen bonds. We have been interested in reaction centers (1) and (3) and have

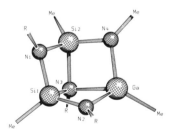

FIG. 6.   The seco norcubane-like polycycle in $(MeSi)_2(MeGa)(NtBu)_4$ **(53)**.

performed several reactions to evaluate their synthetic use. As expected, compounds of types **31–54** are very efficient alkylating reagents. Some selected examples are shown in Eqs. (16)–(18).

The procedures given by Eqs. (16)–(18) may be used either to alkylate silicon halides or tin halides in very high yields (90–100%), or to substitute the alkyl functions in the metallacycles by halogen atoms. The two products in these reactions can be easily separated from one another and compounds **57–60** are isolated in high yields (*63,69*).

The dimers **39–41** are stable toward base addition, and there is no way to transform by coordination of pyridine compound **39** to the monomeric base adduct **61** [Eq. (19)]. The species **61** can only be prepared by other routes, as indicated in Eqs. (20) and (21).

In order to visualize the reaction path of Eq. (21), the reader should be reminded that intramolecular methyl bridges between the two aluminum atoms in **35** need to be invoked (see Section III,D).

The metallacycles **31–54** can also simultaneously function as acceptors and

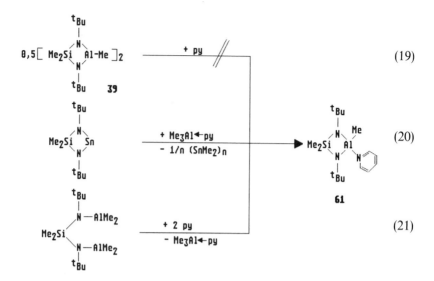

$$
\text{(19)}
$$

$$
\text{(20)}
$$

$$
\text{(21)}
$$

donors and may accommodate different metallic elements. Starting with **29**, a dilithium derivative can first be synthesized [the core structure of this compound has been established to be a distorted $N_4Li_2Si_2$ cube (68)], which then can react with only one equivalent of $ClAlMe_2$ or $ClInMe_2$ to form **62** and **63** [Eq. (22)].

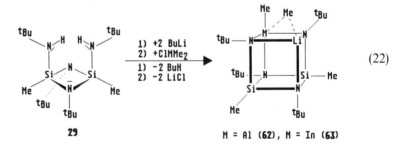

$$
\text{(22)}
$$

The structures of **62** and **63** can be established by ^1H NMR spectra, the bridging methyl group between aluminum or indium and/or lithium showing up as a broad signal at high field [$\delta = -0.35$ ppm (vs. tetramethylsilane for **62**)]. In this series, **62** and **63** are the first compounds that contain two different metals in a silazane. Such poly(metal)silazanes may be prepared by different routes, starting with alkyl metal silazanes that still have a hydrogen atom bonded to one of the nitrogen atoms. Several examples are summarized in Eqs. (23)–(29).

$$
\text{31} \xrightarrow[\text{- 1/2 H}_2\text{ or - RH}]{\text{+ M or + MR}} M^{\oplus}\ \underset{\underset{^tBu}{|}}{\overset{\overset{^tBu}{|}}{\underset{N}{\overset{N}{Me_2Si\ominus AlMe_2}}}} \qquad (23)
$$

64 (M = Li), **65** (M = Na) [64]

$$
\underset{\underset{^tBu\quad H}{|}}{\overset{\overset{^tBu}{|}}{\underset{N}{\overset{N}{Me_2Si\quad AlMe_2}}}}
$$

31

$$
\xrightarrow[\text{- 0.5 H}_2]{\text{+ 0.5 M(activated)}} M^{2\oplus}\left[\ \underset{\underset{^tBu}{|}}{\overset{\overset{^tBu}{|}}{\underset{N}{\overset{N}{Me_2Si\ominus AlMe_2}}}}\ \right]_2 \qquad (24)
$$

66 (M = Mg), **67** (M = Ca) [73,73]

$$
\xrightarrow[\text{- CH}_4]{\text{+ Me}_2\text{Mg}} Me{-}Si\underset{\underset{^tBu}{|}}{\overset{\overset{^tBu}{|}}{\begin{array}{l} N{-}AlMe_2 \\[4pt] N{-}Mg{-}Me \end{array}}} \qquad (25)
$$

68 [64]

$$
\text{64,65} \xrightarrow[\text{- MX}]{\text{+ XMgR}} Me{-}Si\underset{\underset{^tBu}{|}}{\overset{\overset{^tBu}{|}}{\begin{array}{l} N{-}AlMe_2 \\[4pt] N{-}Mg{-}R \end{array}}} \qquad (26)
$$

69 (R = ^tBu), **70** (R = CH_2-Ph)
71 (R = Ph) [64], **72** (R = CPh_3) [73]
73 (R = I) [64]

The poly(metal)silazanes **62–77** are isolated usually as colorless crystals. They are unstable with respect to water and oxygen and are soluble in non-polar solvents (with the exception of compound **77**). The formulas drawn for **64–67** are merely formal and do not imply that these substances have a considerable ionic character. They are soluble in benzene and are monomolecular as found by cryoscopic molecular weight determinations. The ^{1}H NMR spectra of **64** and **65** (*64*) demonstrate that the metal atoms are coordinated by methyl groups, the most reasonable structure being depicted in the formulas **64** and **65**. The trigonal–bipyramidal arrangement of the N_2SiAlM

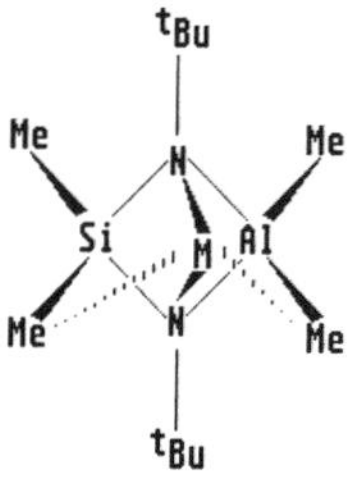

64, 65

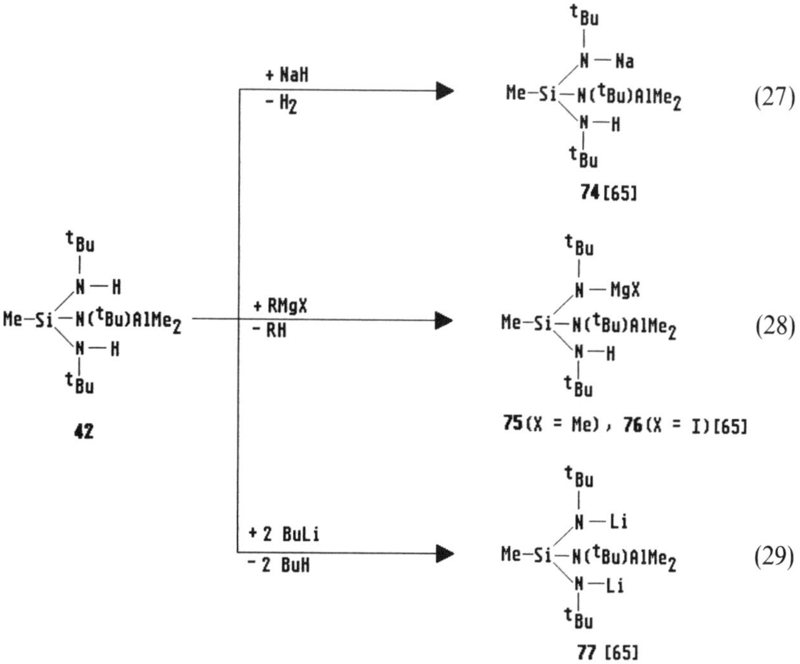

$$\text{Me-Si}-\text{N}(^t\text{Bu})\text{AlMe}_2 \qquad (27)$$

74 [65]

Me-Si—N(^{t}Bu)AlMe$_2$

42

$$\text{Me-Si}-\text{N}(^t\text{Bu})\text{AlMe}_2 \qquad (28)$$

75 (X = Me) , 76 (X = I) [65]

$$\text{Me-Si}-\text{N}(^t\text{Bu})\text{AlMe}_2 \qquad (29)$$

77 [65]

fragment has been found for some compounds in the solid state. In Figs. 7 and 8, $N_2SiAlMg$ and $N_2SiAlCa$ bipyramids, respectively, are observed as the central skeletons. Whereas two of the bipyramids in **73** are linked via iodine bridges, with the magnesium atom situated in an almost square–planar environment, in **67** two bipyramidal $N_2SiAlCa$ cages are connected by the common calcium atom. Compound **67** can also be visualized as a sandwich in which the calcium atom lies between the two SiN_2Al rings. Compared to calcium, magnesium seems to be too small to fit between these four-membered SiN_2Al rings. As is apparent from Fig. 9, the "sandwich" in **66** is more open,

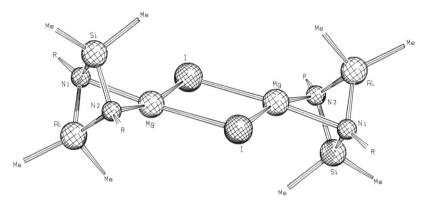

FIG. 7.   The dimeric structure of $(Me_2Si)(Me_2Al)(IMg)(NtBu)_2$ **(73)**. The coordination at the magnesium atom is distorted square–planar.

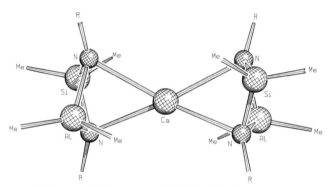

FIG. 8.   The sandwich-structure of $Ca[Me_2Si(NtBu)_2AlMe_2]_2$ **(67)**.

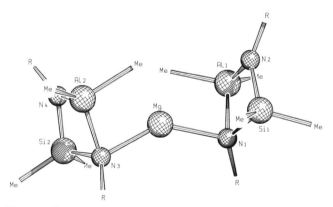

FIG. 9.   Two-coordinate magnesium in $Mg[Me_2Si(NtBu)_2AlMe_2]_2$ **(66)**. The methyl groups at Al1 and Al2 have weak interactions with the magnesium atom (2.60 Å).

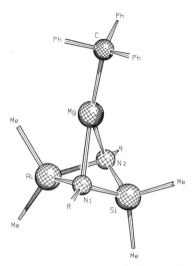

FIG. 10. The trigonal coordinated magnesium atom in $(t\text{BuN})_2(\text{Me}_2\text{Si})(\text{Me}_2\text{Al})\text{MgCPh}_3$ (**72**).

the magnesium atom being almost two coordinate (there is also some intra-molecular methyl interaction (*72,73*). Only the trityl compound in the series **68–72** is suitable for a crystallographic study. Figure 10 shows the skeleton of **72**, which again is a distorted N_2SiAlMg bipyramid (73). The magnesium atom is in trigonal–planar coordination by two nitrogen atoms and a carbon atom.

The distances within the SiN_2Al unit for the compounds **66, 67, 72,** and **73** vary in the range 1.70–1.79 Å for Si–N and 1.85–1.97 Å for Al–N. The aluminum–carbon distances are in the range 1.97–2.06 Å. The magnesium carbon distance in **72** is quite elongated [2.20(1) Å], a feature that might be related to some extent to the triphenylmethyl group.

It was of interest to determine which of the metal–carbon bonds, Mg—C or Al—C, in alkyl poly(metal)silazanes such as **68** would be the more reactive. This was probed by treating **68** with cyclopentadiene or trimethylsilylcy-clopentadiene [Eq. (30)] (*73*).

$$\underset{\text{Me}_2\text{Si}}{}\quad\begin{array}{c}{}^{t}\text{Bu}\quad\text{Me}\\ |\quad\quad\diagup\\ \text{N}-\text{Al}\\ \diagup\quad\quad\diagdown\text{Me}\\ \diagdown\\ \text{N}-\text{Mg}-\text{Me}\\ |\\ {}^{t}\text{Bu}\end{array}\quad\xrightarrow[-\text{CH}_4]{+\ \text{C}_5\text{H}_5\text{X}}\quad\underset{\text{Me}_2\text{Si}}{}\quad\begin{array}{c}{}^{t}\text{Bu}\quad\text{Me}\\ |\quad\quad\diagup\\ \text{N}-\text{Al}\\ \diagup\,\vdots\quad\diagdown\text{Me}\\ \diagdown\,\vdots\\ \text{N}-\text{Mg}\\ |\quad\quad\diagdown\\ {}^{t}\text{Bu}\qquad\diagdown-\text{X}\end{array}\qquad(30)$$

78(X = H),**79**(X = SiMe₃)

NMR spectra showed that in these reactions the $Me_2Si(NtBu)_2AlMe_2$ portion of the molecules remains unchanged, and in each case, the methyl group of the magnesium atom is replaced by the cyclopentadienyl ligand, indicated by a singlet in the ^1H-NMR spectra of **78** and **79** ($\delta = 6.35$ and 6.54 ppm for **78** and **79**, respectively).

The same type of reaction can be used to replace one of the two methyl groups in **51** by a cyclopentadienyl moiety, according to Eq. (31).

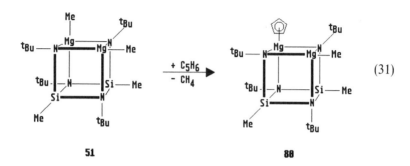

$$(31)$$

51   **80**

Compound **80** was unambiguously characterized by elemental analysis and ^1H NMR. The singlet of the cyclopentadienyl group has a comparable chemical shift ($\delta = 6.52$ ppm) to the same ligand in **78**. Compound **80** is a very rare example of a silazane containing two magnesium atoms with different organic ligands, and it resembles tetrameric cubanelike $(cpMgOEt)_4$ (*74*).

The products of Eqs. (27) and (28) have been studied by single-crystal X-ray diffraction, and the skeleton core structures of the compounds **74–76** are illustrated in Figs. 11–13 (*65*). Surprisingly, **74** is monomeric even in the

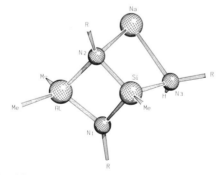

FIG. 11. The two fused four-membered rings in $MeSi(NtBu)_3(H)(Na)AlMe_2$ (**74**). Even in the crystal, the sodium atom has no further contacts.

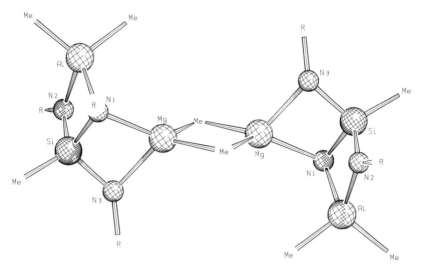

FIG. 12.   Five fused four-membered cycles with magnesium atoms as spiro centers and methyl bridges in the dimeric [(MeSi)(NtBu)$_3$(AlMe$_2$)(H)MgMe]$_2$ (**75**).

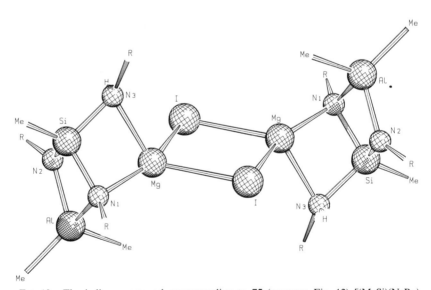

FIG. 13.   The iodine compound corresponding to **75** (compare Fig. 12) [(MeSi)(NtBu)$_3$-(AlMe$_2$)(H)MgI]$_2$ (**76**).

solid state, whereas **75** and **76** are dimeric through Mg–Me–Mg or Mg–I–Mg bridges. In both **75** and **76**, the magnesium atom adopts a distorted tetrahedral coordination.

Compound **74** is one of the few compounds yet characterized by X-ray diffraction that contains a two-coordinate sodium atom (*61,75*). The bonding distances to nitrogen, Na–N2 = 2.476(8) Å and Na–N3 = 2.523(9) Å, are comparatively long, indicating a decreased polar character of the bond. The common skeleton of compounds **74–76** is two edge-sharing four-membered rings that are almost mutually orthogonal. Whereas the methyl bridge in **75** is involved in a three-center two-electron bond [Mg–C = 2.24(2) Å], the iodine bridge in **76** uses two electron pairs [Mg–I = 2.771(1) Å and Mg–I′ = 2.900(1) Å] and is unsymmetrical. The Mg–C distances compare well with the corresponding distances in $Me_8Al_2Mg[2.21$ Å $(76)]$, as does the average Mg–I value (2.84 Å) in **76** with the corresponding Mg–I distance in **73** (2.83 Å).

To close this section, we describe our observations on the reaction of MeSi-(N*t*Bu[H])$_3$ (**26**) with the Grignard reagent CH_3MgI in benzene/tetrahydrofuran (*65*). According to Eq. (32), four equivalents of the Grignard reagent always react with the silazane (**26**), even if the molar ratios are changed.

We have been able to show by spectroscopic methods and by an independent preparation of $MeSi(NtBuMgI)_3 \cdot THF$ that the reaction proceeds via the intermediate compound **82**, which seems to function as a very strong Lewis acid toward CH_3MgI [Eq. (33)] (*65*).

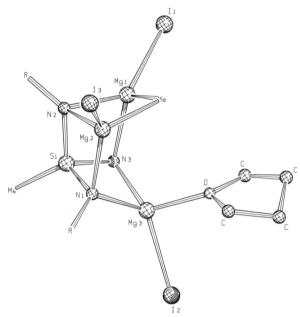

FIG. 14.   The anion MeSi(NtBu)$_3$(MgI) · CH$_3$ · THF$^-$ in compound **81**. The methyl anion lies between the magnesium atoms Mg1 and Mg2. The magnesium atom Mg3 is stabilized by an added THF molecule.

The Grignard reagent is thus cleaved into a cation, IMg(THF)$_5^+$, and the anion **81a**, which contains the methyl anion. We have been able to perform a structure determination on single crystals of **81**. In Fig. 14, the anion is represented as part of this structure. The two magnesium atoms coordinate the methyl group [Me], and the third magnesium in the polycycle is coordinated by a tetrahydrofuran molecule. All metal atoms adopt fourfold distorted tetrahedral coordination. The skeleton of the polycycle is thus derived from a cube with one opened edge. The Mg–C distances in **81** are of the order of 2.32 Å, which is in good agreement with other magnesium complexes that have R$_3$Mg$^-$ or R$_6$Mg$_2^{2-}$ anions and a cation stabilized by a crown ether (*77–80*).

D. *Intramolecular Bond Fluctuations*

The alkyl and aryl metal silazanes and poly(metal) silazanes described so far are representatives of an unusual class of compound containing simultaneously electron-deficient and electron-rich centers that can interact with one another. The interaction may be of the two-electron donor type (Lewis acid/

base interaction) or may be part of a two-electron three-center bond (see the two preceding sections). It is evident that such interactions can be quite labile, and that bond fluctuations, rotation of groups, inversions, etc. need to be considered, especially when several bases are competing for an acidic center in the molecule.

We have used temperature-dependent NMR spectroscopy in order to characterize the "movements" within these molecules. For reference, we generally start with the structures determined in the solid, because they should always represent an energy minimum in a fluctuation process. Such intramolecular rearrangements are not only observed in silazanes of the type described in this article, but are common to all metal-containing silazanes (56,81). The following examples have been selected to illustrate different aspects of these "motions."

The chelates **31–34** have been extensively studied with respect to intramolecular rearrangements (70). From a comparison of the (N–H) frequencies in the different compounds, a series of increasing Lewis acidity of the MMe_2 group can be established: $TlMe_2 < GaMe_2 \approx InMe_2 < AlMe_2$. The λ^4 N—M bond in these compounds is subject to opening, and its frequency can be followed by temperature-dependent ^1H NMR techniques. Intra- and intermolecular hydrogen transfers can be excluded by comparing the rate constants k_H for **31–34** with k_D of the deuterated compounds. The free energy of activation for bond breaking and the rotation of the NHtBu group is of the order 61–84 kJ/mol at 298 K. The process can be described best by the following model [Eq. (34)].

The enantiomeric forms **83a** and **83b** are related by the intermediate **84**, in which rotation around both the N—M and the Si—N bonds may occur. Rotation around the Si—N bond is associated with an inversion at the pyramidal nitrogen atom. In a similar manner, in the chelates **46–48** the breaking of the O—M bond is accompanied by rotation around the respective bonds, and a very rapid transformation of one enantiomeric form (**85a**) into the other (**85b**) is observed [Eq. (35)] (65).

$$(35)$$

85a   **85b**

A very interesting "ligand exchange" at the metal atoms is observed in compounds of types **35–38**. The two metal atoms within the molecule are in competition for the two nitrogen atoms, of which only one is attainable for steric reasons. As a consequence, one aluminum atom becomes electronically unsaturated and has to participate in a three-center two-electron bond (see also Fig. 2). This bonding situation can be rapidly inverted by an exchange of ligands as shown in Eq. (36) for compound **35** (*64*).

$$(36)$$

86a   **86b**

We have been able to demonstrate by synthesizing the unsymmetrical compound **87** and recording its temperature-dependent ^{1}H NMR spectra that, in addition to the equilibrium between the enantiomers **86a** and **86b**, further ligand rearrangements need also to be considered.

87

The model presented so far indicates a difference between the methyl groups Me^1 and Me^4 compared with Me^2 and Me^3. This is the case for compound

35, which has two discrete signals at lower temperatures for the four methyl groups of aluminum. At higher temperatures ($> -40°C$) the two signals collapse to a single resonance, indicating "free rotation" of the $AlMe_2$ groups around the Al—N bond. According to the structures drawn in Eq. (36), the whole process can be described as follows: (1) At high temperatures the aluminum atoms Al^1 and Al^2 exchange ligands N^1 and N^2 and may be bridged by methyl groups Me^1–Me^4. (2) At lower temperatures the exchange of N^1 and N^2 continues, whereas the methyl groups are separated into terminal (Me^1 and Me^4) and bridging (Me^2 and Me^3) groups. (3) At very low temperatures ($-70°C$ in compound **87**) all movement is stopped, and the enantiomers **86a** and **86b** no longer interconvert at rates fast enough to be observed on the ^1H NMR time scale (64).

^{1}H-NMR techniques have also been used to show that the aluminasilazane **42** exists in two different conformers (65). At 308 K in toluene, the spectrum of **42** may be interpreted as a 3:1 mixture of the isomers **88a** and **88b**. If the temperature is raised to 365 K, however, the six signals for tBu groups (three for **88a** and three for **88b**) collapse to two in a ratio 1:2, indicating that the

88a   **88b**

two tBu groups within the molecule have now become chemically equivalent and a mixture of isomers is no longer observed. The best explanation for these observations seems to be that a rapid intramolecular ligand exchange of the tBuNH groups occurs at the aluminum atom, making impossible a distinction between the isomers **88a** and **88b**. Similar ligand exchanges also appear to occur in the metalla derivatives **74–76** (65).

The final example of bond rearrangement in solution is the magnesium amide **66**. Examination of Fig. 9 shows that the nitrogen atoms N1 and N3 are coordinated to the magnesium, whereas the atoms N2 and N4 do not interact with the metal (the molecule has the approximate C_2 point symmetry). The tBu groups at N1 and N3 should thus be different from those at N2 and N4, and therefore two resonances should be observed for tBu in the ^{1}H-NMR spectrum. This, however, is definitely not the case, and the spectrum of **66** shows only one signal in the *tert*-butyl region down to $-88°C$ (the methyl groups of aluminum and silicon are all different, as expected) (73). To explain these observations it is necessary to assume a "rapid rocking" of the four-membered rings with respect to the magnesium, in which the nitrogen atoms N2 and N4 displace the nitrogen atoms N1 and N3 from the coordination sphere of the magnesium [see also Eq. (37)].

66a   **66b**

IV

CONCLUSION

The sophisticated silazanes 25–29 can be used to stabilize (in a coordinative sense) metal atoms that have further organic ligands. There is, of course, no limit to other metallic or nonmetallic elements that can be incorporated in such silazanes (23,56,81). The organic groups attached to the metals should exhibit reactivity patterns different from those of Grignard reagents or aluminum alkyls. Little comparative work in organic synthesis has yet been carried out in this field, but studies are planned for the future. From a structural point of view, these compounds have a very diverse and interesting coordination chemistry. The very prominent bond fluctuations observed for these molecules are often due to competitive reactions between bases for the same acidic center. Multinuclear NMR experiments in solution and in the solid phase are planned for these compounds to complete the investigations.

ACKNOWLEDGMENTS

I am indebted to my co-workers for their help in exploring this field of chemistry. Their names can be found in the references. I also express my thanks to the people of the Fonds der Chemischen Industrie, who have supported our work. Finally I thank Dr. P. G. Harrison, Nottingham, England, for critically reading the manuscript.

REFERENCES

1. J. P. Oliver, *Adv. Organomet. Chem.* **15**, 235 (1977).
2. H. Yamamoto and M. Oki, *Angew. Chem.* **90**, 549 (1978); *Angew. Chem., Int. Ed. Engl.* **17**, 169 (1978).
3. B. B. Snider, D. J. Rodini, M. Karras, T. C. Kirk, E. A. Deutsch, R. Cordova, and R. T. Price, *Tetrahedron* **37**, 3927 (1981).
4. Named in honor of Victor Grignard (1871–1935), who was awarded the Nobel Prize in chemistry in 1912; V. Grignard, *C. R. Hebd. Seances Acad. Sci.* **130**, 1322 (1900).
5. E. C. Ashby, *Q. Rev., Chem. Soc.* **21**, 259 (1967).
6. W. Schlenck and W. Schlenck, Jr., *Chem. Ber.* **62**, 920 (1929).
7. F. W. Walther and E. C. Ashby, *J. Am. Chem. Soc.* **91**, 3845 (1969).
8. R. Benn, H. Lehmkuhl, K. Mehler, and A. Rufinska, *Angew. Chem.* **96**, 521 (1984); *Angew. Chem., Int. Ed. Engl.* **23**, 534 (1984).
9. M. F. Lappert, P. P. Power, A. R. Sanger, and R. C. Srivastava, "Metal and Metalloid Amides." Ellis Horwood, Chichester, U. K., 1980.
10. G. E. Coates and D. Ridley, *J. Chem. Soc. A*, 56 (1967).
11. V. R. Magnuson and G. D. Stucky, *Inorg. Chem.* **8**, 1427 (1969).
12. R. Masthoff, G. Krieg, and C. Vieroth, *Z. Anorg. Allg. Chem.* **364**, 316 (1969).
13. T. Mole and E. A. Jeffery, "Organoaluminium Compounds." Elsevier, Amsterdam, 1972.
14. M. Cesari and S. Cucinella, *in* "The Chemistry of Inorganic Homo- and Hetero-Cycles" (D. B. Sowerby and I. Haiduc, eds.), Vol. 1, p. 167. Academic Press, London, 1987.
15. A. McKillop, J. D. Smith, and I. J. Worrall, "Organometallic Compounds of Aluminium, Gallium, Indium and Thallium." Chapman & Hall, London, 1985.
16. R. B. Hallock, O. T. Beachley, Jr., Y.-J. Li, W. M. Sanders, M. R. Churchill, W. E. Hunter, and J. L. Atwood, *Inorg. Chem.* **22**, 3683 (1983).
17. A. McKillop and E. C. Taylor, *Adv. Organomet. Chem.* **11**, 147 (1973).
18. A. G. Lee, *Q. Rev., Chem. Soc.* **24**, 310 (1970).
19. H. Hess, A. Hinderer, and S. Steinhauser, *Z. Anorg. Allg. Chem.* **377**, 1 (1970).
20. K. Gosling, G. M. McLaughlin, G. A. Sim, and J. D. Smith, *Chem. Commun.*, 1617 (1970).
21. G. M. McLaughlin, G. A. Sim, and J. D. Smith, *J. Chem. Soc., Dalton Trans.*, 2197 (1972).
22. J. L. Atwood and G. D. Stucky, *J. Am. Chem. Soc.* **92**, 285 (1970).
23. M. Veith, *Chem. Rev.* **90**, 3 (1990).
24. V. R. Magnuson and G. D. Stucky, *J. Am. Chem. Soc.* **91**, 2544 (1969).
25. N. Wiberg and W. Baumeister, *J. Organomet. Chem.* **36**, 277 (1972).
26. J. E. Rie and J. P. Oliver, *J. Organomet. Chem.* **133**, 147 (1977).
27. N. Wiberg, W. Baumeister, and P. Zahn, *J. Organomet. Chem.* **36**, 267 (1972).
28. T. Sakakibara, T. Hirabayashi, and Y. Ishii, *J. Organomet. Chem.* **46**, 231 (1972).
29. C. Ya. Zhinkin, G. K. Korneeva, N. N. Korneev, and M. V. Sobolevskii, *J. Gen. Chem. USSR (Engl. Transl.)* **36**, 360 (1966).
30. K. J. Alford, K. Gosling, J. D. Smith, *J. Chem. Soc., Dalton Trans.*, 2203 (1972).
31. B. Walther, A. Zschunke, B. Adler, A. Kolbe, and S. Bauer, *Z. Anorg. Allg. Chem.* **427**, 137 (1976).
32. G. M. Sheldrick and W. S. Sheldrick, *J. Chem. Soc. A*, 2279 (1969).
33. R. Allmann, W. Henke, P. Krommes, and J. Lorberth, *J. Organomet. Chem.* **162**, 283 (1978).
34. H. Hoberg, *Justus Liebigs Ann. Chem.* **746**, 86 (1971).
35. J. R. Jennings and K. Wade, *J. Chem. Soc. A*, 1333 (1967).
36. H. V. Schwering and J. Weidlein, *Chimia* **27**, 535 (1973).
37. Y. Kai, N. Yasuoka, N. Kasai, and H. Kakudo, *J. Organomet. Chem.* **32**, 165 (1971).
38. H. Yasuda, T. Araki, and H. Tani, *J. Organomet. Chem.* **49**, 103 (1973).
39. H. Schmidbaur and G. Kammel, *J. Organomet. Chem.* **14**, 28 (1968).

40. K. Urata, K. Itoh, and Y. Ishii, *J. Organomet. Chem.* **76**, 203 (1974).
41. F. Gerstner and J. Weidlein, *Z. Naturforsch. B: Anorg. Chem., Org. Chem.* **33B**, 24 (1978).
42. H. D. Hausen, F. Gerstner, and W. Schwarz, *J. Organomet. Chem.* **145**, 277 (1978).
43. A. Arduini and A. Storr, *J. Chem. Soc., Dalton Trans.*, 503 (1974).
44. D. F. Rendle, A. Storr, and J. Trotter, *Can. J. Chem.* **53**, 2930 (1975).
45. D. Boyer, R. Gassend, and J. C. Maire, *J. Organomet. Chem.* **215**, 157 (1981).
46. B. Walther, A. Zschunke, B. Adler, A. Kolbe, and S. Bauer, *Z. Anorg. Allg. Chem.* **427**, 24 (1976).
47. H. Köppel, J. Dallorso, G. Hoffmann, and B. Walther, *Z. Anorg. Allg. Chem.* **427**, 24 (1976).
48. O. T. Beachley, Jr. and K. C. Racette, *Inorg. Chem.* **15**, 2110 (1976).
49. K. Urata, T. Yogo, K. Itoh, and Y. Ishii, *Bull. Chem. Soc. Jpn.* **47**, 2709 (1974).
50. H. Schmidbaur, K. Schwirten, and H.-H. Pickel, *Chem. Ber.* **102**, 564 (1969); H. Schrem and J. Weidlein, *Z. Anorg. Allg. Chem.* **465**, 109 (1980).
51. K. Urata, K. Itoh, and Y. Ishii, *J. Organomet. Chem.* **66**, 229 (1974).
52. F. Gerstner, W. Schwarz, H. D. Hausen, and J. Weidlein, *J. Organomet. Chem.* **175**, 33 (1979).
53. H. U. Schwering and J. Weidlein, *J. Organomet. Chem.* **84**, 17 (1975).
54. P. Fischer, R. Gräf, J. J. Stezowski, and J. Weidlein, *J. Am. Chem. Soc.* **99**, 6131 (1977).
55. T. Halder, W. Schwarz, J. Weidlein, and P. Fischer, *J. Organomet. Chem.* **246**, 29 (1983).
56. M. Veith, *Angew. Chem.* **99**, 1 (1987); *Angew. Chem., Int. Ed. Engl.* **26**, 1 (1987).
57. C. Lecompte, J. Protas, P. Cocolios, and R. Guilard, *Acta Crystallogr., Sect. B* **B36**, 2769 (1980).
58. W. Fink, *Helv. Chim. Acta* **47**, 498 (1964).
59. L. Tansjö, *Acta Chem. Scand.* **13**, 35 (1959).
60. M. Veith and R. Rösler, *J. Organomet. Chem.* **229**, 131 (1982).
61. M. Veith and J. Böhnlein, *Chem. Ber.* **122**, 603 (1989).
62. M. Veith, F. Goffing, and V. Huch, *Z. Naturforsch. B: Chem. Sci.* **43B**, 846 (1988).
63. M. Veith, H. Lange, A. Belo, and O. Recktenwald, *Chem. Ber.* **118**, 1600 (1985).
64. M. Veith, H. Lange, O. Recktenwald, and W. Frank, *J. Organomet. Chem.* **294**, 273 (1985).
65. A. Spaniol, Ph.D Thesis, University of Saarbrücken, 1988.
66. M. Veith and J. Pöhlmann, *Z. Naturforsch. B: Chem. Sci.* **43B**, 505 (1988).
67. C. Ruloff, Ph.D. Thesis, University of Saarbrücken, 1988.
68. M. Veith, F. Goffing, and V. Huch, *Chem. Ber.* **121**, 943 (1988).
69. S. Becker, Ph.D Thesis, University of Saarbrücken, 1989.
70. M. Veith and A. Belo, *Z. Naturforsch B: Chem. Sci.* **42B**, 525 (1987).
71. J. F. Malone and W. S. McDonald, *J. Chem. Soc., Dalton Trans.*, 2646 (1972).
72. M. Veith and A. Lengert, to be published.
73. M. Veith and M. Harine, to be published.
74. H. Lehnkuhl, K. Mehler, R. Benn, A. Rufinska, and C. Krüger, *Chem. Ber.* **119**, 1054 (1986).
75. R. Gryning and J. L. Atwood, *J. Organomet. Chem.* **137**, 101 (1977).
76. J. L. Atwood and G. D. Stucky, *J. Am. Chem. Soc.* **91**, 2538 (1969).
77. H. G. Richey and B. A. King, *J. Am. Chem. Soc.* **104**, 4672 (1982).
78. H. G. Richey and D. M. Kushlau, *J. Am. Chem. Soc.* **109**, 2510 (1987).
79. A. D. Pajerski, M. Parnez, and H. G. Richey, *J. Am. Chem. Soc.* **110**, 2660 (1988).
80. E. P. Squiller, R. R. Whitle, and H. G. Richey, *J. Am. Chem. Soc.* **107**, 432 (1985).
81. M. Veith, *Phosphorous, Sulfur and Silicon* **41**, 195 (1989).
82. V. M. Waggoner, H. Hope, P. P. Power, *Angew. Chem.* **100**, 1765 (1988); *Angew. Chem. Int. Ed. Engl.* **27**, 1988 (1988).

ADVANCES IN ORGANOMETALLIC CHEMISTRY, VOL. 31

Heteronuclear Clusters Containing Platinum and the Metals of the Iron, Cobalt, and Nickel Triads

LOUIS J. FARRUGIA

Department of Chemistry
University of Glasgow
Glasgow G12 8QQ, Scotland

I

INTRODUCTION

The subject of heteronuclear cluster compounds of the transition metals remains an active area of research interest, and was reviewed in the early 1980s by Geoffroy *et al.* (*1,2*). Clusters with novel architectures, exemplified by the "star" clusters of Stone and co-workers (*3*), continue to be synthesized. Whereas there is undoubtedly strong academic interest in the structure, bonding, and chemical reactivity of heteronuclear clusters in their own right, additional impetus to this field is given by the important relationship between heteronuclear clusters and bimetallic alloy catalysts. This relationship was the subject of a published symposium (*4*).

This article, which reviews the literature up to December 1989, is intended to be a comprehensive account of one particular subclass of heteronuclear molecular clusters, viz. those containing platinum and at least one Group VIIIA (Groups 8–10) metal atom. This subclass contains many examples of small clusters with three or four metal atoms, and several high-nuclearity clusters e.g., $[Fe_4Pt_6(CO)_{22}]^{2-}$ (*5*), $Os_6Pt_2(CO)_{16}(COD)_2$ (*6,7*), and $[Rh_{11}$-$Pt_2(CO)_{24}]^{3-}$ (*8*), as well as the largest structurally characterized clusters $[Ni_{38}Pt_6(CO)_{48}H_{6-n}]^{n-}$, $n = 4, 5$ (*9*), containing 44 transition metal atoms. Some of the earliest reported (1971) examples of heteronuclear clusters were hetero-Pt clusters of the Fe triad (*10*). The availability of firm structural data is of considerable importance in cluster chemistry, particularly with reference to metal core structures, and a substantial number of hetero-Pt clusters have been crystallographically characterized. For the purpose of this review, only compounds containing polyhedral, linked polyhedral, or planar arrangements of three or more directly bonded metal atoms will be considered as clusters, although some mention will be made of compounds containing more open structures and chain structures. In addition to the compounds discussed here,

301

there are numerous examples of hetero-Pt clusters containing Group 6 transition metals, and also a few Pt–Group 7 clusters. The interested reader is referred to several lead-in papers (*11–15b*).

One underlying reason for the interest in heteronuclear clusters lies in their potential for site-specific reactivity at chemically differing metal atoms. As pointed out by Poë (*16*), our understanding of cluster reactivity in general, and of heteronuclear cluster reactivity in particular, is still at the "toddler" stage. Nevertheless, the available data do suggest that site specificity is an important facet of the reactivity of heteronuclear clusters, particularly in their substitution chemistry. Platinum is well known to exhibit both 16- and 18-electron counts in its stable mononuclear organometallic compounds (*17,18*), and the interconversion between these electron counts is of vital importance in catalysis. The change in electron count between 16 and 18 has geometrical consequences, because 16-electron Pt compounds are almost invariably planar [either square–planar for Pt(II) or trigonal–planar for Pt(0)], whereas the 18-electron configuration is associated with nonplanar coordination spheres, e.g., tetrahedral, square–pyramidal, octahedral, or trigonal–bipyramidal. The consequences of *formal* 16- or 18-electron counts at Pt are evident in the localized coordination geometries of this metal in its heteronuclear clusters, and are responsible for some of the differing core geometries observed. Hetero-Pt clusters might therefore be expected to display a unique reactivity, centering on the ability of the Pt atom to adopt either an 18- or 16-electron count. The subject of homonuclear platinum clusters has been previously reviewed (*19–21*), and the readers' attention is also drawn to recent reviews covering material relevant to this article, including reviews on high-nuclearity carbonyl clusters (*22*), electrochemistry of cluster compounds (*23*), butterfly clusters (*24*), osmium sulfido clusters (*25*), and triosmium clusters (*26*).

II

SYNTHESIS

Rational synthetic routes to heteronuclear clusters are still not as freely available as might be desired. Geoffroy *et al.* (*1,2*) have discussed the general synthetic routes to heteronuclear clusters, and these apply equally to Pt-containing clusters. For a detailed discussion on synthetic strategies and on the methods used in characterizing clusters, the interested reader is directed to earlier reviews (*1,2*). Specific synthetic reactions are discussed under the individual compounds in Section V, but some examples of the most important general synthetic routes are given below.

A. Reaction of Nucleophilic Zero-Valent Platinum Compounds with Metal Carbonyls

Zero-valent Pt compounds such as $Pt(COD)_2$ (COD, cycloocta-1,5-diene), $Pt(C_2H_4)_3$, $Pt(C_2H_4)_{3-n}(PR_3)_n$ ($n = 1,2$), and $Pt(PR_3)_4$ are highly reactive and have been used to synthesize a large number of clusters (11,12,27–29). It is generally assumed that coordinatively unsaturated species, formed by loss of coordinated olefins [or PR_3 in the case of $Pt(PR_3)_4$], are the reactive intermediates. There are no detailed kinetic studies on cluster assembly reactions involving Pt, though Powell and co-workers (30) have recently reported NMR evidence for the sequential synthesis of FePt, $FePt_2$, and $FePt_3$ clusters from $Fe(CO)_4(PR_2H)$ and $Pt(C_2H_4)_2(PCy_3)$.

1. Reactions with Saturated Carbonyl Complexes

Pt(0) compounds are often sufficiently reactive to react with coordinatively saturated metal carbonyl complexes, affording species with new Pt—M bonds [Eqs. (1)–(6)] (7,31–35).

$$[Fe_2(\mu\text{-}H)(CO)_8]^- + 2Pt(C_2H_4)_2(PPh_3) \xrightarrow[0.5 \text{ hr, } 70\%]{Et_2O, RT} [Fe_2Pt_2(\mu\text{-}H)(CO)_8(PPh_3)_2]^- \quad (1)$$

$$Fe_3(\mu\text{-}H)(\mu\text{-}COMe)(CO)_{10} + Pt(C_2H_4)_2(PPh_3) \xrightarrow[50\%]{Et_2O, RT}$$

$$Fe_3Pt(\mu_3\text{-}H)(\mu_3\text{-}COMe)(CO)_{10}(PPh_3) \quad (2)$$

$$Os_6(CO)_{16}(NCMe)_2 + 2Pt(COD)_2 \xrightarrow[18 \text{ hr, } 15\%]{CH_2Cl_2, RT} Os_6Pt_2(CO)_{16}(COD)_2 \quad (3)$$

$$Os_3(\mu_3\text{-}S)(CO)_{10} + Pt(PMe_2Ph)_4 \xrightarrow[RT, 1 \text{ hr}]{THF} \begin{array}{l} Os_3Pt(\mu_3\text{-}S)(CO)_{10}(PMe_2Ph)_2 \quad (15\%) \\ + Os_3Pt(\mu_3\text{-}S)(CO)_9(PMe_2Ph)_3 \quad (20\%) \\ + Os_3Pt(\mu_3\text{-}S)(CO)_9(PMe_2Ph)_2 \quad (2\%) \\ + Os_3Pt(\mu_3\text{-}S)(CO)_8(PMe_2Ph)_3 \quad (26\%) \end{array} \quad (4)$$

$$Ir(CO)_2(\eta\text{-}C_5Me_5) + Pt(C_2H_4)_3 \xrightarrow[0.5 \text{ hr, } 100\%]{Et_2O, 0°C} Ir_3Pt_3(CO)_6(\eta\text{-}C_5Me_5)_3 \quad (5)$$

$$Co_2(\mu\text{-}CH_2)(CO)_2(\eta\text{-}C_5H_5)_2 + Pt(C_2H_4)(PPh_3)_2 \xrightarrow[16 \text{ hr}]{THF \text{ reflux}}$$

$$\begin{array}{l} CoPt(\mu\text{-}CH_2)(CO)(PPh_3)_2(\eta\text{-}C_5H_5) \quad (68\%) \\ + Co_2Pt(CO)_2(PPh_3)_2(\eta\text{-}C_5H_5)_2 \quad (4\%) \end{array} \quad (6)$$

The use of $Pt(PR_3)_4$ reagents usually results in several products being formed
(10,33,36–39a) due to the high reactivity of the liberated phosphine. In addition, with the phosphine(ethylene) Pt reagents, it is often found that a carbonyl
ligand is abstracted from the metal carbonyl complex, so that the final products contain $Pt(CO)PR_3$ groups (31,32,36,37,40–45). The products of these
reactions are not always those expected on a rational basis cf. [Eq. (6)].

2. Reactions with Unsaturated Carbonyl Complexes

The addition of transition metal fragments ML_n (L = two-electron donor
ligand) across formally unsaturated metal–metal or metal–carbon bonds is
a well-developed synthetic route to heteronuclear clusters (1,2,11,12,27) and
has received theoretical justification from Hoffmann's isolobal principle (46).
The addition of a PtL_2 fragment across an M=M double bond may be considered as analogous to the reaction of a carbene with an olefin, resulting
in a cyclopropane. The use of isolobal analogies in the directed synthesis of
heteronuclear clusters has been reviewed (11,12,27).

High-yield syntheses of Os_3Pt clusters are available using this rationale.
Thus treatment of $Os_3(\mu\text{-H})_2(CO)_{10}$, which contains a formal Os=Os double bond, with $Pt(C_2H_4)_2(PR_3)$ affords the tetrahedral 58-electron cluster
$Os_3Pt(\mu\text{-H})_2(CO)_{10}(PR_3)$ (1) (40,40a), whereas with the bis-phosphine reagent

1   **2**   **3** LL'=COD   **5** LL'=$(C_2H_4)(PPh_3)$

4 LL'=$(CO)_2$ **6** LL'=$(CO)(PPh_3)$

$Pt(C_2H_4)(PPh_3)_2$ the 60-electron butterfly cluster $Os_3Pt(\mu\text{-H})_2(CO)_{10}(PR_3)$-
(PR'_3) (**2a**, R = R' = Ph) is formed (47). Similarly, reaction of $Rh_2(\mu\text{-CO})_2$-
Cp^*_2 ($Cp^* = \eta\text{-}C_5Me_5$), which contains a formal Rh=Rh double bond, with
$Pt(COD)_2$, $Pt_2(CO)_2Cp_2$ (Cp = $\eta\text{-}C_5H_5$), $Pt(C_2H_4)_2(PPh_3)$, or $Pt(C_2H_4)$-
$(PPh_3)_2$, affords, respectively, the Rh_2Pt clusters $Rh_2Pt(\mu_3\text{-CO})_2(L)(L')Cp^*_2$
3, (L)(L') = COD; **4**, (L)(L') = $(CO)_2$; **5**, (L)(L') = $(C_2H_4)(PPh_3)$; and **6**,

$(L)(L') = (CO)(PPh_3)]$ *(45,48–50)*. The related unsaturated nitrosyl species $[Rh_2(CO)(NO)Cp_2^*]^+$ reacts with $Pt(COD)_2$ in an analogous fashion *(45)*.

B. *Reactions of Carbonylate Anions with Halide Complexes*

The nucleophilic displacement of halide ions from M—X bonds by carbonylate anions (either mononuclear or polynuclear) is a general synthetic route to metal–metal bonded species *(1,2)*, and numerous hetero-Pt clusters have been obtained in this way. The resulting products are not often those of expected stoichiometry, although under optimized experimental conditions this method can provide very useful syntheses, particularly of high-nuclearity clusters. Some examples are shown in Eqs. (7)–(11) *(5,51–54)*.

$$[Fe_3(CO)_{11}]^{2-} + [PtCl_4]^{2-} \xrightarrow[70°C, 4\ hr]{CH_3CN} [Fe_3Pt_3(CO)_{15}]^{2-} \tag{7}$$

$$[Fe_3(CO)_{11}]^{2-} + [PtCl_4]^{2-} \xrightarrow[4\ hr,\ 70-80\%]{CH_3CN,\ 0°C} [Fe_4Pt(CO)_{16}]^{2-} \tag{8}$$

$$cis\text{- or }trans\text{-}PtCl_2(PPh_3)_2 + [Co(CO)_4]^- \xrightarrow[4\ hr,\ 22\%]{THF,\ 0°C} Co_2Pt_2(CO)_8(PPh_3)_2 \tag{9}$$

$$cis\text{-}PtCl_2(PEt_3)_2 + [Co(CO)_4]^- \xrightarrow[RT,\ 5\ hr]{THF} \begin{array}{l} Co_2Pt_2(CO)_8(PEt_3)_2 \quad (1\%) \\ + Co_2Pt_3(CO)_9(PEt_3)_3 \quad (63\%) \end{array} \tag{10}$$

$$[Pt_{12}(CO)_{24}]^{2-} + Ir(CO)Cl(PPh_3)_2 \xrightarrow[3\ hr,\ 50\%]{CH_3CN,\ RT} Ir_2Pt_2(CO)_7(PPh_3)_3 \tag{11}$$

Related synthetic routes include the "redox condensation" reaction of carbonylate anions with neutral carbonyls, e.g., Eq. (12) *(55)*, metal exchange reactions between carbonylate anions, e.g., Eq. (13) *(56)*, and direct reductive carbonylation of metal halides, e.g., Eq. (14) *(57)*. The stoichiometry of the products are not rational, and the mechanisms clearly are very complicated, though once again these reactions, under experimental optimization, can provide very useful synthetic routes.

$$[Rh_4Pt(CO)_{14}]^{2-} + Rh_4(CO)_{12} \xrightarrow[48\ hr,\ 34\%]{acetone,\ RT} [Rh_8Pt(CO)_{19}]^{2-} \tag{12}$$

$$[Pt_{12}(CO)_{24}]^{2-} + [Rh_{12}(CO)_{30}]^{2-} \xrightarrow[RT]{THF} [Rh_5Pt(CO)_{15}]^- \tag{13}$$

$$[PtCl_6]^{2-} + IrCl_3 + OH^- + CO \xrightarrow[24\ hr,\ 45\%]{MeOH,\ RT} [Ir_4Pt(CO)_{14}]^{2-} \tag{14}$$

III

BONDING AND ELECTRON COUNTING

Due to the difficulties of obtaining realistic *ab initio* MO calculations on even small to medium-sized transition metal clusters, efforts to understand their electronic structure have focused on more intuitive approaches based on symmetry arguments and semiempirical MO methods. Various electron counting schemes (*58*), such as the skeletal electron pair theory (SEPT) (*59–63*) and the topological electron counting theory (TECT) (*64–68*), have been proposed; these relate the number of cluster valence electrons (CVEs) to the metal skeletal geometry. Within the fragment formalism, the important cluster bonding and antibonding MOs can be envisaged as arising from symmetry-restricted interactions between the frontier orbitals of individual ML_n units. The generalized interactions between the radial and tangential MOs of conical ML_n units resulting in deltahedral structures have been elegantly analyzed by Stone (*69–71*) using tensor surface harmonic theory. This approach has been extended by Mingos and co-workers to include capped polyhedral structures (*72*), four-connected polyhedra (*63*), bispherical clusters (*73*), and, recently, clusters with cylindrical topologies (*74*).

The most common "building blocks" in hetero-Pt clusters are the nonconical fragments PtL, C_{2v}-PtL_2, and C_{2v}-PtL_3^+. The frontier orbitals of these fragments have been described by Mingos and co-workers (*75–77*) and are shown in Fig. 1. For the PtL unit, those orbitals of predominantly d character are filled, and the bonding characteristics are mainly determined by the s–z hybrid and the higher lying degenerate e set (p_x and p_y) (*75*). These latter orbitals are generally too high-lying to contribute significantly to cluster bonding, so that the PtL fragment may be considered as isolobal with H^+. For the PtL_2 unit, the high-lying b_1 orbital is likewise not usually involved in cluster bonding, so that this fragment may be considered as isolobal with CH_2 (*76*). When one of the donor ligands L is a π acid such as in the common fragment $Pt(CO)(PR_3)$, an additional low-lying pseudo-b_1 acceptor orbital is also present (*77a*). The C_{2v}-PtL_3^+ moiety is isolobal with H or CH_3, and the b_2 orbital is not involved in cluster bonding (*77*).

An important consequence of the nonutilization of tangential orbitals is that platinum clusters often do not obey the normal electron counting rules and appear to be electron deficient (*19,21,29,58,75,76*). Electron counts are usually intermediate between those found in "normal" transition metal clusters (*58–68*) and those observed in gold clusters (*58,78*), but no satisfactory general electron counting theory has been developed for Pt-containing clusters. In small Pt clusters constructed from PtL_2 units, theoretical studies have shown that the total electron count depends on the relative orientation of the

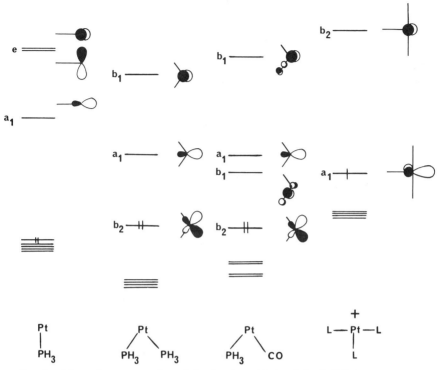

FIG. 1. Schematic representation of the frontier orbitals of the $Pt(PH_3)$, C_{2v}-$Pt(PH_3)_2$, $Pt(CO)(PH_3)$, and C_{2v}-PtL_3^+ fragments.

fragments (76), and may also depend on the nature of any bridging ligands (77a). Rules for electron counts in condensed polyhedral clusters have been adapted for Pt-containing clusters, assuming the Pt atom utilizes only eight valence orbitals (79).

Although the vast majority of homonuclear Pt clusters appear electron deficient according to SEPT or TECT, there are a significant number of hetero-Pt clusters in which the Pt atom appears to adopt an 18-electron configuration and the expected CVE count according to SEPT is actually observed. Examples include 60-electron tetrahedral clusters, e.g., $Fe_3Pt(\mu_3$-H)-$(\mu_3$-COMe$)(CO)_{10}(PPh_3)$ (7) (32) and $Os_3Pt(\mu$-H$)_2(\mu$-CO$)(CO)_9(COD)$ (8) (80), 62-electron butterfly clusters, e.g., $Os_3Pt(\mu_3$-S$)(CO)_{10}(PMe_2Ph)_2$ (9) (33), 64-electron spiked-triangular clusters, e.g., $Os_3Pt(\mu_3$-S$)_2(CO)_9(PMe_2Ph)_2$ (10) (39), 72-electron trigonal–bipyramidal clusters, e.g., $[M_4Pt(CO)_{12}]^{2-}$ [11a, M = Rh (81) and 11b, M = Ir (57)], and 86-electron octahedral clusters, e.g.,

7   **8**   **9**

10   **11a**

$Fe_5Pt(\mu_6-C)(CO)_{15}(PPh_3)$ (**12**) (see later, Fig. 12, for **12**) (*82*). In all these clusters the coordination geometry about the Pt center is nonplanar, assuming *nonbridged* Pt–M vectors are included in the coordination sphere. Several observations about electron counts in hetero-Pt clusters can be made:

1. In triangular clusters in which the Pt atoms have planar coordination geometries, they behave as 16-electron centers, giving rise to 46 CVEs for clusters containing one Pt atom and 44 CVEs for those with two Pt atoms. When face-capping ligands are present, resulting in nonplanar coordination geometries at Pt, then both electron-precise structures,

13   **14**   **15**

e.g., in the 48-CVE cluster $WFePt(\mu_3\text{-}CR)(\mu\text{-}CO)(CO)_5(PMe_2Ph)_2Cp$
(**13**) (*83,83a*), or electron-deficient structures, e.g., in the 46-CVE cluster
$WFePt(\mu_3\text{-}CR)(CO)_6(PEt_3)Cp$ (**14**) (*83,83a*), may be observed.

2. In tetrahedral clusters the Pt atoms have, of necessity, nonplanar coordination geometries and usually obey SEPT. Exceptions include cluster 1
(*40,40a*), which is formally electron deficient, and $FePt_3(\mu\text{-}H)(\mu\text{-}PnPr_2)$-
$(\mu\text{-}CO)_2(PtBu_2Ph)_3$, a 56-CVE tetrahedron that is Pt rich (*30*). On the
other hand, butterfly clusters with planar Pt centers give rise to 60 CVEs
for clusters with one Pt atom, e.g., **2a** (*47*), and 58 CVEs for those with
two Pt atoms, e.g. $Os_2Pt_2(\mu\text{-}H)_2(CO)_8(PR_3)_2$ (**15**) (*41,84*).

3. In high-nuclearity clusters containing deltahedral or condensed polyhedral skeletons, the CVE count predicted by SEPT is often observed.
Some unusual core geometries may be found, e.g., in $Os_6Pt_2(\mu\text{-}H)_2(\mu\text{-}$
$C_8H_{10})(CO)_{16}(COD)$ (*85*), which are less straightforward to rationalize.

4. When a nonplanar PtL_n unit such as the conical fragment $Pt(\eta^5\text{-}C_5H_5)$
is present, available evidence suggests that all tangential orbitals are
utilized, and "normal" electron counts are observed, e.g., in the 60-CVE
tetrahedral clusters $Co_3Pt(\mu\text{-}CO)_3(CO)_6Cp^*$ (**16**) (*86*) and $[Pt_4(\mu_3\text{-}CO)\text{-}$
$(CO)_2Cp_3^*]^+$ (*87*). Similar effects are seen for the isolobal PdCp* frag-
ment; the cluster $Pd_2Co_2(\mu\text{-}CO)_4(CO)_4Cp_2^*$ (*86*) has a 64-CVE
count and a planar butterfly skeleton with four metal–metal bonds (*24*).

16   **17a**   **17b**

18   **19**   **22**

The factors responsible for the apparent utilization of all tangential orbitals by nonconical PtL_2 and PtL_3^+ fragments in certain clusters, and not in others, are not well understood. The nature of the ancilliary ligands is evidently of crucial importance. Recent MO calculations (88) suggest that π-acceptor ligands that bridge the M—M bond in 60-CVE M_3Pt clusters may stabilize the tetrahedral geometry. This is borne out by the observed structures of the 60-CVE clusters **8** (80), both isomers of $Os_3Pt(\mu\text{-H})_2(\mu\text{-CH}_2)(CO)_{10}(PCy_3)$ (**17a,b**) ($89,89a$) and $Os_3Pt(\mu\text{-H})_2(\mu\text{-SO}_2)(CO)_{10}(PCy_3)$ (**18**) (90). Nevertheless, closely related 60-CVE Os_3Pt clusters such as $Os_3Pt(\mu\text{-H})_2(CO)_{11}(PCy_3)$ (**19**) ($89a,90a$) and cluster **2** ($47,91$) are found to possess butterfly metal frameworks. It is likely that the potential energy surface connecting the tetrahedral and butterfly Os_3Pt skeletal geometries in 60-CVE clusters is shallow, and, as discussed in Section IV,D, the fluxional behavior of **19** is consistent with a facile interchange between such geometries. These phenomena may also be related to the "slip" distortion of the PtL_2 unit observed in several platina-carbaboranes (92).

Rather few MO studies on specific hetero-Pt systems have been reported. The anionic raft clusters $[Fe_3Pt_3(CO)_{15}]^{n-}$ (**20a**, $n = 2$; **20b**, $n = 1$) (5) shown in Fig. 2 may be treated as derivatives of the well-known $[Pt_3(\mu\text{-L})_3(L')_3]$ system ($93,94$). Calculations by Hoffmann et al. (94) and Evans and Mingos (95) show that the HOMO of the 86-electron dianion **20a** is of a_2' symmetry. This a_2' orbital is Pt–Pt antibonding (Fig. 2), but is stabilized by extensive Fe–Pt bonding through the bridging $Fe(CO)_4$ groups. ESR and diffuse reflectance studies on the paramagnetic monoanion **20b** (96) are consistent with a SOMO of a_2' symmetry primarily derived from Pt atomic orbitals, and the X-ray structures (5) show shorter Pt–Pt separations in the monoanion relative

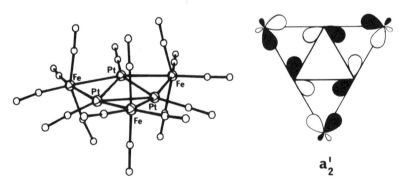

a_2'

Fig. 2.   Structure of the anion $[Fe_3Pt_3(CO)_{15}]^{n-}$, $n = 2, 1, 0$, and the a_2' frontier orbital.

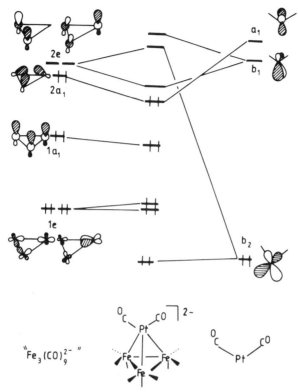

FIG. 3. EHMO interaction diagram for the hypothetical dianion $[Fe_3Pt(CO)_{11}]^{2-}$.

to the dianion [average Pt–Pt distances are 2.656(1) and 2.750(1) Å, respectively]. Recently Adams *et al.* (97) have synthesized the 84-electron neutral analogue (**20c**; $n = 0$) and the average Pt–Pt distance of 2.590(2) Å is shorter still, consistent with this analysis.

Schilling and Hoffmann (98) have described the interaction of the PtL_2 fragment with $[Fe_3(CO)_9]^{2-}$ (Fig. 3) that affords the hypothetical dianion $[\{PtL_2\}Fe_3(CO)_9]^{2-}$, a model for cluster **1**. A low-lying LUMO was observed, consistent with the chemical unsaturation of **1** (Section V,C,2). Because the main interactions involve one of a_1 symmetry, which is invariant to rotation, and a b_2 orbital on the PtL_2 unit, which can interact with either component of the cluster $2e$ set, a low barrier (~ 25 kJ mol^{-1}) to rotation of the PtL_2 unit was predicted. Evidence for this rotation is discussed in Section IV,A. In a related MO study, Ewing and Farrugia (88) have shown that the principal result of protonation of the dianion (affording a dihydride) is the reversal in the relative energies of the HOMO and LUMO. The LUMO in **1** is predicted to be localized about the hydride-bridged Os—Os bond, and to be strongly Os—Os antibonding.

The 76- and 72-CVE trigonal–bipyramidal anionic clusters $[M_4Pt(CO)_{14}]^{2-}$ and $[M_4Pt(CO)_{12}]^{2-}$ [M = Rh (*81*) and M = Ir (*57*)] provide interesting examples of the variable CVE count for this polyhedron. Simple SEPT predicts that the trigonal bipyramid, a *closo* deltahedron, should have 72 CVEs, although there are many examples of trigonal–bipyramidal transition metal clusters with CVE counts of 76 (*58*). Theoretical studies by Johnston and Mingos (*99*) show that for bipolar deltahedra such as the trigonal bipyramid, an elongation along the threefold axis results in stabilization of the doubly degenerate $\bar{D}^\pi$, so that an extra four electrons may be accommodated. Essentially similar results are obtained by Teo's TEC theory (*64–66*), wherein a variable value of the adjustment parameter X (either 0 or 2) also allows for CVE counts of 72 or 76. Occupation of the $\bar{D}^\pi$ orbitals in the 76-CVE clusters implies longer basal–apical M—M bonds than for 72 CVE systems, because these orbitals are antibonding between the basal and apical atoms (*99*). The crystal structures of the 76- and 72-CVE clusters (M = Ir) (*57*), which show mean Ir_{ap}–Ir_{bas} distances of 3.025(1) and 2.840(5) Å, respectively, are consistent with this analysis.

IV

FLUXIONAL BEHAVIOR AND ISOMERIZATION REACTIONS

Stereochemical nonrigidity is a fundamental property of transition metal clusters (*100,101*), and is normally only manifest on the nuclear magnetic resonance (NMR) time scale, though recently carbonyl exchange on the infrared time scale in $Os_4(CO)_{14}$ has been proposed (*102*). NMR studies on Pt-containing clusters are greatly aided by the presence of the spin-1/2 Pt isotope [195]Pt (33.8% natural abundance), and its high relative receptivity (3.4×10^{-3} relative to ^{1}H) enables easy direct observation. [195]Pt chemical shifts in homonuclear Pt clusters have been previously tabulated by Mingos and Wardle (*19*), and those for hetero-Pt clusters are given in Table 1. They cover a wide range ($\sim +1700$ to -1400 ppm for the hetero clusters), but are unfortunately of relatively little diagnostic use. To date direct observation of [195]Pt has played little role in studies on the dynamic behavior of hetero-Pt clusters, but has proved useful for homo-Pt clusters (*102a*). However, J couplings to [195]Pt (and also to nuclei such as [187]Os and [103]Rh) have proved more useful.

TABLE I
^{195}Pt Chemical Shift Data for Hetero-Platinum Clusters

Complex	$\delta(Pt)^a$	$J(Pt-P)$	$J(Pt-X)^b$	Ref.
Fe$_3$Pt(μ_3-H)(μ_3-COMe)(CO)$_{10}$(PPh$_3$)	-116.2	2598	—	32
FePt$_2$(μ-dppm)$_2$(CO)$_4$	790	3186	(Pt) 1730	136
	497	4387, 2923	—	
Ru$_2$Pt(μ-CH$_2$)(CO)$_3$(PR$_3$)Cp$_2$				116
R = Cy	794.4	3433	—	
R = iPr	770.5	3462	—	
Ru$_2$Pt$_2$(μ-H)(μ_4-CH)(CO)$_3$(PR$_3$)$_2$Cp$_2$				116
R = Cy	330.5	5659, 281	(Pt) 2080	
	572.7	3964, 342	—	
R = iPr	298.6	5650, 287	(Pt) 2075	
	588.2	4023, 347	—	
Ru$_2$Pt$_2$(μ-H)$_2$(μ_4-C)(CO)$_2$(PR$_3$)$_2$Cp$_2$				116
R = Cy	-723.9	5186	—	
R = iPr	-763.4	5227	(Pt) 365	
Os$_3$Pt(μ-H)$_2$(μ-CH$_2$)(CO)$_{10}$(PCy$_3$)				89a
(C_s isomer)	1045	2513	—	
(C_1 isomer)	172	2661	—	
Co$_2$Pt(μ-dppm)(CO)$_7$	386	3294, 47	—	135
Rh$_2$Pt(CO)$_4$Cp$_2^*$	51.8	—	(Rh) 15	45
Rh$_2$Pt(CO)$_3$(PPh$_3$)Cp$_2^*$	-5.8	4004	(Rh) 27, 6	45
Rh$_2$Pt(CO)$_2$(COD)Cp$_2^*$	934	—		45
[Rh$_2$Pt(CO)(NO)(COD)Cp$_2^*$]$^+$	988.7	—	(Rh) 21	45
Rh$_4$Pt(CO)$_4$Cp$_4^*$	288	—	(Rh) 68	50
[Rh$_4$Pt(CO)$_{14}$]$^{2-}$	255.6	—	(Rh) 55, 69	81
[Rh$_4$Pt(CO)$_{12}$]$^{2-}$	421	—	—	81
[Rh$_5$Pt(CO)$_{15}$]$^-$	36.2	—	(Rh) 24, 73	81
[Ir$_4$Pt(CO)$_{14}$]$^{2-}$	799	—	(C) 2383, 479, 90	57
[Ir$_4$Pt(CO)$_{12}$]$^{2-c}$	950	—	(C) 2327	57
Ir$_3$Pt$_3$(CO)$_6$Cp$_3^*$	-874.9	—	—	34
[Ni$_3$Pt$_3$(CO)$_{12}$]$^{2-}$	20.7	—	—	102a
[Ni$_9$Pt$_3$(CO)$_{21}$]$^{4-}$	185	—	—	120
[Ni$_9$Pt$_3$(CO)$_{21}$H]$^{3-}$	-222	—	(H) 260.4	120
[Ni$_9$Pt$_3$(CO)$_{21}$H$_2$]$^{2-}$	-323	—	(H) 201.1	120
[Pd$_2$Pt(μ-Cl)(μ-PPh$_2$)$_2$(PPh$_3$)$_3$]$^+$	-1406.7	4037, 2621, 212.6	—	195
FeWPt(μ_3-CR)(CO)$_6$(PMe$_2$Ph)$_2$Cp^d	-390.7	3339	—	83,83a
	-477.3^e	3646, 2991	—	
FeWPt(μ_3-CR)(CO)$_6$(PMe$_3$)Cp^d	169.6	3650	—	83,83a
RuW$_2$Pt(μ-CR)(μ_3-CR)(CO)$_7$Cp$_2^d$	1057	—	(W) 75	210
Ru$_2$W$_2$Pt(μ_3-CR)$_2$(CO)$_{11}$Cp$_2^d$	1273	—	(Pt) 1787	210
	893	—	—	

(continued)

TABLE I (*continued*)

Complex	$\delta(\text{Pt})^a$	$J(\text{Pt–P})$	$J(\text{Pt–X})^b$	Ref.
$Co_2PdPt(\mu\text{-dppm})_2(CO)_7$	379	3164, 3939	—	198
$CoPdPt(\mu\text{-dppm})_2(CO)_3I$	379	3227, 3845	—	198
$Co_2Rh_2Pt(CO)_4Cp_4^*$	199.8	—	(Rh) 69	110
$Ir_2Rh_2Pt(CO)_4Cp_4^*$	194.5	—	(Rh) 86	110
$NiW_2Pt(\mu\text{-CR})(\mu_3\text{-CR})(CO)_4$ $(COD)Cp_2^d$	457	—	—	206
$Ni_2W_4Pt_2(\mu_3\text{-CR})_2(\mu_3\text{-CR}')_2$ $(CO)_8Cp_4^h$	1383 1289	— —	(W) 69, 165 (W) 88, 136	15a
$Ni_2Mo_2W_2Pt_2(\mu_3\text{-CR})_4(CO)_8Cp_4^d$	1241	—	—	15a
$Ni_2Mo_2W_2Pt_2(\mu_3\text{-CR})_2(\mu_3\text{-CR}')_2$ $(CO)_8Cp_4^g$	1252 1242	— —	(W) 156 —	15a
$Ni_2Mo_2W_2Pt_2(\mu_3\text{-CR})_2(\mu_3\text{-CR}')_2$ $(CO)_8Cp_2Cp_2^{*h}$	1304	—	(W) 158	15a
$Ni_2W_4Pt_2(\mu\text{-CR})(\mu_3\text{-CR}')_3(CO)_8Cp_4^d$ Isomer a	1726 1138	— —	(Pt) 410 (W) 136 (Pt) 410 (W) 117	
Isomer b	1641 1002	— —	(Pt) 234 (W) 420 (Pt) 234 (W) 108	
$Ni_2W_4Pt_2(\mu_3\text{-CR})_4(CO)_8Cp_4^d$	1236	—	—	
$Ni_2W_4Pt_2(\mu\text{-CR})(\mu_3\text{-CR})_3(CO)_8Cp_4^i$ Isomer a	1740 1138	— —	(Pt) 413 (W) 126 (Pt) 413 (W) 100	208
Isomer b	1648 1004	— —	(Pt) 242 (W) 410 (Pt) 242 (W) 121	
$NiW_2Pt(\mu\text{-CR})(\mu_3\text{-CR})(CO)_4$ $(COD)Cp_2^{*d}$	439 481^f	— —	(W) 142 (W) 136	206
$NiMo_2Pt(\mu\text{-CR})(\mu_3\text{-CR})(CO)_4$ $(COD)Cp_2^d$	482	—	—	15a
$NiW_4Pt_2(\mu\text{-CMe})_2(\mu_3\text{-CMe})_2$ $(CO)_8Cp_4^*$	1532 1389	— —	(W) 176 (W) 166, 126	207
$NiMo_2W_2Pt_2(\mu\text{-CR})_2(\mu_3\text{-CR}')_2$ $(CO)_8Cp_4^g$	1652 1355	— —	— —	15a
$NiMo_2W_2Pt_2(\mu\text{-CR})_2(\mu_3\text{-CR}')_2$ $(CO)_8Cp_2Cp_2^{*h}$	1407 1249	— —	— —	15a
$Ni_2Mo_4Pt_2(\mu\text{-CR})(\mu_3\text{-CR})_3(CO)_8Cp_4^d$ Isomer a	1752 1242	— —	(Pt) 430 (Pt) 430	15a

(*continued*

TABLE I (*continued*)

Complex	$\delta(Pt)^a$	$J(Pt-P)$	$J(Pt-X)^b$	Ref.
Isomer b	1595	—	(Pt) 426	
	1147	—	(Pt) 426	
$Ni_2Mo_4Pt_2(\mu_3\text{-}CR)_4(CO)_8Cp_4{}^d$	1245	—	—	15a
$Ni_2W_4Pt_2(\mu_3\text{-}CR)_4(CO)_8Cp_4{}^i$	1245	—	(W) 73, 105	208
$NiW_4Pt_3(\mu\text{-}CR)(\mu_3\text{-}CR)_3(CO)_8Cp_4{}^d$	1604(A)	—	(W) 137	208
	1062(B)	—	(Pt) AB 361	
	904(C)	—	AC 1338, BC 1553	

^{a} The δ values are to high frequency of $\Xi(^{195}Pt) = 21.4$ MHz; data at ambient temperatures unless otherwise stated.
^{b} The X nucleus is given in parentheses.
^{c} At $-40°C$.
^{d} $R = C_6H_4Me\text{-}4$.
^{e} At $-95°C$.
^{f} Minor isomer.
^{g} $R = Ph, R' = C_6H_4Me\text{-}4$.
^{h} $R = Me, R' = C_6H_4Me\text{-}4$.
^{i} $R = Ph$.

Three major classes of fluxional behavior are commonly observed in clusters:

1. Metal localized ligand scrambling, most commonly seen in the $M(CO)_3$ or $M(CO)_2L$ groups. For Pt clusters some examples of PtL_2 localized scrambling have also been observed.
2. Intermetallic ligand migrations, most commonly involving CO or hydrides, and usually thought to proceed via bridging intermediates. Ligands that do not display ground-state bridging modes, e.g., tertiary phosphines, do not generally migrate easily.
3. Metal framework rearrangements.

These three classes are not mutually exclusive, and all have been observed in hetero-Pt clusters. Whereas most examples of fluxional behavior seen in hetero-Pt clusters are common to other systems, there are several interesting examples that appear to be intimately connected with formal 16 ↔ 18-electron conversions at Pt centers.

Modern NMR techniques such as quantitative analysis of multisite exchange using either 1D magnetization transfer experiments (*103*) or the 2D exchange spectroscopy (EXSY) method (*104,105*) promise to be of great help in unraveling the complex stereochemical exchange networks involved in cluster fluxionality. The usefulness of EXSY in the context of this article is illustrated by the phase-sensitive $^{13}C\{^1H\}$ EXSY spectrum (255 K, $t_m = 0.5$ sec)

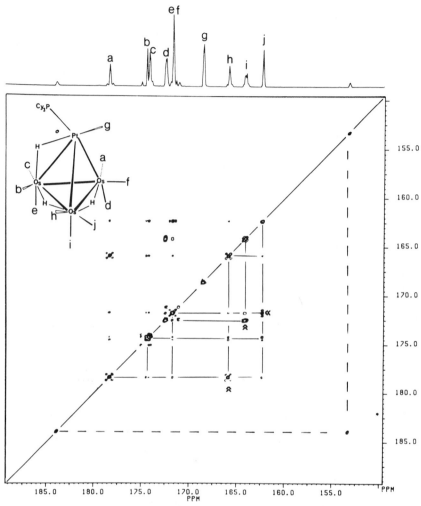

FIG. 4. $^{13}C\{^{1}H\}$ EXSY spectrum of $[Os_3Pt(\mu\text{-}H)_3(CO)_{10}(PCy_3)]^+$ (21) in the carbonyl region.

of $[Os_3Pt(\mu\text{-}H)_3(CO)_{10}(PCy_3)]^+$ (21) shown in Fig. 4 (106). Two exchange processes are manifest: (1) a more rapid process ($k \sim 1.8\ \text{sec}^{-1}$) that gives rise to the more intense cross-peaks (marked by a chevron) and exchanges carbonyls a/h, b/c (not marked), d/i, and f/j; and (2) a slower process ($k \sim 0.15\ \text{sec}^{-1}$) that gives rise to weaker cross-peaks and exchanges all the pseudo-equatorial COs a, b, c, f, h, and j with each other, and the axial COs d, e, and i with each other. Process (1) involves racemization of the cluster by migration of an Os(μ-H)Os hydride to an adjacent unbridged Os—Os edge, and process (2) involves a rotation of the Pt(H)(CO)(PR$_3$) unit, which coupled with the lower energy hydride migration results in the observed CO exchanges. EXSY spectra in general also show exchange between ^{195}Pt satellites (dashed line, Fig. 4), which is due to the relaxation of the ^{195}Pt nucleus.

A. *Rotation and Localized Exchange in PtL $_2$ and PtL $_3$ Groups*

The prediction by Schilling and Hoffmann (*98*) that the barrier to rotation of the PtL$_2$ unit in clusters of the type [{PtL$_2$}M$_3$(CO)$_9$]$^{2-}$ would be low (Section III) has prompted a number of studies into the fluxionality of Os$_3$Pt clusters. Several derivatives of the 58-electron tetrahedral cluster 1 have been studied, and evidence is found for a concerted rotation of the Pt(H)(L)(PR$_3$) group. Unambiguous evidence for the rotation of the Pt(H)(CNCy)(PCy$_3$) unit in Os$_3$Pt(μ-H)$_2$(CO)$_9$(CNCy)(PCy$_3$) (22) has been recently obtained from the dynamic behavior of ^{187}Os J couplings to the hydride ligands, and a value of $\Delta G^{\ddagger} = 56(1)$ kJ mol^{-1} has been estimated (*107*). The observed CO scrambling in 1a (R$_3$ = Cy$_3$) is consistent with a similar process, though exchange between the hydrides complicates the quantitative interpretation (*108*). As stated above, EXSY studies on 21 also indicate that a similar process occurs, with $\Delta G^{\ddagger} = 66(2)$ kJ mol^{-1}.

Variable-temperature ^{13}C studies on Os$_3$Pt(μ-H)$_2$(CO)$_9$(PCy$_3$)(PMe$_2$Ph) (23) (Fig. 5) show that phosphine substitution at an Os center has a marked effect on the fluxional behavior of the system (*106*). In contrast to 1a (*108*), 21 (*106*), and 22 (*107*), cluster 23 shows a low-energy exchange between the Pt—CO and carbonyls in the Os(CO)$_3$ groups. The similarities in the activation barriers to exchange of the diastereotopic methyl groups in the PMe$_2$Ph ligand ($\Delta G^{\ddagger}_{260} = 53.7(7)$ kJ mol^{-1}) and of the diastereotopic carbonyls in the Os(CO)$_2$(PMe$_2$Ph) group ($\Delta G^{\ddagger}_{268} = 53.5(7)$ kJ mol^{-1}) suggest a concerted process involving restricted rotation of the Pt(H)(CO)(PCy$_3$) and Os(CO)$_2$(PMe$_2$Ph) groups and a hydride migration. No convincing evidence for rotation of the Pt(H)(L)(PCy$_3$) group was found in Os$_3$Pt(μ-H)$_2$(μ-CH$_2$)-(CO)$_9$(CNCy)(PCy$_3$) (24) (*107*), or in 18 (*90*), but in these systems a degenerate

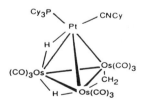

24

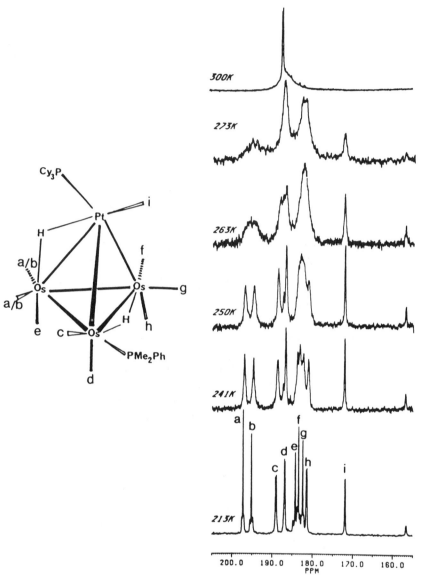

FIG. 5. Variable-temperature $^{13}C\{^1H\}$ NMR spectrum of $Os_3Pt(\mu\text{-}H)_2(CO)_9(PCy_3)$-$(PMe_2Ph)$ (23) in the carbonyl region.

exchange would also require a concerted movement of the bridging CH_2 or SO_2 group to adjacent Os–Os edges, which is presumably a high-energy process. The cluster $Fe_3(\mu_3\text{-}H)(\mu_3\text{-COMe})(CO)_{10}(PPh_3)$ **(7)** shows a single signal for the Fe-bound CO ligands, consistent with rapid rotation of the $Pt(H)(CO)(PPh_3)$ group *(32)*. The orientation of this latter unit in the crystal structure *(32)* corresponds to the postulated transition state for $Pt(H)L_2$ rotation *(107)*, with a μ_3-hydride and the Pt–CO vector eclipsing an Fe–Pt edge. The rotation of the $Pt(H)(L)(PR_3)$ in all these systems appears to require other lower energy hydride or carbonyl migrations.

Similar dynamic behavior is observed for the $Pt(H)(L)(L')$ unit in the triangular clusters $[Rh_2Pt(\mu\text{-}H)(\mu\text{-CO})_2(L)(L')Cp_2^*]^+$ [**25**, $(L)(L') = (CO)_2$; **26**, $(L)(L') = (CO)(PPh_3)$; and **27**, $(L)(L') = COD$] *(48,109)*, with $\Delta G^\ddagger$ values in the range 49.3–58.1 kJ mol^{-1}. Exchange was detected through averaging of $^{103}Rh\text{-}^1H$ J couplings. Interestingly, rotation of the PtL_2 unit is not observed

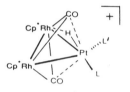 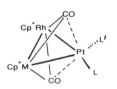

25 LL'=(CO)₂   **26** LL'=(CO)(PPh₃)   **28** M=Co;LL'=COD   **29** M=Ir;LL'=COD

27 LL'=COD   **30** M=Ir;LL'=(ᵗBuNC)₂

in the related clusters **6** *(45)*, $MRhPt(\mu\text{-CO})_2L_2Cp_2^*$ [**28**, M = Co, L_2 = COD *(110)*; **29**, M = Ir, L_2 = COD *(110)*; **30**, M = Ir, L_2 = $(t\text{BuNC})_2$ *(110)*; or $Ir_2\text{-}Pt(CO)_3(PCy_3)(\eta\text{-}C_9H_7)_2$ **(31)** *(111)*]. EHMO calculations by Hoffmann *et al.* *(112)* on $[Rh_3(\mu\text{-CO})_2(CO)_2Cp_2]^-$ indicate that the high barrier (~ 88 kJ mol^{-1}) to rotation of the $Rh(CO)_2$ group arises from an energy mismatch involving the b_2 orbital of the d^{10} $Rh(CO)_2^-$ unit and the $2b_2$ orbital of the

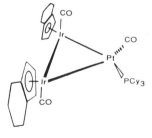

31   **32**

$Rh_2(CO)_2(\eta-C_5H_5)_2$ moiety. Similar considerations should also apply to the above clusters containing the PtL_2 unit, because this is also a d^{10} ML_2 unit. The $Pt(H)L_2$ unit is, however, formally a d^9 ML_3 unit, isolobal with CH_3 (77), and hence rotation barriers involving this group are expected to be much lower. Electronic factors may mitigate against a high barrier to PtL_2 rotation in some systems; for instance, the COD ligand in $[Rh_2Pt(\mu_3-CO)(\mu_3-NO)(COD)Cp_2^*]^+$ (32) is dynamic (45).

Rotation of PtL_2 units has been observed in other systems. The inequivalent olefinic environments of the COD ligands in 8 (80) and $Ru_3Pt(\mu-H)(\mu_4-\eta^2-C\equiv CtBu)(CO)_9(COD)$ (33) (113) undergo exchange, while the inequivalent phosphine environments in 34 also show a slow exchange at ambient

33 LL = COD   34 LL = dppe   35   36a L=PPh₃   36b L=CO

temperatures ($\Delta G_{295}^\ddagger = 63.2(5)$ kJ mol^{-1}) (113). Exchange between the two Pt-bound PMe_2Ph groups in $Os_3Pt(\mu_3-S)(CO)_8(PMe_2Ph)_3$ (35) (33) and in cluster 13 (83a) ($\Delta G_{248}^\ddagger = 46(2)$ kJ mol^{-1}) is presumed to arise from PtL_2 rotation about the pseudo-threefold axes. When the two ligands L are chemically distinct, e.g., in $Pt(CO)(PR_3)$, then PtL_2 rotation is a nondegenerate process. Low-temperature ^{31}P NMR data on cluster 14 show the presence of two isomers, which are in rapid exchange at 25°C due to facile rotation of the $Pt(CO)(PEt_3)$ unit (33).

B. Localized Exchange in other ML_n Groups

Localized exchange in $M(CO)_3$ groups is a very common fluxional process in tri- and tetranuclear clusters of the iron and cobalt triads and is usually lower in energy than intermetallic CO migration. The barrier to exchange generally increases with increasing atomic weight of the metal M. Studies by Rosenberg et al (114) and Hawkes et al. (115) indicate that a concerted pseudo-C_3 (tripodal) rotation is the most likely process. Tripodal rotation of $Os(CO)_3$ groups occurs in $OsPt_2(\mu_3-MeC\equiv CMe)(CO)_5(PPh_3)_2$ (36a) (47), and it is the lowest energy CO migration observed in clusters 1a (108) and 22 (107), with

$\Delta G^{\ddagger} = 49.7(5)$ and $52(1)$ kJ mol^{-1}, respectively. The presence of bridging ligands cis to $M(CO)_3$ groups often raises the barrier to tripodal exchange, and this is seen in several examples. Protonation of **1a** affords **21**, in which tripodal rotation of $Os(CO)_3$ groups was not detectable by either EXSY (*106*) nor DANTE (*108*) experiments. The barrier to exchange for the two hinge $Ru(CO)_3$ groups in $Ru_3Pt(\mu_4-\eta^2-C{=}C(H)t Bu)(CO)_9(dppe)$ (**37**) is lower than in the protonated cluster **38** (*113*) (see later, Scheme 2). Two independent $Os(CO)_3$ tripodal rotations in **24** are detected by EXSY experiments (*107*), but these are relatively high-energy processes ($\Delta G^{\ddagger}_{318} = 81(1)$ and $88(2)$ kJ mol^{-1}). Tripodal rotation of the $Fe(CO)_3$ group in **14** is rapid even at $-100°C$, and a similar facile process scrambles the $Fe(CO)_3$ and $Fe(\mu-CO)Pt$ carbonyls in **13** (*83a*).

Localized "rocking" motions of the $Ir(CO)_2Cp^*$ groups in the raft cluster $Ir_3Pt_3(CO)_6Cp_3^*$ (**39**) (Fig. 6) are presumably responsible for the observed equivalence of the CO and Cp^* environments at ambient temperatures (*34*). At $-90°C$, the NMR data indicate effective C_s symmetry. Cluster **39** provides a clear example of the utility of ^{195}Pt couplings in determination of a fluxional mechanism. The retention of long-range $^{195}Pt-^{13}C$ couplings on the carbonyls at all temperatures implies that intermetallic CO migration does not occur, and that the CO groups remain bonded to the same Ir atom and bridge only to adjacent Pt centers (*34*).

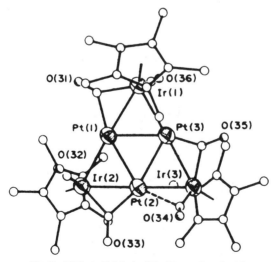

FIG. 6. Structure of $Ir_3Pt_3(CO)_6(\eta-C_5Me_5)_3$ (**39**). [Reproduced with permission from Freeman *et al.* (*34*).]

C. Intermetallic Ligand Migrations

Migrations of ligands, particularly CO and hydrides, between metal centers are also commonly observed, and several hetero-Pt clusters show such behavior. Intermetallic CO migration is generally more common in anionic clusters, in which the barriers to CO exchange are lower. Complete CO scrambling has been reported for $[Fe_3Pt(\mu\text{-}H)(CO)_{11}(PPh_3)]^-$ (**40**) at $-90°C$ (*31*); for $[Rh_8Pt(CO)_{19}]^{2-}$ (**41**) (*55*), $[Rh_5Pt(CO)_{15}]^-$ (**42**) (*81*), and $[M_4Pt(CO)_{12}]^{2-}$ (**11a**, M = Rh) (*81*) at ambient temperatures; and for **1a** above 60°C (*108*). Clusters **42**, $[M_4Pt(CO)_{14}]^{2-}$ (**43**, M = Rh, Ir), and **11a** also show lower energy dynamic processes involving localized intermetallic CO exchange. At $-50°C$ the carbonyls b, c, and e in **42** (Fig. 7) are exchanging rapidly, as are

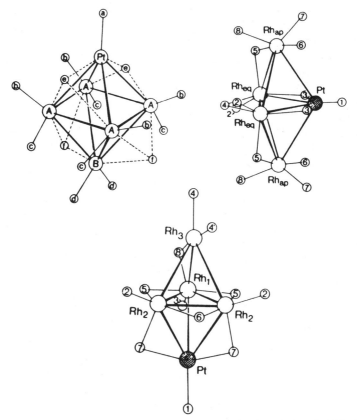

FIG. 7. Carbonyl arrangements in the anions $[Rh_5Pt(CO)_{15}]^-$ (**42**), $[M_4Pt(CO)_{14}]^{2-}$ (**43**, M = Rh, Ir), and $[Rh_4Pt(CO)_{12}]^{2-}$ (**11a**), respectively. [Reproduced with permission from Fumagalli *et al.* (*81*).]

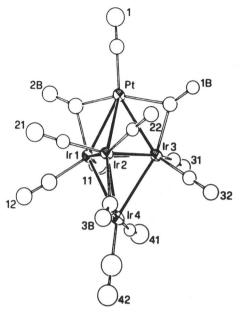

FIG. 8. Structure of the anion $[Ir_4Pt(CO)_{12}]^{2-}$ (11b). [Reproduced with permission from Fumagalli *et al.* (57).]

the carbonyls 5, 6, 7, and 8 in 43 [at $-75°C$ for 43a, M = Rh (81), and at ambient temperature for 43b, M = Ir (57)]. These processes involve scrambling between terminal COs and those COs that bridge to the same metal center. However, the anion 11a (Fig. 7) shows exchange between all COs except 5 and 6 at $-30°C$, implying rapid migration of carbonyls between the unconnected metal centers Rh3 and Pt (81). The iridium analogue 11b has a slightly different carbonyl arrangement (Fig. 8) and shows intermetallic exchange processes, which up to $-10°C$ at least, do not involve the Pt–carbonyl (57). From reported NMR data in the literature, intermetallic CO exchange presumably also occurs in the clusters $Ru_2Pt(\mu\text{-}CH_2)(CO)_3(PCy_3)Cp_2$ (116), Co_3Pt_3-$(\mu\text{-}H)(\mu_6\text{-}C)(CO)_9(PR_3)_3$ (117), and $Os_3Pt(\mu\text{-}H)_2(CO)_{10}(CNCy)(PCy_3)$ (118), and is indeed likely to be common in many other systems for which ^{13}C data are presently unavailable.

Clusters with chemically inequivalent hydrides often show hydride exchange. In cases in which hydrides are chemically equivalent, J couplings to, e.g., ^{195}Pt or ^{31}P may allow observation of "hidden" fluxional exchange. In the cluster $Ru_2Pt_2(\mu\text{-}H)_2(\mu_4\text{-}C)(\mu\text{-}CO)_2(PR_3)_2Cp_2$ (44, R = Cy, iPr), the two hydrides are chemically equivalent (Fig. 9) but are clearly non-fluxional, because there are two distinct ^{195}Pt–hydride couplings of ~ 600 and 23 Hz

(116). A recent suggestion by Nevinger and Keister *(119)*, that low-energy hydride exchange requires adjacent unbridged M–M edges, may be relevant here, because there are no unbridged M–M edges in **44**. However, the corollary may not be true, because the cluster $Ru_2Pt_2(\mu\text{-}H)(\mu_4\text{-}CH)(\mu\text{-}CO)(CO)_2$-$(PR_3)_2Cp_2$ (**45**, R = Cy, *i*Pr), with an unbridged Ru–Ru edge adjacent to the hydride (Fig. 9), is also nonfluxional at ambient temperatures, as shown by two distinct ^{195}Pt couplings ($\sim$625 and 75 Hz) *(116)*. In contrast, the clusters $Os_2Pt_2(\mu\text{-}H)_2(CO)_8(PR_3)_2$ (**15**), which contain adjacent unbridged M–M edges and chemically equivalent hydrides, do show hydride exchange *(41)*. The second-order multiplet (the AA′ part of the AA′XX′ pattern) seen for the hydrides at low temperatures collapses to a simple triplet at higher temperatures, indicating magnetic equivalence of the hydrides, while the value of $J(^{195}Pt\text{--}^1H)$ decreases from 520 to 260 Hz. The PMe_3 derivative shows a lower activation energy than does the PPh_3 derivative *(41)*. Several triosmium–platinum clusters that also possess unbridged M–M edges show hydride mobility *(40a,42,89a,90,106–108,118)*, and the free energy of activation is strongly dependent on the ancilliary ligands. Details are given in Table II. At the present time it is not possible to elucidate the relative importance of steric and electronic factors in determining activation parameters.

The magnitudes of $J(^{195}Pt\text{--}^1H)$ and the splitting patterns observed in $[Ni_9Pt_3H_{n-4}(CO)_{21}]^{n-}$ ($n = 2, 3$) suggest that the interstitial hydrides in these

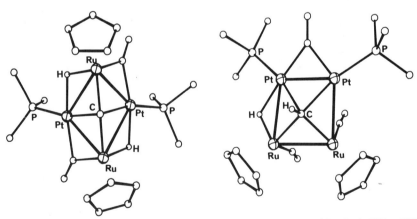

FIG. 9.   Structures of $Ru_2Pt_2(\mu\text{-}H)_2(\mu_4\text{-}C)(\mu\text{-}CO)_2(PR_3)_2Cp_2$ (**44**) and $Ru_2Pt_2(\mu\text{-}H)(\mu_4\text{-}CH)$-$(\mu\text{-}CO)(CO)_2(PR_3)_2Cp_2$ (**45**).

TABLE II

Free Energies of Activation for Hydride Exchange in Triosmium–Platinum Clusters

Complex	$\Delta G^{\ddagger}$ (kJ mol^{-1})	Temp. (K)	Ref.
$Os_3Pt(\mu\text{-}H)_2(CO)_{10}(PCy_3)$	63.1(0.5)[a]	298	108
$Os_3Pt(\mu\text{-}H)_2(CO)_{10}(PPh_3)$	59.4(0.6)[b]	306	106
$Os_3Pt(\mu\text{-}H)_2(CO)_{10}\{P(tBu)_2Me\}$	58(2)[b]	303	40a
$Os_3Pt(\mu\text{-}H)_2(CO)_9(CNCy)(PCy_3)$	>74[c]	333	107
$Os_3Pt(\mu\text{-}H)_2(CO)_9(PMe_2Ph)(PCy_3)$	57.1(0.6)[b]	303	106
$[Os_3Pt(\mu\text{-}H)_3(CO)_{10}(PCy_3)]^{+d}$	62.7(0.5)[a]	293	108
$[Os_3Pt(\mu\text{-}H)_3(CO)_9(CNCy)(PCy_3)]^{+d}$	67.3(1.1)[b]	330	118
$Os_3Pt(\mu\text{-}H)_2(CO)_{10}(PCy_3)_2$	Rigid up to 373[e]		106
$Os_3Pt(\mu\text{-}H)_2(CO)_{11}(PPh_3)$	46(1)[b]	259	106
$Os_3Pt(\mu\text{-}H)_2(\mu\text{-}CH_2)(CO)_9(CNCy)(PCy_3)$	85.6(0.8)[a]	373	107

[a] Values for k determined from line shape analysis.

[b] Values for k determined from coalescence temperature using approximation $k = \pi \delta \nu / \sqrt{2}$.

[c] Determined from line width at 333 K.

[d] Only the two $Os(\mu\text{-}H)Os$ hydrides exchange.

[e] No detectable line broadening at this temperature.

high-nuclearity anions (see later, Fig. 31) migrate rapidly between the tetra-hedral and trigonal prismatic holes, but the data on $[Ni_9Pt_3{}^1H^2H(CO)_{21}]^{2-}$ imply that intercluster proton migration is slow (120).

D. Metal Framework Rearrangements

Several fluxional processes have been reported that imply some metal framework rearrangement. The general restrictions that apply to the pathways for such rearrangements have been discussed in several theoretical papers (121–124). Adams and co-workers (39,39a) have reported ^1H and ^{31}P NMR data on $Os_3Pt(\mu_3\text{-}S)_2(CO)_9(PMe_2Ph)_2$ (10) indicating that the molecule ac-quires a time-averaged mirror plane. The proposed mechanism (Scheme 1) involves two possible intermediates. The bond making path B is preferred on the grounds of the very low energy ($\Delta G^{\ddagger}_{167} \sim 32$ kJ mol^{-1}), though the inter-mediate A, with a planar 16-electron Pt center, cannot be ruled as a permu-tationally feasible route. No evidence was found for a similar dynamic process in the related cluster $Os_3Pt(\mu_3\text{-}S)_2(CO)_9(=C=C_6H_6Me_4)(PCy_3)$ (46) (125).

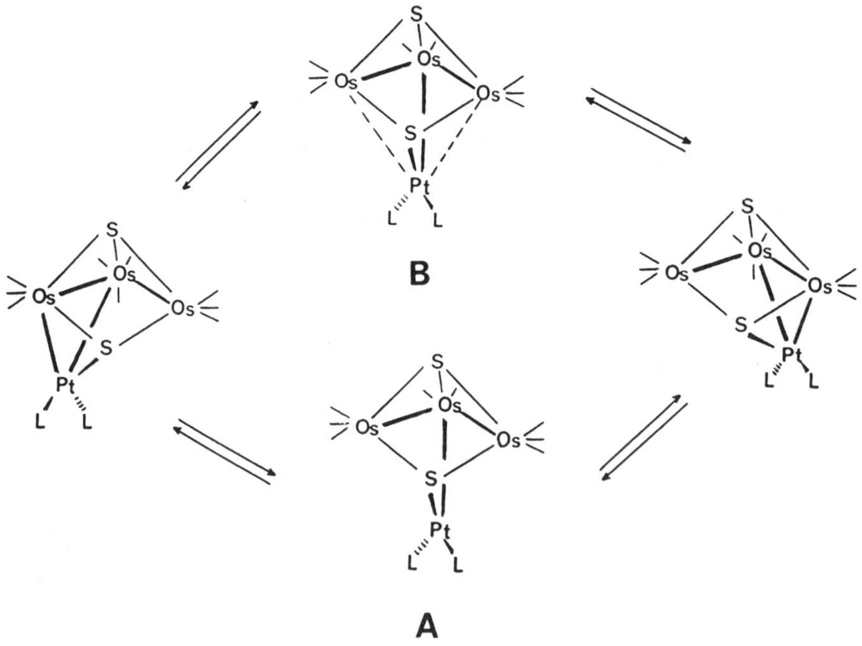

SCHEME 1

46

Studies on some 60-CVE Os_3Pt clusters, which have butterfly core geometries in the solid state, indicate that this framework is quite flexible in solution. The cluster $Os_3Pt(\mu-H)_2(CO)_{11}(PCy_3)$ (**19**) shows a low-energy fluxional process involving an exchange of CO ligands between the wing-tip Os and Pt centers (*108*). This is postulated to occur via a closo-tetrahedral intermediate or transition state. Such a process involves a change in coordination of the Pt center from a planar to a pseudotrigonal–bipyramidal environment, and implies a 16 ↔ 18-electron interchange. Venanzi et al. (*126*) have proposed a similar butterfly ↔ tetrahedron interconversion to explain the fluxional behavior of $Pt_4(\mu-CO)_5(PR_3)_4$, and EHMO calculations indicate a soft potential energy surface for this process (*122*). A ^{13}C EXSY spectrum on the closely related species $Os_3Pt(\mu-H)_2(CO)_{10}(PR_3)(PR'_3)$ (**2b**, R = R' = Cy) shows a strikingly different CO exchange network, which implies cluster enantiomerization (*108*). A mechanism is proposed that involves a planar–butterfly transition state. Further studies on the derivative (**2c**, R = Cy, R'_3 = Me_2Ph), with the prochiral phosphine PMe_2Ph bonded to Os, confirm that enantiomerization is taking place, though this species shows additional fluxional processes (see later, Scheme 3) that complicate a quantitative interpretation (*106*).

The single ^{13}C signal for the CO ligands in $[Rh_8Pt(CO)_{19}]^{2-}$ (**41**) indicates complete scrambling over the metal skeleton (see later, Fig. 29), while the multiplicity suggests either accidental similarities of $^{103}Rh-^{13}C$ coupling constants or a fluxional process of the metal skeleton that renders the six Rh atoms in the outer Rh_3 triangles magnetically equivalent by rotation about the pseudothreefold axis (*55*). ^{195}Pt data show that a similar rotation of Rh_2Pt triangles is probably occurring in the anion $[Rh_2Pt(CO)_x]_2^-$ (*56*), but that the octahedral core in $[Rh_5Pt(CO)_{15}]^-$ (**42**) is rigid (*56,81*).

E. Metal Core Isomerizations

Isomerizations involving metal core rearrangements that are not fast on the NMR time scale have also been observed. Couture and Farrar (*127*) have

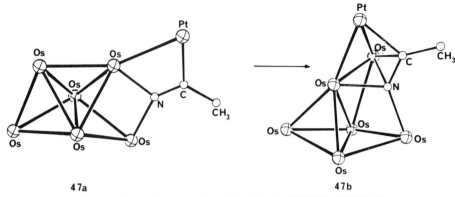

47a   47b

FIG. 10.   Skeletal isomerization of $Os_6Pt(\mu\text{-MeCN})(CO)_{17}(COD)$ (47).

reported that the complex $Os_6Pt(\mu\text{-MeCN})(CO)_{17}(COD)$ (47, Fig. 10) undergoes an isomerization that involves a substantial core rearrangement. This is
coupled with a change in the bonding mode of the MeCN ligand from μ_3 to
μ_4, and in the coordination sphere of the Pt atom (see Fig. 10). Kinetic studies
show the process 47a → 47b obeys a first-order rate law, with activation pa-
rameters $\Delta H^{\ddagger} = 120.5(5)\,\text{kJ mol}^{-1}$ and $\Delta S^{\ddagger} = 55(17)\,\text{J K}^{-1}\,\text{mol}^{-1}$, which are
typical of a slow process involving substantial reorganization in the transition
state. Isomer 47b is the more thermodynamically stable product. Replacement
of the COD ligand by dppe or $P(OMe)_3$ results in structure 47a being exclusively favored. Ewing and Farrugia have shown that the hydrido alkynyl clus-
ter $Ru_3Pt(\mu\text{-H})(\mu_4\text{-}\eta^2\text{-C}\equiv C t Bu)(CO)_9(dppe)$ (34) reversibly isomerizes via
an intramolecular hydride transfer to the vinylidene species $Ru_3Pt(\mu_4\text{-}\eta^2\text{-}$
$C=C(H)tBu)(CO)_9(dppe)$ (37), with a concomitant core rearrangement from
a spiked triangle to a butterfly (see later, Scheme 2) (113). Studies using iso-
topic labeling of the hydride site show an inverse deuterium kinetic isotope
effect ($k_H/k_D = 0.65$), consistent with a transition state engendering significant C—H bonding (128).

V

STRUCTURAL AND REACTIVITY ASPECTS
OF HETERO-PT CLUSTERS

It is convenient to discuss the structures and chemical reactivity of hetero-
Pt clusters under the individual metals, because there is little homogeneity of
structural type, particularly when the details of ligand coordination are taken
into account. There are remarkably few examples common to more than one

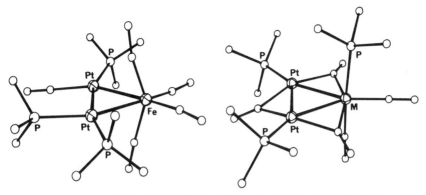

FIG. 11. Structures of $Fe_2Pt(CO)_5(PR_3)_3$ (**48a**) and $M_2Pt(\mu\text{-}CO)_3(CO)_2(PR_3)_3$ (**48b,c**, M = Ru, Os).

metal atom of a given triad. This stems in part from their differing synthetic routes; very often a crucial starting complex is only known for one of the metals of that triad. For example, the triosmium complex $Os_3(\mu\text{-}H)_2(CO)_{10}$ is an important precursor to many Os_3Pt clusters, but its iron and ruthenium analogues are not as yet described. Only two series of structurally character-ized clusters are known for all metals of a triad:

1. $MPt_2(CO)_5(PR_3)_3$ [**48a**, M = Fe, R_3 = $(OPh)_3$ (*129*); **48b**, M = Ru, R_3 = Ph_2Me (*130*), or R_3 = Ph_3 (*131*); and **48c**, M = Os, R_3 = Ph_3 (*132*)].
2. $M_2Rh_2Pt(\mu_3\text{-}CO)_4Cp_2^*$ [M = Co, Rh, and Ir (*50,110*)].

In the case of (1), even the details of ligand coordination vary, because the Fe cluster has a structure with all terminal carbonyls, whereas the iso-structural Ru and Os cogenors have three bridging carbonyls and a different phosphine distribution (Fig. 11). Important metrical details for all the crys-tallographically characterized Fe–Pt, Ru–Pt, Os–Pt, Co–Pt, Rh–Pt, and Ir–Pt clusters are given in Tables III–VIII respectively. Due to the paucity of examples, crystallographically characterized Ni–Pt and Pd–Pt clusters are not tabulated separately but are discussed in the individual sections, as are clusters containing more than two different metals.

A. Iron–Platinum Clusters

1. Trinuclear Clusters

The earliest examples were provided by the work of Stone *et al.* on reactions of zero-valent Pt complexes with carbonyls of the iron triad (*10,36,37*). Thus

TABLE III

CRYSTALLOGRAPHICALLY CHARACTERIZED Fe–Pt CLUSTERS

Cluster	CVE count	Polyhedral shape	Fe–Pt[a]	Fe–Fe[a]	Pt–Pt[a]	Ref.
1. $Fe_2Pt(CO)_9(PPh_3)$	46	Triangle	2.597(5), 2.530(5)	2.758(8)	—	133
2. $Fe_2Pt(CO)_8(COD)$	46	Triangle	2.561(2), 2.553(3)	2.704(4)	—	31
3. $Fe_2Pt(CO)_8(PEt_3)_2$	46	Triangle	2.640(1)	—	—	134
4. $Fe_2Pt(\mu_3\text{-}P\equiv C\mathit{t}Bu)(CO)_6(dppe)$	46	Triangle	2.671(1), 2.669(1)	2.518(1)	—	140,140a
5. $FePt_2(CO)_5[P(OPh)_3]_3$	44	Triangle	2.550(5), 2.583(6)	—	2.633(1)	129
6. $Fe_3Pt(\mu_3\text{-}H)(\mu_3\text{-}COMe)(CO)_{10}(PPh_3)$	60	Tetrahedron	2.617(2), 2.696(2), 2.739(2)	2.556(3), 2.543(2), 2.694(3)	—	32
7. $Fe_3Pt(\mu_4\text{-}C=C(H)Ph)(CO)_9(dppe)$	60	Butterfly	No metrical details available			143
8. $Fe_3Pt(\mu\text{-}H)(CO)_{11}(PCy_3)$	59	Tetrahedron	2.680(1), 2.714(1), 2.692(1)	2.569(1), 2.576(1), 2.588(1)	—	142
9. $[Fe_2Pt_2(\mu\text{-}H)(CO)_8(PPh_3)_2]^-$	58	Butterfly	2.756(2), 2.626(2), 2.555(2), 2.562(2)	2.522(2)	2.966(1)	31
10. $Fe_2Pt_2(\mu\text{-}H)_2(CO)_8(PPh_3)_2$	58	Butterfly	2.604(3), 2.696(3) 2.694(3), 2.631(3)	2.567(4)	2.998(2)	31
11. $[Fe_4Pt(CO)_{16}]^{2-}$	76	Bow-tie	2.601(1) (av)	2.708(2) (av)	—	51

12. $Fe_5Pt(\mu_6\text{-}C)(CO)_{15}(PPh_3)$	86	Octahedron	2.650(2), 2.912(2), 2.886(2)	2.622(2), 2.771(2), 2.786(2), 2.626(2), 2.658(2)	—	*82*
13. $[Fe_5Pt(\mu_6\text{-}C)(CO)_{15}]^{2-}$	86	Octahedron	No metrical details available			*144*
14. $[Fe_3Pt_3(CO)_{15}]^{2-}$	86	Raft	2.596(4) *(av)*	—	2.750(1) *(av)*	*5*
15. $[Fe_3Pt_3(CO)_{15}]^-$	85	Raft	2.587(4) *(av)*	—	2.656(1) *(av)*	*5*
16. $Fe_3Pt_3(CO)_{15}$	84	Raft	2.571(2), 2.580(2), 2.584(2)	—	2.592(1), 2.587(2)	*97*
17. $FePt_5(CO)_9(PEt_3)_4$	84	See Fig. 14	2.603(4), 2.570(4)	—	2.709(2), 2.727(2), 2.722(2), 2.767(2), 2.762(2), 2.666(2), 2.869(2), 2.895(2)	*134*
18. $Fe_2Pt_5(CO)_{12}(COD)_2$	98	See Fig. 14	2.546(3), 2.582(3), 2.602(3), 2.591(3) *(av)*	—	2.640(1), 2.643(1), 2.824(1), 2.827(1), 2.663(1), 2.775(1), 2.802(1), 2.958(1)	*97*
19. $[Fe_4Pt_6(CO)_{22}]^{2-}$	138	See Fig. 13	2.597(4), 2.540(4) *(av)*	—	2.677(1), 2.790(1) *(av)*	*5*

[a] In angstroms; estimated standard deviations in parentheses. Mean Fe−Pt = 2.632 Å.

TABLE IV
CRYSTALLOGRAPHICALLY CHARACTERIZED Ru–Pt CLUSTERS

Cluster	CVE count	Polyhedral shape	Ru–Pt[a]	Ru–Ru[a]	Pt–Pt[a]	Ref.
1. $Ru_2Pt(\mu\text{-}CH_2)(CO)_3(PCy_3)(Cp)_2$	46	Triangle	2.609(1), 2.613(1)	2.813(1)	—	116
2. $RuPt_2(CO)_5(PPh_2Me)_3$	44	Triangle	2.707(2), 2.729(2)	—	2.647(2)	130
3. $RuPt_2(CO)_5(PPh_3)_3$	44	Triangle	2.734(1)[b], 2.741(1)[b], 2.739(2)[c], 2.721(2)[c]	—	2.651(1)[b] 2.661(2)[c]	131
4. $Ru_3Pt(\mu\text{-}H)\{\mu_4\text{-}C{\equiv}C(tBu)\}(CO)_9(COD)$	62	Spiked triangle	2.645(1)	2.810(1), 2.815(1), 2.791(1)	—	113
5. $Ru_3Pt(\mu\text{-}H)\{\mu_4\text{-}C{\equiv}C(tBu)\}(CO)_9(dppe)$	62	Spiked triangle	2.681(1)	2.824(1), 2.792(1), 2.796(1)	—	113
6. $Ru_3Pt(\mu\text{-}H)\{\mu_4\text{-}C{\equiv}C(tBu)\}(CO)_9(dppet)$[d]	62	Spiked triangle	2.700(3)	2.793(4), 2.799(4), 2.799(4)	—	128
7. $Ru_3Pt\{\mu_4\text{-}C{\equiv}C(H)tBu\}(CO)_9(dppe)$	60	Butterfly	2.730(1), 2.792(1)	2.708(2), 2.799(2), 2.823(2)	—	113
8. $Ru_3Pt\{\mu_4\text{-}C{\equiv}C(H)tBu\}(CO)_9(S,S\text{-}dppb)$[e]	60	Butterfly	2.723(3), 2.790(4)	2.718(4), 2.775(4), 2.826(4)	—	128
9. $[Ru_3Pt(\mu\text{-}H)\{\mu_4\text{-}C{\equiv}C(H)tBu\}(CO)_9(dppe)]^+$	60	Butterfly	2.784(1), 2.782(1)	2.774(1), 2.835(1), 2.830(1)	—	113
10. $Ru_2Pt_2(\mu\text{-}H)(\mu_4\text{-}CH)(CO)_3(PiPr_3)_2(Cp)_2$	60	Square	2.803(1), 2.820(1)	2.809(1)	2.662(1)	116
11. $Ru_2Pt_2(\mu\text{-}H)_2(\mu_4\text{-}C)(CO)_2(PiPr_3)_2(Cp)_2$	60	Butterfly	2.707(1), 2.858(1), 2.850(1), 2.707(1)	—	3.132(1)	116
12. $Pt[Ru_3(\mu\text{-}H)\{\mu_4\text{-}C{\equiv}C(tBu)\}(CO)_9]_2$	106	See Fig. 15	2.695(1), 2.688(1)	2.818(2), 2.824(2), 2.797(2), 2.807(2), 2.820(2), 2.812(2)	—	152

[a] In angstroms; estimated standard deviations in parentheses. Mean Ru–Pt = 2.734 Å.

[b] Unsolvated structure.

[c] Benzene solvate.

[d] cis-1,2-(diphenylphosphino)ethene.

[e] (2S,3S)-bis(diphenylphosphino)butane.

TABLE V
Crystallographically Characterized Os–Pt Clusters

Cluster	CVE count	Polyhedral shape	Os–Pt[a]	Os–Os[a]	Pt–Pt[a]	Ref.
1. $Os_2Pt(CO)_{10}$	46	Triangle	2.669(2), 2.689(2)	2.860(2)	—	153
2. $OsPt_2(\mu_3\text{-MeC}\equiv\text{CMe})(CO)_5(PPh_3)_2$	46	Triangle	2.664(2), 2.669(2)	—	3.033	47
3. $OsPt_2(CO)_5(PPh_3)_3$	44	Triangle	2.735(1), 2.744(1)	—	2.650(1)	132
4. $Os_3Pt(\mu_3\text{-}S)_2(CO)_{10}(PPh_3)$	64	Spiked triangle	2.858(1), 2.905(1), 3.401(1)	2.826(1), 2.990(1), 3.585(1)	—	44
5. $Os_3Pt(\mu_3\text{-}S)_2(CO)_9(PPh_3)_2$	64	Spiked triangle	2.878(1), 2.904(1), 3.336(1)	2.819(1), 2.998(1), 3.612(1)	—	44
6. $Os_3Pt(\mu_3\text{-}S)_2(CO)_9(PMe_2Ph)_2$	64	Spiked triangle	2.936(1), 2.789(1), 3.771(1)	3.027(1), 2.798(1)	—	39,39a
7. $Os_3Pt(\mu_3\text{-}S)_2(CO)_9(\eta^1\text{-}C{=}C_6H_6Me_4)(PCy_3)$	64	Spiked triangle	2.848(1), 2.862(1), 3.656(1)	2.810(1), 3.030(1), 3.618(1)	—	125
8. $Os_3Pt(\mu\text{-}H)_2(\mu_4\text{-}C)(CO)_{10}(PCy_3)$	62	Spiked triangle	2.764(1), 3.486(1), 3.834(1)	2.906(1), 2.864(1), 2.828(1)	—	42
9. $Os_3Pt(\mu\text{-}CH_2)(CO)_{11}(PPh_3)_2$	62	Spiked triangle	2.772(4)	2.877(4), 2.882(4), 2.857(3)	—	157
10. $Os_3Pt(\mu\text{-}H)(\mu_4\text{-}C{\equiv}CPh)(CO)_{10}(PCy_3)$	62	Spiked triangle	2.712(1)	2.845(1), 2.824(1), 2.838(1)	—	155
11. $Os_3PtSn(\mu\text{-}H)_2(Cl)(OEt_2)(SnCl_3)(CO)_{10}(PCy_3)$	62	Spiked triangle	2.782(3), 2.865(3)	3.003(3), 3.092(3)	—	88
12. $Os_3Pt(\mu_3\text{-}S)(CO)_{10}(PMe_2Ph)_2$	62	Butterfly	2.768(1), 2.836(1)	2.869(1), 2.867(1), 2.870(1)	—	33
13. $Os_3Pt(\mu_3\text{-}S)(CO)_9(PMe_2Ph)_3$	62	Butterfly	2.789(1), 2.835(1), 2.782(1), 2.803(1)	2.891(1), 2.888(1), 2.884(1), 2.870(1), 2.833(1), 2.907(1)	—	33
14. $Os_3Pt(\mu\text{-}H)_2(CO)_{10}(PPh_3)_2$	60	Butterfly	2.717(1), 2.848(2), 3.530(2)	3.034(2), 2.914(2), 2.773(2)	—	47

(continued)

333

TABLE V (continued)

Cluster	CVE count	Polyhedral shape	Os–Pt[a]	Os–Os[a]	Pt–Pt[a]	Ref.
15. $Os_3Pt(\mu\text{-}H)_2(CO)_{10}(PCy_3)_2$	60	Butterfly	2.853(1), 2.732(1), 3.506(1)	2.923(1), 3.023(1), 2.760(1)	—	91
16. $Os_3Pt(\mu\text{-}H)_2(CO)_{11}(PCy_3)$	60	Butterfly	2.729(1), 2.914(1), 3.775(1)	2.877(1), 2.882(1), 2.869(1)	—	89a
17. $Os_3Pt(\mu\text{-}H)_4(CO)_{10}(PCy_3)^b$	60	Tetrahedron	2.714(4), 2.967(4), 2.977(4)	2.867(4), 2.991(4), 2.983(4)	—	89a
18. $Os_3Pt(\mu_3\text{-}S)(CO)_8(PMe_2Ph)_3$	60	Tetrahedron	2.853(1), 2.740(1), 2.766(1)	2.816(1), 2.770(1), 3.064(1)	—	33
19. $Os_3Pt(\mu\text{-}H)(\mu\text{-}PMe_2C_6H_4)(\mu_3\text{-}S)(CO)_7(PMe_2Ph)$	60	Tetrahedron	2.816(1), 2.684(1), 2.806(1)	2.920(1), 2.837(1), 2.972(1)	—	33
20. $Os_3Pt(\mu\text{-}H)_2(\mu\text{-}CH_2)(CO)_{10}(PCy_3)^b$ (red C_1 isomer)	60	Tetrahedron	2.867(1), 2.854(1), 2.777(1)	2.975(1), 2.866(1), 2.808(1)	—	89a
21. $Os_3Pt(\mu\text{-}H)_2(\mu\text{-}CH_2)(CO)_{10}(PCy_3)$ (yellow C_s isomer)	60	Tetrahedron	2.730(1), 2.814(1), 2.826(1)	2.954(1), 2.941(1), 2.826(2)	—	89a
22. $Os_3Pt(\mu\text{-}H)_2(\mu\text{-}CH_2)(CO)_9(PCy_3)(CNCy)$	60	Tetrahedron	2.865(1), 2.858(1), 2.787(2)	2.965(2), 2.796(2), 2.841(2)	—	118
23. $Os_3Pt(\mu\text{-}H)_2(\mu\text{-}SO_2)(CO)_{10}(PCy_3)$	60	Tetrahedron	2.906(1), 2.821(1), 2.792(1)	2.795(1), 2.988(1), 2.976(1)	—	90
24. $Os_3Pt(\mu\text{-}H)_2(\mu\text{-}CH_2)(CO)_9(COD)$	60	Tetrahedron	2.712(1), 2.788(1), 2.785(1)	2.936(1), 2.940(1), 2.833(1)	—	156
25. $Os_3Pt(\mu\text{-}H)_2(\mu\text{-}CO)(CO)_9(COD)$	60	Tetrahedron	2.730(1), 2.811(1), 2.714(1)	2.837(1), 2.919(1), 2.952(1)	—	80
26. $Os_3Pt(\mu\text{-}H)_2(CO)_{10}(PCy_3)$	58	Tetrahedron	2.863(1), 2.791(1), 2.832(1)	2.777(1), 2.741(1), 2.789(1)	—	40,40a
27. $Os_3Pt(\mu\text{-}H)_2(CO)_9(CNCy)(PCy_3)$	58	Tetrahedron	2.844(1), 2.773(1), 2.781(1)	2.745(1), 2.742(1), 2.752(1)	—	118

No.	Compound	Geometry				Ref.
28.	[Os$_3$Pt(μ-H)$_3$(CO)$_{10}$(PCy$_3$)]$^+$	Tetrahedron	2.812(1), 2.819(1), 2.845(1)	2.747(1), 2.891(1), 2.790(1)	—	108
29.	Os$_2$Pt$_2$(μ-H)$_2$(CO)$_8$(PPh$_3$)$_2$	Butterfly	2.863(1), 2.709(1)	2.781(1)	3.206(1)	41
30.	Os$_2$Pt$_2$(μ-H)$_2$(CO)$_8$(PCy$_3$)$_2$	Butterfly	2.854(1), 2.826(1), 2.710(1), 2.714(1)	2.774(1)	3.230(1)	84
31.	Os$_4$Pt(μ_3-S)$_2$(CO)$_{11}$(PMe$_2$Ph)$_2$	Bridged butterfly	2.762(1), 2.967(1)	2.665(1), 2.882(1), 2.809(1), 3.106(1), 2.760(1)	—	39a
32.	Os$_4$Pt(μ_4-S)(CO)$_{13}$(PPh$_3$)	Square–pyramidal	2.900(1), 2.875(1), 2.749(1)	2.802(1), 2.799(1), 2.862(1), 2.770(1), 2.911(1)	—	43a
33.	Os$_3$Pt$_2$(μ-H)$_2$(μ_5-C)(CO)$_{10}$(PCy$_3$)$_2$	Bridged square	2.929(1), 2.935(1), 3.693(1), 3.628(1)	2.855(1), 2.916(1), 2.917(1)	—	42
34.	Os$_3$Pt$_2$(μ-H)(μ-OMe)(μ_5-C)(CO)$_{10}$(PCy$_3$)$_2$	Spiked square	2.891(1), 2.941(1)	2.853(1), 2.813(1), 3.328(1)	2.668(1)	42
35.	Os$_5$Pt(μ_4-S)(CO)$_{15}$(PPh$_3$)	Capped square pyramid	2.986(1), 2.791(1), 2.646(1)	2.847(1), 2.804(1), 2.830(1), 2.859(1), 2.914(1), 2.703(1), 2.847(1), 2.919(1)	—	43a
36.	Os$_5$Pt(μ_4-S)(CO)$_{15}$(PPh$_3$)$_2$	Bridged square pyramid	3.030(1), 2.862(1), 2.818(1)	2.878(1), 2.791(1), 2.984(1), 2.896(1), 2.823(1), 2.719(1), 2.761(1)	—	43a
37.	Os$_7$Pt$_2$(CO)$_{18}$	Puckered raft	2.652(3), 2.916(4), 2.816(4), 2.849(4), 2.869(4), 2.658(4)	2.906(2), 2.876(3)	2.667(2)	153
38.	Os$_6$Pt(μ_3-NCMe)(CO)$_{17}$(COD)	See Fig. 10	2.783(1)	2.728(1), 2.804(1), 2.774(1), 2.829(1), 2.806(1), 2.890(1), 2.828(1), 2.815(1), 2.853(1), 2.856(1), 2.802(1)	—	127
39.	Os$_6$Pt(μ_4-NCMe)(CO)$_{17}$(COD)	See Fig. 10	2.696(2), 2.903(2)	2.876(3), 2.870(2), 2.821(3), 2.617(2), 2.754(3), 2.777(2), 2.881(2), 2.760(2), 2.842(2), 3.175(2)	—	127

(continued)

TABLE V (Continued)

Cluster	CVE count	Polyhedral shape	Os–Pt[a]	Os–Os[a]	Pt–Pt[a]	Ref.
40. $Os_6Pt_2(CO)_{17}(COD)_2$	110	Bicapped octahedron	2.645(3), 2.787(4), 2.966(4), 2.659(3), 3.068(3), 2.762(2)	2.783(3), 2.815(4), 2.788(4), 2.796(3), 2.922(3), 2.875(4), 2.855(3), 2.909(3), 2.923(3), 2.846(3), 2.900(4), 2.935(3)	—	158
41. $Os_6Pt_2(CO)_{16}(COD)_2$	108	See Fig. 19	2.624(1), 2.805(1), 2.901(1)	2.647(1), 2.537(1), 2.807(1), 2.831(1), 2.807(1), 2.872(1), 2.768(1), 2.708(1)	—	7
42. $Os_6Pt_2(CO)_{16}\{P(OMe)_3\}_2$	108	See Fig. 20	2.647(3), 3.083(3), 2.703(3), 2.930(3), 2.667(3), 2.836(3)	2.818(3), 2.761(3), 2.698(3), 3.014(3), 2.936(3), 2.685(3), 2.804(3), 2.659(3), 2.765(3), 2.782(3), 2.865(3), 2.738(3)	—	7
43. $Os_6Pt_2(\mu\text{-}H)_2(\mu\text{-}C_8H_{10})(CO)_{16}(COD)$	112	See Fig. 21	2.669(2), 2.733(3)	2.741(3), 2.879(3), 2.665(3), 2.828(3), 2.867(3), 2.788(3), 2.865(3), 2.868(3), 2.870(4), 2.788(3), 2.856(4), 2.836(3)	2.859(3)	85
44. $[Os_3Pt(\mu\text{-}H)(\mu_3\text{-}S)(\mu_4\text{-}S)(\mu\text{-}PPh_2C_6H_4)(CO)_8]_2$	128	See Fig. 23	2.680(1), 2.744(1)	2.956(1), 2.930(1)	3.313(1)	160

[a] In angstroms; estimated standard deviations in parentheses. Mean Os–Pt = 2.806 Å.
[b] Average values for crystallographically independent molecules.

336

TABLE VI

CRYSTALLOGRAPHICALLY CHARACTERIZED Co–Pt CLUSTERS

Cluster	CVE count	Polyhedral shape	Co–Pt[a]	Co–Co[a]	Pt–Pt[a]	Ref.
1. $Co_2Pt(CO)_8(PPh_3)$	46	Triangle	2.519(1), 2.533(1)	2.507(2)	—	163
2. $Co_2Pt(CO)_2(PPh_3)_2(Cp)_2$	46	Triangle	2.560(2), 2.566(2)	2.372(2)	—	35
3. $CoPt_2(\mu\text{-}PPh_2)(CO)_3(PPh_3)_3$	44	Triangle	2.540(2), 2.574(2)	—	2.664(1)	164
4. $Co_2Pt_2(CO)_8(PPh_3)_2$	58	Butterfly	2.540(2), 2.579(2), 2.554(2), 2.528(3)	2.498(3)	2.987(4)	52a
5. $Co_3Pt(\mu\text{-}CO)_3(CO)_6Cp^*$	60	Tetrahedron	2.526(2), 2.527(2), 2.527(3)	2.465(2), 2.493(2), 2.479(2)	—	86
6. $Co_2Pt_3(CO)_9(PEt_3)_3$	72	Trigonal–bipyramidal	2.67(1), 2.87(1), 2.51(1), 2.50(1), 2.85(1), 2.64(1)	2.49(1)	2.885(4), 2.892(4)	53
7. $Co_3Pt_3(\mu\text{-}H)(\mu_6\text{-}C)(CO)_9(PiPr_3)_3$	86	Distorted octahedron	2.649(2), 2.652(2), 2.937(2), 2.959(2), 3.060(2), 2.599(2), 2.842(2), 2.661(2)	2.692(3), 2.693(3)	3.260(1), 2.868(1)	117

[a] In angstroms; estimated standard deviations in parentheses. Mean Co–Pt = 2.647 Å.

337

TABLE VII

CRYSTALLOGRAPHICALLY CHARACTERIZED Rh–Pt CLUSTERS

Cluster	CVE count	Polyhedral shape	Rh–Pt[a]	Rh–Rh[a]	Pt–Pt[a]	Ref.
1. $Rh_2Pt(CO)_3(PPh_3)Cp^*_2$[b]	46	Triangle	2.618(2), 2.691(2)	2.647(2)	—	45
2. $[Rh_2Pt(\mu\text{-}H)(CO)_3(PPh_3)Cp^*_2]^+$	46	Triangle	2.705(1), 2.805(1)	2.667(1)	—	109
3. $Rh_4Pt(CO)_4Cp^*_4$	74	Bow-tie	2.620(2)[b], 2.617(2)[b], 2.617(2)[c], 2.617(2)[c], 2.616(2)[c]	2.617(3)[b], 2.620(3)[b], 2.617(2)[c], 2.622(3)[c]	—	50
4. $Rh_4Pt(\mu_3\text{-}CF_3C{\equiv}CCF_3)_2(CO)_2Cp_4$	78	Bow-tie	2.647, 2.712, 2.654, 2.703	2.697, 2.690	—	169
5. $[Rh_4Pt(CO)_{12}]^{2-}$	72	Trigonal–bipyramidal		No metrical details available		81
6. $[Rh_4Pt(CO)_{14}]^{2-}$	76	Trigonal–bipyramid	No metrical details available			81
7. $[Rh_5Pt(CO)_{15}]^-$	86	Octahedron	2.776(2), 2.780(2), 2.811(2), 2.794(1)	2.827(2), 2.698(2), 2.725(2), 2.805(2), 2.746(2), 2.755(2), 2.758(2), 2.768(2)	—	56
8. $[Rh_6Pt(CO)_{16}]^{2-}$	98	Capped octahedron	2.698(2), 3.167(2), 2.771(2)	2.830(3), 2.834(3), 2.730(3), 2.751(3), 2.820(3), 2.814(3), 2.810(3), 2.844(3), 2.795(3), 2.857(3), 2.746(3), 2.781(3)	—	174

No.	Compound		Structure				Ref.
9.	$[Rh_8Pt(CO)_{19}]^{2-}$	122	Face-shared bioctahedron	2.741(2), 2.717(2), 2.732(2)	2.760(2), 2.730(2), 2.767(2), 2.754(2), 2.860(4), 2.833(3), 2.752(2), 2.778(3), 2.806(3)	—	55
10.	$[Rh_{10}Pt(\mu_5\text{-}N)(CO)_{21}]^{3-}$	150	See Fig. 30	2.752(2), 2.700(2), 2.681(2), 3.180(2)	2.755(2), 2.728(2), 2.912(2) 2.797(2), 2.984(2), 2.816(2) 3.088(2), 2.828(2), 2.789(2) 2.755(2)	—	176
11.	$[Rh_9Pt_2(CO)_{22}]^{3-}$	148	Edge-fused trioctahedron	2.666(3), 2.735(3), 2.720(3), 2.743(3), 2.709(3), 2.698(3), 2.687(3), 2.679(3), 2.654(3), 2.669(3), 2.761(3), 2.765(3)	2.721(4), 2.911(4), 2.792(4), 2.834(4), 2.944(4), 2.734(4), 2.997(4), 2.854(4), 2.826(4), 2.954(4), 2.781(4), 2.820(4), 2.893(3), 2.893(4), 2.824(3)	2.812(2)	9
12.	$[Rh_{12}Pt(CO)_{24}]^{4-}$	170	Centered cuboctahedron	2.776 (av), range 2.743(4)–2.812(6)	2.778 (av), range 2.713(8)–2.849(7)	—	8
13.	$[Rh_{11}Pt_2(CO)_{24}]^{3-}$	170	Centered cuboctahedron (disordered)	No metrical details available		—	8
14.	$[Rh_{18}Pt_4(CO)_{35}]^{4-}$	276	See Fig. 30	No metrical details available		—	177a

[a] In angstroms; estimated standard deviations in parentheses. Mean Rh–Pt = 2.730 Å.
[b] Orthorhombic form.
[c] Monoclinic form.

TABLE VIII

Crystallographically Characterized Ir–Pt Clusters

Cluster	CVE count	Polyhedral shape	Ir–Pt[a]	Ir–Ir[a]	Pt–Pt[a]	Ref.
1. $Ir_2Pt_2(CO)_7(PPh_3)_3$	58	Butterfly	2.675(1), 2.681(1), 2.704(1), 2.714(1)	2.741(1)	2.976(1)	54
2. $[Ir_4Pt(CO)_{12}]^{2-}$	72	Trigonal-bipyramidal	2.701(4), 3.024(4), 2.709(4)	2.716(4), 2.794(4), 2.870(4), 2.712(5), 2.780(5), 2.871(4)	—	57
3. $[Ir_4Pt(CO)_{14}]^{2-}$	76	Trigonal-bipyramidal	2.697(1), 2.692(1), 3.106(1), 3.165(1)	2.718(1), 3.062(1), 3.007(1), 2.987(1), 3.044(1)	—	57
4. $Ir_3Pt_3(CO)_6Cp^*_3{}^b$	84	Raft	2.682(3), 2.642(3), 2.645(3), 2.695(3), 2.635(3), 2.702(3)	—	2.701(3), 2.709(3), 2.701(3)	34

[a] In angstroms; estimated standard deviations in parentheses. Mean Ir–Pt = 2.757 Å.
[b] Average values quoted for two crystallographically independent molecules.

49 L=CO   **50** L=L'   **51**   **52**

53   **54**

reaction of $Fe_2(CO)_9$ with PtL_4 affords (10,36) the 46-CVE triangular clusters $Fe_2Pt(CO)_9L$ (**49**, L = PPh_3, PPh_2Me, $PPhMe_2$, PMe_3, or $AsPh_3$) and $Fe_2Pt(CO)_8L_2$ (**50**, L = PPh_3, PPh_2Me, $PPhMe_2$, PMe_3, $PPh(OMe)_2$, or $P(OPh)_3$; L_2 = dppe, diars). Crystal structures on **49** [L = PPh_3 (133) and **50** [L = PEt_3 (134) and L_2 = COD (31)] have been reported. The reaction of $Fe_2(CO)_9$ with $Pt\{P(OPh)_3\}_4$ affords in addition the 44-CVE diplatinum species $FePt_2(CO)_5\{P(OPh)_3\}_3$ (**48a**) (10,36,129). Related clusters $Fe_2Pt(CO)_8$-(μ-dppm) (**51**), $Fe_2Pt(CO)_6(\mu$-dppm$)_2$ (**52**), $FePt_2(CO)_6(\mu$-dppm) (**53**), and $FePt_2(CO)_4(\mu$-dppm$)_2$ (**54**) have been prepared by Braunstein et al. (135) from the reaction of $PtCl_2$(dppm) with either $Fe_2(CO)_9$ or $[Fe_2(CO)_8]^{2-}$. The synthesis of **54** by reaction of $Pt_2Cl_2(\mu$-dppm$)_2$ with $[Fe(CO)_4]^{2-}$ had been previously described by Grossel et al. (136), who also briefly mentioned the synthesis of Ru and Os analogues. Russian workers have shown that the reaction of **49** (L = PPh_3) with the donor ligands L = PPh_3, $AsPh_3$, $SbPh_3$, $P(OEt)_3$, and $P(OPh)_3$ results in site-specific carbonyl substitution exclusively at the Pt center (137). Polarographic studies on the products of these reactions show a completely reversible first reduction but an irreversible second reduction (138). The authors conclude the LUMO is an M–M σ^* orbital rather than a vacant p orbital on the Pt center. An ESR spectrum of the relatively stable radical anion $[\{(PhO)_3P\}_2PtFe_2(CO)_8]^-$ has been obtained with g = 2.054 (139).

Reaction of the phosphaalkyne complex $Pt(dppe)(\eta^2\text{-}P\equiv CtBu)$ with either $Fe_2(CO)_9$ or $Fe_3(CO)_{12}$ affords high yields of the red cluster $Fe_2Pt\text{-}(\mu_3\text{-}(\eta^2\text{-}\perp)\text{-}P\equiv CtBu)(CO)_6(dppe)$ (55), which contains a novel $\mu_3\text{-}(\eta^2\text{-}\perp)$ phosphaalkyne ligand (140,140a). This bonding mode is similar to the alkyne coordination observed in some alkyne clusters, such as $Fe_3(\mu_3\text{-}(\eta^2\text{-}\perp)\text{-}PhC\equiv CPh)(CO)_9$ (141), and the low value of $J(^{195}Pt-^{31}P) = 128$ Hz indicates that the P atom of the phosphaalkyne is not coordinated to the Pt

(CO)₃Fe structure with ᵗBuC≡P, Pt, (CO)₃Fe, PPh₂, PPh₂

55

(CO)₃Fe structure with ᵗBuC≡P, Pt, (CO)₃Fe, PPh₂, PPh₂, Ph₂P, Fe(CO)₄

56

center via its lone pair. A related complex, $Fe_3Pt(\mu_3\text{-}(\eta^2\text{-}\perp)\text{-}P\equiv CtBu)\text{-}(CO)_{10}\text{-}(triphos)$ (56) contains a similar Fe_2Pt(phosphaalkyne) cluster with an $Fe(CO)_4$ group coordinated to the dangling arm of the triphos ligand (140a). Powell et al. (30) have reported spectroscopic evidence for $FePt_2(\mu\text{-}H)(\mu\text{-}CO)\text{-}(\mu\text{-}PR_2)(CO)_3(PCy_3)_2$, which exists as a mixture of interconverting isomers.

2. Tetranuclear Clusters

Reaction of $[Fe_3(\mu\text{-}H)(CO)_{11}]^-$ with $Pt(C_2H_4)_2(PPh_3)$ affords high yields of the anion $[Fe_3Pt(\mu\text{-}H)(\mu\text{-}CO)(CO)_{10}(PPh_3)]^-$ (40) (31). The isolated NEt_3H^+ salt was not amenable to X-ray analysis, but recent studies (142) on the PCy_3 analogue have shown that the 60-electron monoanion is unstable in solution and is easily oxidized to the 59-electron neutral paramagnetic species $Fe_3Pt(\mu\text{-}H)(\mu\text{-}CO)(CO)_{10}(PCy_3)$ (57). The crystal structure (142) of 57 reveals a tetrahedral Fe_3Pt core, and in view of the ready one-electron reversible reduction to the monoanion and the similarity of their IR spectra, it is highly likely that 40 and 57 have analogous structures. Reaction of $[Fe_3\text{-}(\mu\text{-}H)(CO)_{11}]^-$ with the cation $[Pt(COD)(C_8H_{13})]^+$ afforded poor yields of $Fe_2Pt(CO)_8(COD)$, but no Fe_3Pt clusters were isolated (31). Treatment of the methoxymethylidyne species $Fe_3(\mu\text{-}H)(\mu\text{-}COMe)(CO)_{10}$ with $Pt(C_2H_4)_2\text{-}(PPh_3)$ gave the tetrahedral cluster $Fe_3Pt(\mu_3\text{-}H)(\mu_3\text{-}COMe)(CO)_{10}(PPh_3)$ (7) (32).

Treatment of the anion $[Fe_2(\mu\text{-}H)(CO)_8]^-$ with $Pt(C_2H_4)_2(PPh_3)$ does not afford any isolable Fe_2Pt clusters but gives instead the tetranuclear diplatinum species $[Fe_2Pt_2(\mu\text{-}H)(CO)_8(PPh_3)_2]^-$ (58). This anion is easily protonated giving the neutral species $Fe_2Pt_2(\mu\text{-}H)_2(CO)_8(PPh_3)_2$ (59), and X-ray structures on both clusters have been reported (31). They have butterfly metal cores with a long (~ 2.9 Å) nonbonding $Pt\cdots Pt$ interaction, but the mono-

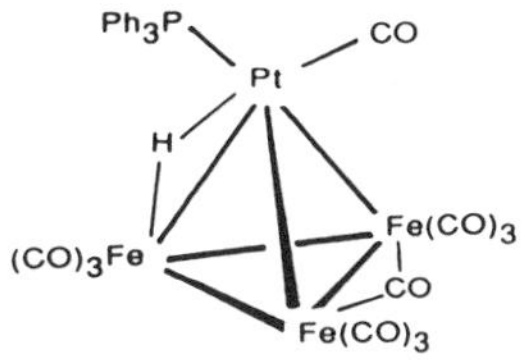

57

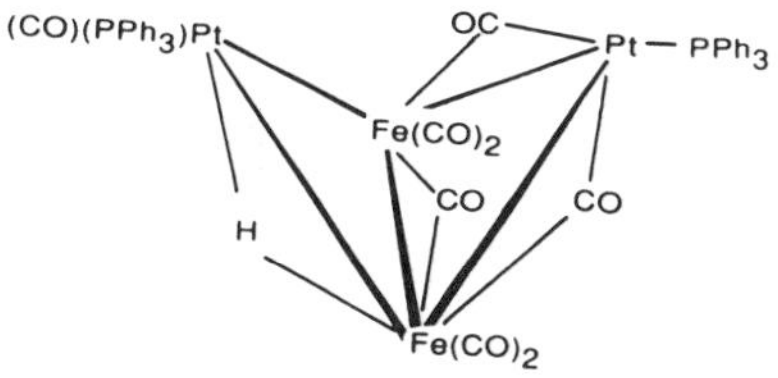

58

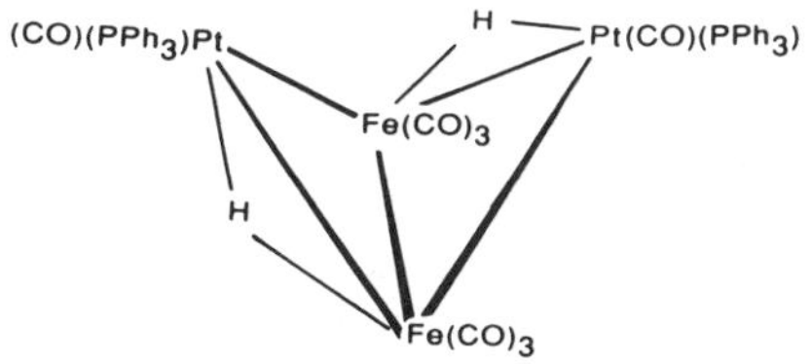

59

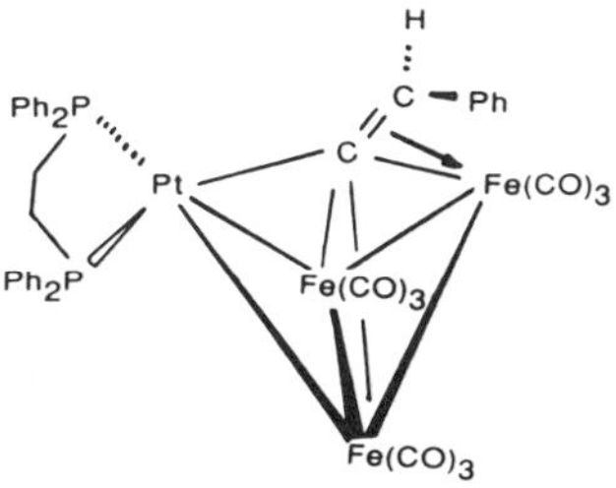

60

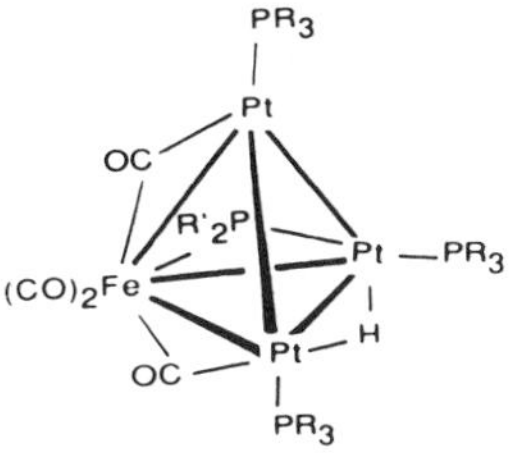

61 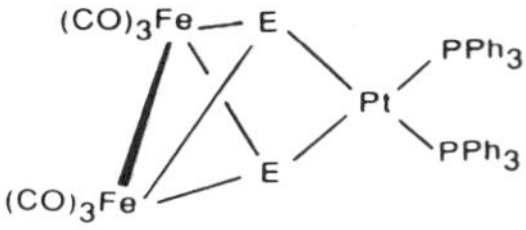   **66** 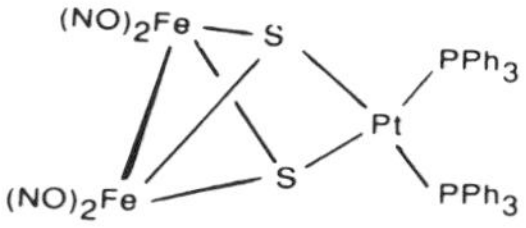   **67**

anion has three bridging carbonyl ligands, whereas the neutral cluster has all terminal COs. The osmium analogue of **59** has been prepared by a different route (see Section V,C,2). The synthesis and crystal structure of the vinylidene cluster $Fe_3Pt(\mu_4-\eta^2C=C(H)Ph)(CO)_9(dppe)$ (**60**) have been briefly reported (*143*). This cluster is closely related to Ru_3Pt cogenors discussed in Section V,B,2. An X-ray study on the tetrahedral 56-CVE cluster $FePt_3(\mu-H)(\mu-CO)_2(\mu-PnPr_2)(CO)_2(PPhtBu_2)_3$ (**61**) has been briefly noted (*30*).

3. High-Nuclearity Clusters

The 86-CVE carbido clusters $[Fe_5Pt(\mu_6-C)(CO)_{14}(PPh_3)]^{2-}$ (*82*), $Fe_5Pt-(\mu_6-C)(CO)_{15}(PPh_3)$ (**12**) (*82*), and $[Fe_5Pt(\mu_6-C)(CO)_{15}]^{2-}$ (*144*) have been reported, and X-ray structures on the latter two confirm the presence of an

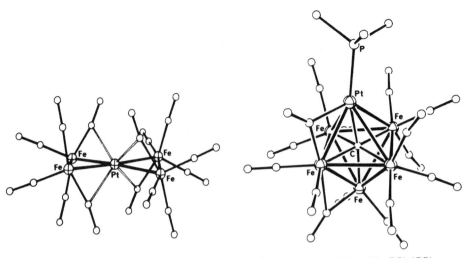

FIG. 12.   Structures of the anion $[Fe_4Pt(CO)_{16}]^{2-}$ (**62**) and $Fe_5Pt(\mu_6\text{-}C)(\mu\text{-}CO)_2(CO)_{13}$-(PPh$_3$) (**12**).

octahedral metal core, which is shown for **12** in Fig. 12. The Pt center is thus behaving as an 18-electron metal and the clusters conform to normal SEPT rules. Lopatin *et al.* (*82*) attribute this to the high stability of the skeletal "superaromatic" electronic system of octahedral carbonyl clusters. Several other clusters with more unusual core structures have been reported. The "bow-tie" anion $[Fe_4M(CO)_{16}]^{2-}$ (**62**, M = Pt, Pd), shown in Fig. 12, has been characterized by Longoni *et al.* (*51*). It is formed by treatment of Pt^{2+} (or Pd^{2+}) salts with various anionic iron clusters such as $[Fe_3(CO)_{11}]^{2-}$, $[Fe_2(CO)_8]^{2-}$, or $[Fe_4(CO)_{13}]^{2-}$. The molar proportions are critical, and increasing Pt^{2+} concentrations leads to the higher nuclearity species $[Fe_3Pt_3$-$(CO)_{15}]^{2-}$ (**20a**) and $[Fe_4Pt_6(CO)_{22}]^{2-}$ (**63**) (*5*). The CVE counts in **20a**, **62**, and **63** may be rationalized by Mingos's rules for condensed polyhedra containing Pt atoms (*79*). The dianion **20a** is easily oxidized to a stable monoanion **20b**, and both species have approximately planar raft structures, shown in Fig. 2. There is a slight distortion from idealized D_{3h} symmetry, because one Fe vertex lies above the Pt_3 plane and the other two Fe vertices lie below (*5*). The recently characterized neutral cogenor **20c** has a similar structure (*97*). Theoretical studies on these systems have been undertaken (see Section III), and the stability of the paramagnetic species is attributed (*96*) to a high degree of delocalization of the odd electron and to the steric difficulties associated with any potential dimerization through the Pt atoms. ^{57}Fe Mössbauer spectra have been recorded (*144a*) for **62** and **20a**. The isomer shift (δ 0.21 mm sec^{-1}) and quadrupole splitting (Δ 0.26 mm sec^{-1}) for **20a** are consistent with the

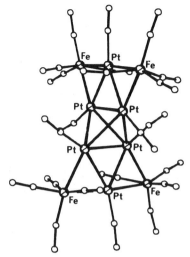

FIG. 13. Structure of the anion $[Fe_4Pt_6(CO)_{22}]^{2-}$ (**63**).

electron-rich distorted $Fe(CO)_4$ units found in the crystal structure (5), but variable-temperature data for **62** suggest that a mixture of isomers exists in the solid phase.

Several clusters formally derived from the planar Fe_3Pt_3 framework have been reported. Removal of one $Fe(CO)_4$ vertex, and orthogonal condensation along the *pseudo-C*$_2$ axis of two resulting trapezoidal $[Fe_2Pt_3(CO)_{11}]^-$ moieties, produce the metal geometry found in the dianion **63** (5) (Fig. 13). This contains a central tetrahedron of Pt atoms, and a similar feature is found in the skeletal structures of the recently reported clusters $FePt_5(CO)_9(PEt_3)_4$ (**64**) (*134*) and $Fe_2Pt_5(CO)_{12}(COD)_2$ (**65**) (*97*) (Fig. 14). These are formally related to **63** by removal of one $PtL\{Fe(CO)_4\}_2$ unit, and in the case of **64**, an additional $Fe(CO)_4$ vertex. These examples illustrate possible steps in the buildup of high-nuclearity heterometallic frameworks via sequential addition of mononuclear metal fragments.

4. Chalcogenide Clusters

A few Fe–Pt clusters containing chalcogenides have been reported, and, although these do not contain direct Fe–Pt bonds, they are included here because the synthetic routes are very similar to those that afford Os–Pt sulfido clusters (Sections V,C,2 and V,C,3). Rauchfuss *et al.* (*145,145a*) report that $Pt(C_2H_4)(PPh_3)_2$ reacts with $Fe_2(\mu\text{-}E_2)(CO)_6$ (E = S, Se, Te) to insert a $Pt(PPh_3)_2$ unit into the E—E bond, affording $Fe_2Pt(\mu_3\text{-}E)_2(CO)_6(PPh_3)_2$

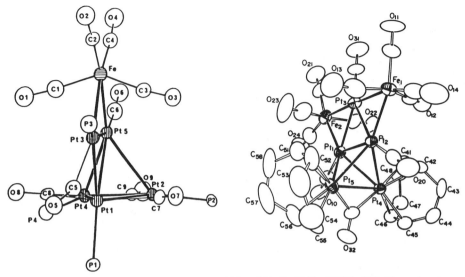

FIG. 14.   Structures of FePt$_5$(CO)$_9$(PEt$_3$)$_4$ (**64**) and Fe$_2$Pt$_5$(CO)$_{12}$(COD)$_2$ (**65**). [Reproduced with permission from Bender *et al.* (*134*) and Adams *et al.* (*97*), respectively.]

(**66a**, E = S; **66b**, E = Se; **66c**, E = Te). Seyferth *et al.* have independently reported the preparation of **66a** either from Fe$_2$(μ-S$_2$)(CO)$_6$ and Pt(PPh$_3$)$_4$ (*146*) or from [Fe$_2$(μ-S)$_2$(CO)$_6$]$^{2-}$ and *cis*-PtCl$_2$(PPh$_3$)$_2$ (*147*). The crystal structure of **66b** has been determined (*145a*). In addition, clusters **66a** and **66c** may be prepared using Pt(C$_2$H$_4$)(PPh$_3$)$_2$ and Fe$_3$(μ_3-E)$_2$(CO)$_9$ (*148*), and Mathur *et al.* (*149,150*) report that the latter reagent also reacts with Pt(PPh$_3$)$_4$, affording **66a–c**. Seyferth and co-workers (*151*) have prepared and structurally characterized Fe$_2$Pt(μ_3-S)$_2$(NO)$_4$(PPh$_3$)$_2$ (**67**) from the reaction of Roussin's red salt [Fe$_2$(μ-S)$_2$(NO)$_4$]$^{2-}$ with *cis*-PtCl$_2$(PPh$_3$)$_2$. The Fe—Fe distance of 2.802(5) Å in **67** is considerably longer than that in **66b** [2.533(2) Å] (*145a*), though both species require a single Fe—Fe bond in order to conform to the 18-electron rule.

B. *Ruthenium–Platinum Clusters*

1. *Trinuclear Clusters*

The reaction of Ru$_3$(CO)$_{12}$ with zero-valent Pt reagents is rather slow. Using PtL$_4$ or Pt(PPh$_3$)$_2$(stilbene), the clusters RuPt$_2$(μ-CO)$_3$(CO)$_2$(L)$_3$ [**48b**, L = PPh$_2$Me, PMe$_2$Ph, PPh$_3$, P(OPh)$_3$, PPh(OMe)$_2$, and AsPh$_3$] have been prepared (*10,37*) and the PPh$_2$Me (*130*) and PPh$_3$ (*131*) derivatives have been structurally characterized. These clusters (Fig. 11) have the unusual feature of an axially bound phosphine ligand on the Ru center. Presumably steric factors, which usually mitigate against phosphine ligation in this crowded site, are favorable in this case, as the other axial sites on the Pt centers are unoccupied. For L = PMe$_2$Ph, the diruthenium–platinum cluster M$_2$Pt(μ-CO)$_3$(CO)$_4$(PR$_3$)$_3$ (**68a**, M = Ru, R$_3$ = Me$_2$Ph) is also formed, whereas for

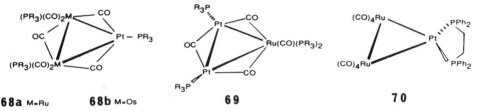

68a M=Ru   **68b** M=Os   **69**   **70**

L = P(OPh)$_3$ and PPh(OMe)$_2$, the tetra-substituted RuPt$_2$ species RuPt$_2$-(μ-CO)$_3$(CO)(L)$_4$ (**69**) are also observed (*37*). With the reagent Pt(dppe)$_2$, phosphine transfer between Pt and Ru does not occur, and the cluster Ru$_2$Pt(CO)$_8$(dppe) (**70**), structurally analogous to the Fe–Pt species **50**, is obtained (*37*).

In closely related, work·Bruce *et al.* (*131*) have studied the much faster reaction of Ru$_3$(CO)$_{11}$(CNtBu) with Pt(C$_2$H$_4$)$_2$(PPh$_3$). This reaction is complex, and an intermediate crimson species, tentatively identified as "RuPt$_2$(CO)$_4$-(PPh$_3$)$_3$" is initially formed at $-20°$C. This intermediate decomposes on warming to give at least 13 complexes, including the Ru–Pt clusters **68a** (R$_3$ = Ph$_3$), **48b** (R$_3$ = Ph$_3$), and Ru$_2$Pt$_2$(CO)$_9$(CNtBu)(PPh$_3$) (**71**). Reflux of the reaction mixture leads to the hexanuclear clusters Ru$_2$Pt$_4$(CO)$_{9-n}$-(CNtBu)(PPh$_3$)$_n$ (**72**, $n = 4$, 5), which were identified spectroscopically. The clusters RuPt$_2$(μ-CO)$_3$(CO)(CNtBu)(PPh$_3$)(L)$_2$ [(L)$_2$ = (CO)$_2$, (PPh$_3$)$_2$, (PPh$_3$)(P{OMe}$_3$)], and **69** [(L)$_4$ = (PPh$_3$)$_3$\{P(OMe)$_3$\}] and Ru$_2$Pt(μ-CO)$_3$-(CO)$_2$(CNtBu)(PMe$_3$)(PPh$_3$)$_3$, derived from the crimson intermediate, are also reported, as well as several other tentatively characterized derivatives (*131*).

Treatment of Ru$_2$(μ-CH$_2$)(μ-CO)(CO)$_2$Cp$_2$ with Pt(C$_2$H$_4$)$_2$(PR$_3$)(R = Cy, iPr) affords the trinuclear cluster Ru$_2$Pt(μ-CH$_2$)(μ-CO)(CO)$_2$(PR$_3$)Cp$_2$ (**73**)

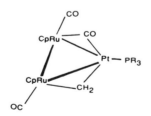

73

(*116*). An X-ray structure shows that the two Ru—Pt bonds are bridged by a carbonyl and methylene group, and hence the methylene group has migrated from the Ru—Ru edge during the reaction.

2. Tetranuclear Clusters

A few examples of tetranuclear species have been reported. The previously mentioned reaction of $Pt(C_2H_4)_2(PR_3)$ with $Ru_2(\mu\text{-}CH_2)(\mu\text{-}CO)(CO)(L)$-$Cp_2$ (L = CO or MeCN) yields, in addition to cluster 73, the methylidyne species $Ru_2Pt_2(\mu\text{-}H)(\mu_4\text{-}CH)(\mu\text{-}CO)(CO)_2(PR_3)_2Cp_2$ (45) (for L = CO) or the carbido cluster $Ru_2Pt_2(\mu\text{-}H)_2(\mu_4\text{-}C)(\mu\text{-}CO)_2(PR_3)_2Cp_2$ (44) (for L = MeCN) (*116*). The structures of both these complexes (Fig. 9) are of some interest, in that oxidative addition into the C—H bonds of the CH_2 ligand occurs; 45 has the unique feature of a symmetric μ_4-methylidyne group [NMR chemical shifts $\delta(^1H) \sim 14.9$, $\delta(^{13}C) \sim 248$]. The Ru_2Pt_2 core is not exactly planar, and this cluster is thought to model the binding of a C_1 fragment on the (100) surface of a body-centered cubic metal. Cluster 44 has a folded pseudo-butterfly Ru_2Pt_2 core with a nonbonding Pt···Pt separation, and the coordination geometry of the carbido ligand is similar to that observed in several other butterfly $M_4(\mu_4\text{-}C)$ clusters. The spiked-triangular cluster $Ru_3Pt(\mu\text{-}H)$-$(\mu_4\text{-}\eta^2\text{-}C{\equiv}CtBu)(CO)_9(COD)$ (33), prepared from the reaction of $Pt(COD)_2$ with $Ru_3(\mu\text{-}H)(\mu_3\text{-}\eta^2\text{-}C{\equiv}CtBu)(CO)_9$, has a labile COD ligand easily displaced by diphosphines such as dppe, giving 34. As mentioned in Section IV,E, cluster 34 isomerizes reversibly to the vinylidene species 37, with a concomitant change in the metal core geometry from spiked-triangular to butterfly. Protonation of 34 with $HBF_4 \cdot Et_2O$ occurs at the α-alkynyl carbon, affording the butterfly alkyne cluster 74, whereas protonation of 37 occurs at the hinge Ru—Ru vector, giving the metal hydride species 38. Cluster 74 irreversibly isomerizes to 38 over a period of 12 hr (*113*). The reactions are summarized in Scheme 2.

3. High-Nuclearity Clusters

To date only one such cluster has been reported. The tetranuclear cluster 33 is unstable in solution and over a period of several days eliminates the elements of $Pt(COD)_2$, giving the Pt bridged cluster $Pt[Ru_3(\mu\text{-}H)(\mu_4\text{-}\eta^2\text{-}C{\equiv}CtBu)(CO)_9]_2$ (75) in reasonable yields (40–50%) (*152*). The C_2 metal core (Fig. 15) is chiral, and a variable-temperature ^{13}C NMR study showed that the cluster undergoes enantiomerization, with $\Delta G^{\ddagger}_{266} = 57$ kJ mol^{-1}. The enantiomerization presumably proceeds via an intermediate or transition state with a planar coordination at the Pt atom.

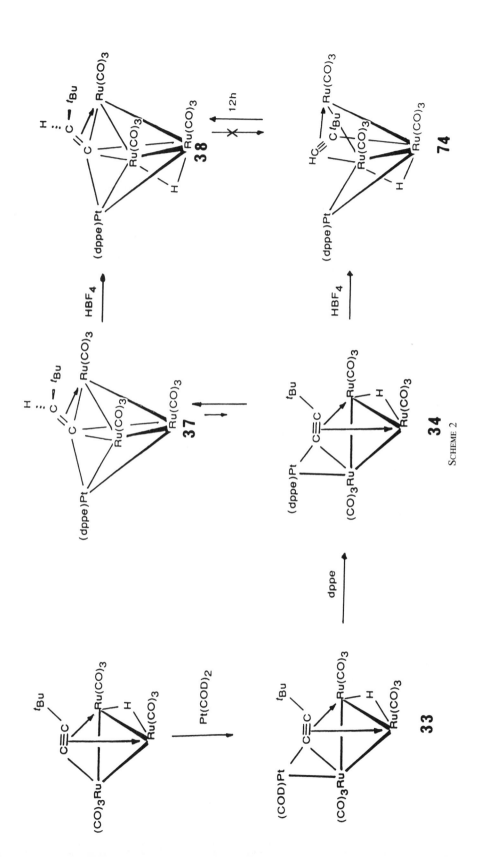

SCHEME 2

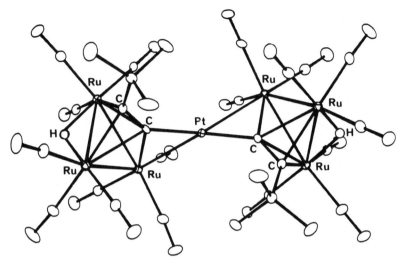

FIG. 15.   Structure of $Pt[Ru_3(\mu\text{-}H)(\mu_4\text{-}\eta^2\text{-}C\equiv CtBu)(CO)_9]_2$ (**75**).

C. *Osmium–Platinum Clusters*

These form the largest single group of hetero-Pt clusters, with more than 40 crystal structures (Table V) reported. Their chemistry, particularly with regard to triosmium–platinum species, is the most developed of any hetero-Pt clusters.

1. *Trinuclear Clusters*

$Os_3(CO)_{12}$ is too unreactive to afford any mixed-metal species with zero-valent Pt complexes, in contrast to $Fe_3(CO)_{12}$ and $Ru_3(CO)_{12}$. However, treatment of the more reactive *cis*-$OsH_2(CO)_4$ with $Pt(C_2H_4)(PPh_3)_2$ affords $OsPt_2(\mu\text{-}CO)_3(CO)_2(PR_3)_3$ (**48c**, $R_3 = Ph_3$) (*10,37*), which has been shown (*132*) to be isostructural with the Ru analogue. Reaction of *cis*-$OsH_2(CO)_4$ with $Pt(PPh_2Me)_4$ gave **48c** ($R_3 = Ph_2Me$) and $M_2Pt(\mu\text{-}CO)_3(CO)_4(PR_3)_3$ (**68b**, M = Os, $R_3 = Ph_2Me$). The latter species is presumed to be isostructural with the Ru analogue on the basis of IR data, although no crystal structures of either cogenor have been reported (*10,37*). It is interesting to note that for the heavier metals Ru and Os, the phosphine-substituted clusters in the series $M_2Pt(CO)_{10-n}L_n$ generally have structure **68** with bridging carbonyls, with the exception of the dppe derivative **70** (*37*). However, for M = Fe, structures **49** and **50**, with all terminal COs, are observed (*36,133,134*). Nevertheless, the homoleptic Os–Pt cluster $Os_2Pt(CO)_{10}$ (**76**) is found (*153*)

76   **77**   **78**

to have only terminal carbonyls. This behavior is indicative of the small energy differences between structures with bridged and unbridged carbonyls.

The diplatinum–osmium clusters $OsPt_2(\mu_3-(\eta^2-\|)MeC\equiv CMe)(CO)_6-$
$(PPh_3)(L)$ (**36a**, L = PPh_3 and **36b**, L = CO) are formed on treatment of $Os_3Pt(\mu-H)_2(CO)_{10}(PPh_3)$ with but-2-yne, together with other homonuclear Os species (47). Cluster **36** is unusual in that it has two Pt atoms but a CVE count of 46. However, the crystal structure (47) of **36a** shows a long Pt···Pt distance of 3.033(2) Å, which together with the small value of $J(Pt-Pt) =$ 57 Hz implies only a weak direct Pt–Pt interaction.

2. Tetranuclear Clusters

The first reported tetranuclear cluster was $Os_2Pt_2(\mu-H)_2(CO)_8(PR_3)_2$ (**15**, $R_3 = Ph_3$) (37), prepared in poor yield from the reaction of cis-$OsH_2(CO)_4$ with $Pt(C_2H_4)(PPh_3)_2$. An improved synthesis (41) using the bis-ethylene complexes $Pt(C_2H_4)_2(PR_3)$ (R = Ph, Cy, Me) led to structure determinations on the PPh_3 (41) and PCy_3 (84) derivatives some years later. The butterfly metal core is similar in both derivatives, with a nonbonded Pt···Pt wing-tip distance of 3.206(1) Å (PPh_3) and 3.230(1) Å (PCy_3), consistent with two 16-electron Pt centers, and the 58-CVE count. The chemically equivalent hydrides undergo mutual exchange (see Section IV,C).

Numerous triosmium–platinum clusters have been prepared, many of them derived from $Os_3Pt(\mu-H)_2(CO)_{10}(PR_3)$ (**1**) (40,40a). Although, as discussed in Section III, the participation in cluster bonding of the tangential orbitals of nonconical PtL_n units may be difficult to predict a priori, the observed tetrahedral skeleton of **1** (40,40a) and nonplanar coordination geometry at the Pt atom suggest that the Pt atom should behave as an 18-electron center. The 58-CVE count therefore implies that the cluster is unsaturated, and this is consistent with EHMO studies (88,98) and the high chemical reactivity of **1**. The crystal structures of **1a** (40,40a) and the 58-CVE derivatives **21** (108) and **22** (118) all show unusually short Os(μ-H)Os distances [2.789(1),

2.752(1),and 2.747(1) Å, respectively], suggesting localization of unsaturation at this bond. Localized unsaturation at the hydride-bridged Os—Os
bond is also indicated by an EHMO study (88).

At ambient-temperatures dark green **1** reacts immediately with two electron donor ligands L, affording the yellow or orange 60-CVE adducts
$Os_3Pt(\mu-H)_2(CO)_{10}(PR_3)(L)$ (47,89a,90,118). Depending on the nature of
the ligand L, these adducts may either have a butterfly metal core, e.g., in **2**
(47,91) or **19** (89a,90a), or tetrahedral cores, e.g., in $Os_3Pt(\mu-H)_4(CO)_{10}(PCy_3)$
(**77**) (89,89a), **17** (89,89a), or **18** (90). The high reactivity of **1** is reminiscent
of that of the unsaturated cluster $Os_3(\mu-H)_2(CO)_{10}$ (26), but there are two important differences that mean that Os_3Pt chemistry is less well developed
than Os_3 chemistry. First, under more demanding reaction conditions, the
Os_3Pt core is not robust and fragmentation products are observed (47,154).
Second, a mixture of isomers is often formed, whose relative stabilities may
be markedly dependent on the ancilliary ligands. Thus at ambient temperatures the reaction of **1a** with RNC (R = Cy, tBu) results initially in three
spectroscopically observed isomers of the adduct $Os_3Pt(\mu-H)_2(CO)_{10}$-
$(PCy_3)(CNR)$, which are thought to differ in the dispositions of the isocyanide
and hydride ligands. On thermal equilibration, one major isomer **78** is formed
(118), and this is also the principal product from the reaction at 195 K. Likewise, the products of reaction of **1** with SO_2 are dependent on the reaction
temperature (90). For **1a**, low-temperature NMR data indicate that the major
species formed has a Pt-coordinated SO_2 ligand and a butterfly metal core,
and the spectroscopic data suggest the structure **79**. On warming, this cluster

79   **80**   **81**

rearranges to a mixture of complexes, the major species being the C_1 tetrahedral isomer of $Os_3Pt(\mu-H)_2(\mu-SO_2)(CO)_{10}(PCy_3)$ (**18**) with an SO_2 ligand
bridging an Os—Os bond. However, reaction of **1b** ($R_3 = Ph_3$) with SO_2 at
ambient temperature gives primarily the C_s tetrahedral isomer **80** (90).

The butterfly phosphine adducts $Os_3Pt(\mu-H)_2(CO)_{10}(PR_3)(PR_3')$ (**2**) may
also exist as isomers in solution, depending on the phosphine substituents.

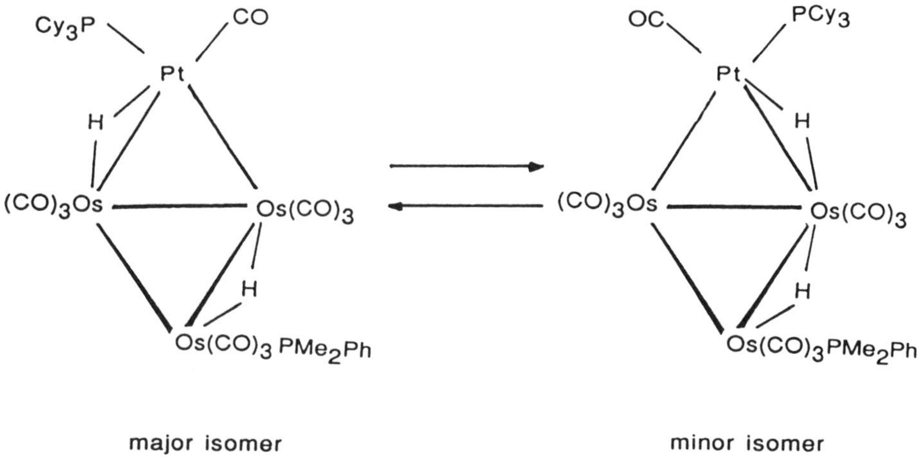

major isomer   minor isomer

SCHEME 3

For $R = R' = Cy$, only one solution species is detectable over a wide temperature range according to ^1H and ^{31}P NMR (106,108), whereas for $R = R' = Ph$ a small amount ($\sim 1-2\%$) of a second isomer is observed at low temperatures, which exchanges with the main species at higher temperatures (47). For the mixed species $PR_3 = PCy_3$, $PR'_3 = PMe_2Ph$, a second isomer is present at $\sim 10\%$, which also exchanges with the main species (106). The probable structures are shown in Scheme 3, and it is likely that exchange occurs via a rotation of the $Pt(H)(CO)(PCy_3)$ group (see Section IV,A). The methylene adduct $Os_3Pt(\mu\text{-}H)_2(\mu\text{-}CH_2)(CO)_{10}(PCy_3)$ (17) exists as a mixture of C_s and C_1 isomers ($\sim 1:4$ ratio) that are in thermodynamic equilibrium, and are formally related by a rotation of the $Pt(CO)(PCy_3)$ group (89,89a). In contrast, the isocyanide derivative 24 exists solely as the C_1 isomer in solution (107,118).

Several of the 60-CVE adducts of 1 are unstable to loss of a ligand, reverting to 58-CVE unsaturated clusters. The reactions of 1 with CO and H_2 are readily reversible (47,89,89a), and the adducts are only stable in solution in the presence of excess reagent. The adducts $Os_3Pt(\mu\text{-}H)_2(CO)_{10}(PCy_3)(CNR)$ (78, R = Cy, tBu) readily lose CO, giving cluster 22, in which the isocyanide is ligated to Pt (118). Interestingly, treatment of the 22 with CO, even at 200 K, affords the same isomer 78 as results from reaction of 1 with isocyanides (118). This indicates that the opening of the tetrahedron to give a butterfly is accompanied by a migration of the isocyanide from the Pt atom to an Os center. These results, in conjunction with the low-temperature reaction of 1a with SO_2 (90), indicate that the initial site of nucleophilic attack on 1 may be at the Pt center rather than at an Os center, as would be predicted from orbital considerations. Reactions of 1 are summarized in Scheme 4.

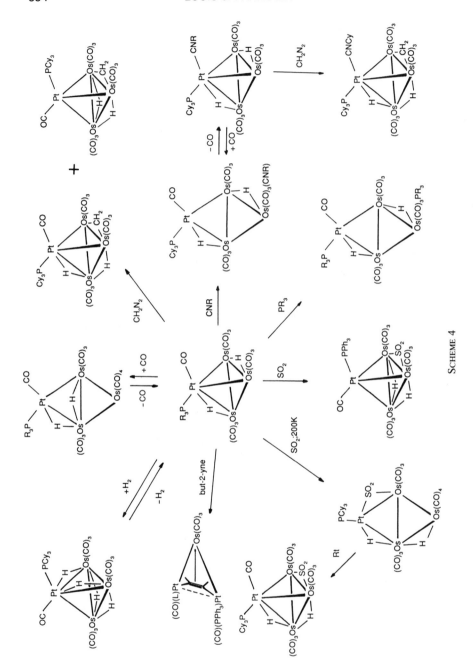

SCHEME 4

In addition to the tetrahedral and butterfly adducts discussed above, there are some Os_3Pt clusters prepared from **1a** that have more open structures. Treatment of **1a** with $Li^+C\equiv CPh^-$ followed by protonation affords low yields of the spiked triangular $Os_3Pt(\mu\text{-}H)(\mu_4\text{-}\eta^2\text{-}C\equiv CPh)(CO)_{10}(PCy_3)$ **(81)** (*155*), a 62-CVE cluster analogous to several Ru_3Pt species (Scheme 2). Reaction with the thioketene 1,1,3,3-tetramethyl-2-thiocarbonylcyclohexane yields the 64-CVE cluster $Os_3Pt(\mu_3\text{-}S)_2(CO)_9(PCy_3)(\eta^1\text{-}C{=}C_6H_6Me_4)$ **(46)** as the major characterized product (*125*). This cluster contains an unusual η^1-vinylidene ligand coordinated to an Os atom, and is formally related to several disulfido Os_3Pt clusters such as **10** described by Adams *et al.* (*39,39a,44*). Reaction of **1a** with 2 mol of $SnCl_2$ affords high yields of $Os_3PtSn(\mu\text{-}H)_2(Cl)(OEt_2)(SnCl_3)(CO)_{10}(PCy_3)$ **(82)**. This 62-CVE cluster (Fig.16) has a central Os_2SnPt tetrahedron, with one Os–Sn edge bridged by an $Os(CO)_4$ moiety, resulting in a spiked triangular Os_3Pt core (*88*).

Several Os_3Pt clusters have been prepared by other routes. Adams and co-workers (*33,38*) report that reaction of $Os_3(\mu_3\text{-}S)(CO)_{10}$ with $Pt(PMe_2Ph)_4$

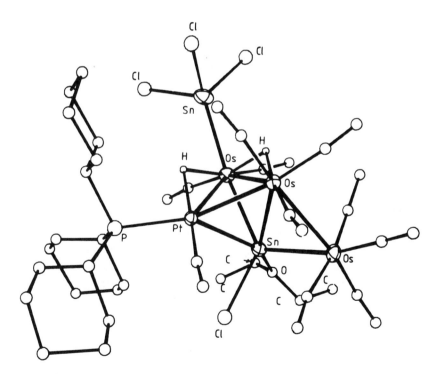

FIG. 16.   Structure of $Os_3PtSn(\mu\text{-}H)_2(Cl)(OEt_2)(SnCl_3)(CO)_{10}(PCy_3)$ **(82)**.

83

84

85

affords the clusters $Os_3Pt(\mu_3\text{-}S)(CO)_{10}(PMe_2Ph)_2$ (9), $Os_3Pt(\mu_3\text{-}S)(CO)_9$-$(PMe_2Ph)_3$ (83), $Os_3Pt(\mu_3\text{-}S)(CO)_8(PMe_2Ph)_3$ (35), and $Os_3Pt(\mu_3\text{-}S)(CO)_9$-$(PMe_2Ph)_2$ (84). Clusters 9 and 83 are 62-CVE butterfly clusters with nearly planar metal cores, and 35 and 84 are tetrahedral 60-CVE clusters. The pairs of clusters with the same number of phosphine ligands, i.e., 9 and 84, and 83 and 35, can be readily interconverted by addition or loss of carbon monoxide. UV photolysis of 35 results in further loss of one CO ligand and consequent orthometallation of a phenyl group at an osmium center forming Os_3Pt-$(\mu\text{-}H)(\mu\text{-}PMe_2C_6H_4)(\mu_3\text{-}S)(CO)_7(PMe_2Ph)$ (85) (33). This reaction also involves a migration of a CO and PMe_2Ph ligand between Pt and Os atoms.

86 L-CO   **87** L=PPh₃   **89** LL=COD   **90** LL=(CO)₂   **91**

Reaction of $Os_3(\mu_3\text{-}S)_2(CO)_9$ with $Pt(C_2H_4)(PPh_3)_2$ or $Pt(PMe_2Ph)_4$ yields $Os_3Pt(\mu_3\text{-}S)_2(CO)_{11-n}(PPh_3)_n$ (86, $n = 1$ and 87, $n = 2$) (44) or cluster 10 (39), respectively. These clusters have the same $Os_3Pt(\mu_3\text{-}S)_2$ core, and an interesting skeletal rearrangement observed for 10 (39,39a) is discussed in Section IV,D. For these and related sulfido clusters it is believed (33,38,39,39a,43,43a,44) that initial coordination of the Pt atom to the cluster via a lone pair on a sulfur atom plays a pivotal role in the synthetic pathway.

Reaction of the semibridging methylidyne cluster $Os_3(\mu\text{-}H)(\mu_3\text{-}CH)(CO)_{10}$ with $Pt(C_2H_4)_2(PCy_3)$ affords the tetranuclear 62-CVE carbido species $Os_3Pt(\mu\text{-}H)_2(\mu_4\text{-}C)(CO)_{10}(PCy_3)$ (88), together with a pentanuclear Os_3Pt_2

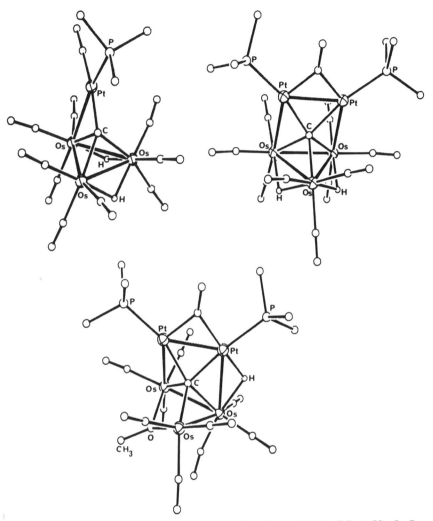

FIG. 17. Structures of the carbido clusters $Os_3Pt(\mu\text{-}H)_2(\mu_4\text{-}C)(CO)_{10}(PCy_3)$ (**88**), Os_3Pt_2-$(\mu\text{-}H)_2(\mu_5\text{-}C)(\mu\text{-}CO)(CO)_9(PCy_3)_2$ (**103**), and $Os_3Pt_2(\mu\text{-}H)(\mu_5\text{-}C)(\mu\text{-}OMe)(\mu\text{-}CO)(CO)_9(PCy_3)_2$ (**104**).

cluster discussed later (*42*). Cluster **88** contains a spiked triangular metal core, with a relatively exposed carbide (Fig. 17), though no chemistry at this center has been reported. The 60-CVE tetrahedral clusters $Os_3Pt(\mu\text{-}H)_2(\mu\text{-}X)$-$(CO)_9(COD)$ have been prepared by reaction of $Os_3(\mu\text{-}H)_2(CO)_{10}$ with either $PtMe_2(COD)$ (**89**, $X = \mu\text{-}CH_2$) (*156*) or $Pt(COD)_2$ (**8**, $X = \mu\text{-}CO$) (*80*), though a better route to the latter cluster is to use **33** as a reagent to transfer a Pt(COD) unit (80). The COD ligand is easily displaced by CO in both species, affording $Os_3Pt(\mu\text{-}H)_2(\mu\text{-}X)(CO)_{11}$ (**90**, X = CH_2) (156) or **91** (X = CO) (80). Both **90** and **91** have only been identified spectroscopically but presumably also contain tetrahedral Os_3Pt cores. Geoffroy et al. (157) have reported that treatment of $Os_3(\mu\text{-}CH_2)(CO)_{11}$ with $Pt(C_2H_4)(PPh_3)_2$ affords the spiked triangular cluster $Os_3Pt(\mu\text{-}CH_2)(CO)_{11}(PPh_3)_2$ (**92**), which has the $\mu\text{-}CH_2$

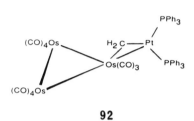

92

group bridging the exo-Os—Pt bond. Unlike $Os_3(\mu\text{-}CH_2)(\mu\text{-}CO)(CO)_{10}$, cluster **92** reacts with CO to replace a PPh_3 ligand, rather than form a $\mu\text{-}CH_2CO$ moiety (157).

3. High-Nuclearity Clusters

Farrar et al. (6,7,85,127,158) have described the reactions of $Os_6(CO)_{18-n}$-$(NCMe)_n$ (n = 0, 1, 2) with $Pt(COD)_2$, which affords several high-nuclearity Os–Pt clusters with unusual core variations. $Os_6(CO)_{18}$ reacts with $Pt(COD)_2$ under ethylene atmosphere to give $Os_6Pt_2(CO)_{17}(COD)_2$ (**93**) (158). This 110-CVE cluster (Fig. 18) has a bicapped octahedral framework, in line with SEPT adapted for capped structures (58,72), and is electronically saturated. Using the "lightly stabilized" cluster $Os_6(CO)_{17}(NCMe)$, in addition to **93** (7), two other crystallographically characterized products are formed (127) that retain the coordinated acetonitrile ligand $Os_6Pt(\mu_3\text{-}NCMe)(CO)_{17}(COD)$ (**47a**) and $Os_6Pt(\mu_4\text{-}NCMe)(CO)_{17}(COD)$ (**47b**) (127). Cluster **47a** rearranges to **47b** (see Fig. 10), and both clusters have unusual open core structures that are difficult to rationalize by electron counting rules. Treatment of either **47a** or **47b** with CO results in cluster degradation, involving loss of the Pt atom and formation of $Os_6(\mu_3\text{-}NCMe)(CO)_{19}$ (127).

Reaction of the bis(acetonitrile) derivative $Os_3(CO)_{16}(NCMe)_2$ with $Pt(COD)_2$ leads to the formation of two structurally characterized clusters, $Os_6Pt_2(CO)_{16}(COD)_2$ (**94**) (6,7) and $Os_2Pt_2(\mu\text{-}H)(\mu_3\text{-}H)(\mu\text{-}C_8H_{10})(CO)_{16}$-(COD) (**95**) (85). Cluster **94** (Fig. 19) has a Pt-bicapped tetrahedron of Os atoms linked to an Os_4 tetrahedron via an Os–Os edge. The 108-CVE count can only be rationalized within the condensed polyhedral approach (58,61,63),

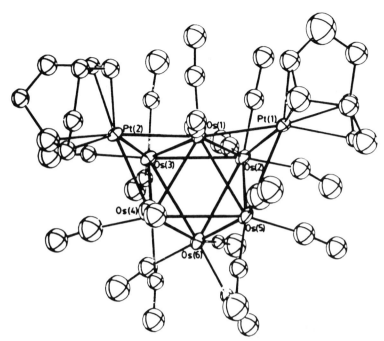

FIG. 18. Structure of $Os_6Pt_2(CO)_{17}(COD)_2$ (**93**). [Reproduced with permission from Couture *et al.* (*158*).]

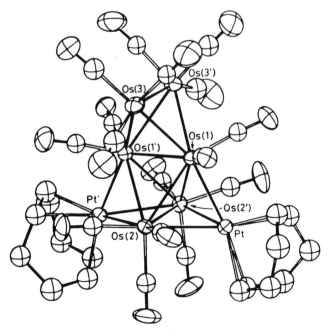

FIG. 19. Structure of $Os_6Pt_2(CO)_{16}(COD)_2$ (**94**). [Reproduced with permission from Couture and Farrar (*6*).]

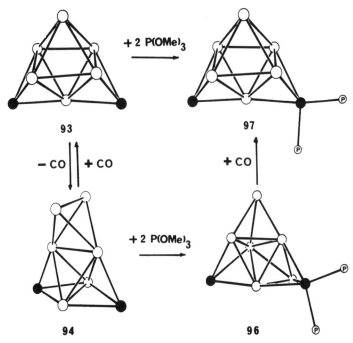

FIG. 20. The relationship between the metal skeletal cores in $Os_6Pt_2(CO)_{17}(COD)_2$ (93), $Os_6Pt_2(CO)_{16}(COD)_2$ (94), $Os_6Pt_2(CO)_{16}[P(OMe)_3]_2(COD)$ (96), and $Os_6Pt_2(CO)_{17}$-$[P(OMe)_3]_2(COD)$ (97).

assuming the cluster is unsaturated, and there is some evidence that this is the case. There are two unusually short Os—Os bonds in 94 [Os(1)—Os(1') = 2.537(1) Å and Os(3)—Os(3') = 2.647(1) Å], and 94 reacts immediately with CO to form 93. Reaction of 94 with $P(OMe)_3$, however, leads to displacement of one COD ligand only, giving the 108-CVE species $Os_6Pt_2(CO)_{16}$-$[P(OMe)_3]_2(COD)$ (96) (7). The core structure of 96 (bicapped tetrahedron of Os atoms with two capping Pt atoms) is different from that found in 94, and is consistent with SEPT adapted for condensed polyhedra. Nevertheless, 96 reacts immediately with CO, affording the heptadecacarbonyl derivative $Os_6Pt_2(CO)_{17}[P(OMe)_3]_2(COD)$ (97) (7), which has the same core structure as 93. These cluster core rearrangements (Fig. 20), in which the central Os_6 unit changes from an octahedron to a bicapped tetrahedron, involve the diamond–square–diamond process that is often postulated in skeletal reorganizations (121–123). Cluster 95 has a very unusual core geometry (Fig. 21), with an Os_6 bicapped tetrahedron edge bridged by a Pt_2 unit through one Pt atom (85). The Pt atoms are bridged by a μ-C_8H_{10} unit, which is bonded to

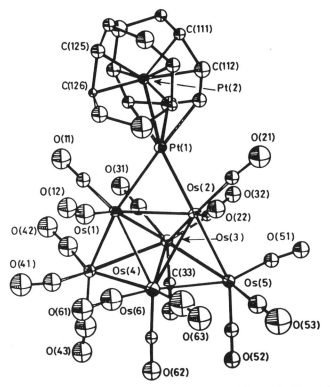

FIG. 21. Structure of $Os_6Pt_2(\mu\text{-}H)(\mu_3\text{-}H)(\mu\text{-}C_8H_{10})(CO)_{16}(COD)$ (**95**). [Reproduced with permission from Couture and Farrar (*85*).]

Pt(1) via an allylic link and to Pt(2) via an eneyl link. Potential energy minimization calculations (*159*) indicate that the chemically distinct hydride ligands [which were originally assigned terminal sites on Os(1) and Os(2) (*85*)] bridge the Os(1)–Os(2) edge and the Os(1)–Os(2)–Os(6) face.

Adams and co-workers (*39a,43,43a,160*) have prepared several high-nuclearity species by reaction of osmium sulfido clusters with mononuclear Pt reagents. Treatment of $Os_4(\mu_3\text{-}S)_2(CO)_{12}$ with $Pt(PMe_2Ph)_4$ afforded $Os_4Pt\text{-}(\mu_3\text{-}S)_2(CO)_{11}(PMe_2Ph)_2$ (**98**) together with **10** (*39a*). The metal framework of **98** (Fig. 22) consists of a Pt-bridged Os_4 butterfly. Reaction of Pt-$(C_2H_4)(PPh_3)_2$ with the square–pyramidal $Os_5(\mu_4\text{-}S)(CO)_{15}$ affords three structurally characterized Os–Pt clusters, $Os_4Pt(\mu_4\text{-}S)(CO)_{13}(PPh_3)$ (**99**), $Os_5Pt(\mu_4\text{-}S)(CO)_{15}(PPh_3)_n$ (**100**, $n = 1$), and **101** ($n = 2$) (*43,43a*). All three contain nonplanar Pt centers and are based on square-pyramidal metal arrays (Fig. 22). The CVE counts for clusters **98–101** are all consistent with SEPT

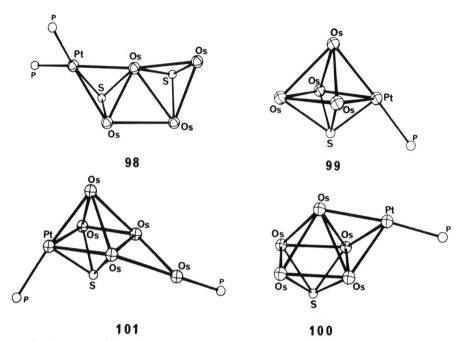

FIG. 22. Metal skeletal structures in $Os_4Pt(\mu_3\text{-}S)_2(CO)_{11}(PMe_2Ph)_2$ (**98**), $Os_4Pt(\mu_4\text{-}S)$-$(CO)_{13}(PPh_3)$ (**99**), $Os_5Pt(\mu_4\text{-}S)(CO)_{15}(PPh_3)$ (**100**), and $Os_5Pt(\mu_4\text{-}S)(CO)_{15}(PPh_3)_2$ (**101**).

adapted for condensed polyhedra. The metal–metal separations show considerable variation, and this has been attributed to the asymmetrical ligand distribution (*43a*). An interesting dimeric structure is produced on decarbonylation of $Os_3Pt(\mu_3\text{-}S)_2(CO)_{10}(PPh_3)$ with $Me_3NO \cdot 2H_2O$ (*160*). The two trigonal–prismatic subunits of the resulting cluster $[Os_3Pt(\mu\text{-}H)(\mu_3\text{-}S)(\mu_4\text{-}S)$-$(\mu\text{-}PPh_2C_6H_4)(CO)_8]_2$ (**102**) are linked by μ_4-sulfido groups with "inverted" tetrahedral geometry (Fig. 23).

Reaction of the methylidyne cluster $Os_3(\mu\text{-}H)(\mu_3\text{-}CH)(CO)_{10}$ with Pt-$(C_2H_4)_2(PCy_3)$ affords a mixture of two carbido clusters, the tetranuclear **88** and the pentanuclear $Os_3Pt_2(\mu\text{-}H)_2(\mu_5\text{-}C)(CO)_{10}(PCy_3)_2$ (**103**) (*42*). The core of **103** (Fig. 17) consists of a nearly planar–square Os_2Pt_2 unit with the Os–Os edge bridged by an Os atom. The coordination of the Pt centers is approximately planar, and the CVE count of 74 indicates that the Pt atoms are 16 electron, and the cluster as a whole does not obey SEPT. The same mixture of **88** and **103** is formed using the triosmium ketenylidene cluster $Os_3(\mu\text{-}H)_2(\mu_3\text{-}CCO)(CO)_9$, whereas reaction with $Os_3(\mu\text{-}H)(\mu\text{-}COMe)(CO)_{10}$ yields $Os_3Pt_2(\mu\text{-}H)(\mu_5\text{-}C)(\mu\text{-}OMe)(CO)_{10}(PCy_3)_2$ (**104**) (*42*), which has one

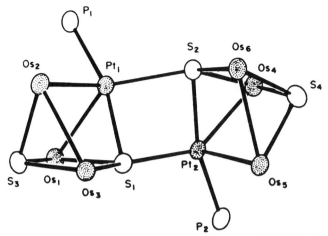

FIG. 23.   Core structure of $[Os_3Pt(\mu\text{-}H)(\mu_3\text{-}S)(\mu_4\text{-}S)(\mu\text{-}PPh_2C_6H_4)(CO)_8]_2$ (**102**). [Reproduced with permission from Adams and Hor (*160*).]

less Os—Os bond than **103** (Fig. 17). A methoxide moiety bridges the open Os···Os vector. The carbide and μ-OMe ligands are thought to arise from an unusual C—O bond cleavage of the methoxymethylidyne ligand. Reaction of $PtMe_2(CO)_2$ with $Os_3(\mu\text{-}H)_2(CO)_{10}$ proceeds via elimination of methane and

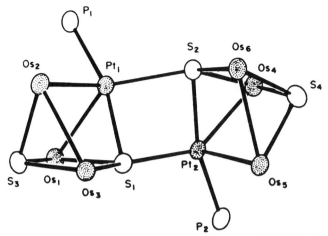

105

affords the unusual puckered raft cluster $Os_4Pt_2(CO)_{18}$ (**105**), which reacts with CO to afford $Os_2Pt(CO)_{10}$ (*153*). The synthesis of $Os_4Pt(\mu\text{-}H)_2(CO)_{12}$-(dppe) from a similar elimination reaction between $Os_4(\mu\text{-}H)_4(CO)_{12}$ and $PtMe_2(dppe)$ has been briefly noted (*161*).

D. Cobalt–Platinum Clusters

1. Tri- and Tetranuclear Clusters

Dehand and Nennig (*162*) reported in 1974 the first Co–Pt cluster, Co_2Pt-$(\mu\text{-}CO)(CO)_6(L_2)$ (**106a**, L_2 = dppe, diars), which was prepared by the reaction

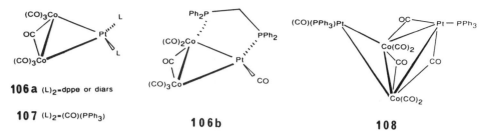

106a (L)₂=dppe or diars

107 (L)₂=(CO)(PPh₃)

106b

108

of $PtCl_2(L_2)$ with 2 mol of $[Co(CO)_4]^-$. When L_2 = dppm, isomer **106b**, with a bridging diphosphine, is formed (135). A crystal structure on the mono-phosphine species $Co_2Pt(\mu\text{-}CO)(CO)_7(PPh_3)$ (**107**) has been published (163). The 58-CVE cluster $Co_2Pt_2(\mu\text{-}CO)_3(CO)_5(PPh_3)_2$ (**108**) is formed in the reaction of *cis* or *trans*-$PtCl_2(PPh_3)_2$ with $[Co(CO)_4]^-$ (52,52a), and the crystal structure shows a butterfly core with a nonbonding $Pt\cdots Pt$ separation of 2.987(4) Å. The ligand distribution closely resembles that found in the isoelectronic Fe_2Pt_2 anion **58** (31). Pyrolysis of solid **108** at 145°C affords the phosphido-bridged cluster $CoPt_2(\mu\text{-}PPh_2)(\mu\text{-}CO)_2(CO)(PPh_3)_3$ (**109**) (164). The reaction of $Co_2(\mu\text{-}CH_2)(CO)_2Cp_2$ with $Pt(C_2H_4)(PPh_3)_2$ leads to Co_2Pt-$(\mu\text{-}CO)_2(PPh_3)_2Cp_2$ (**110**) (35). This cluster may be viewed as the $Pt(PPh_3)_2$ "adduct" of the unsaturated cobalt dimer $Co_2(\mu\text{-}CO)_2Cp_2$, and so is closely

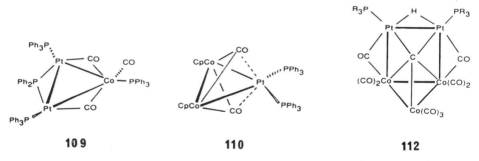

109

110

112

related to other clusters prepared from the 32-electron complexes $MRh(\mu\text{-}CO)_2Cp_2^*$ (M = Co, Rh, Ir) (see Section V,E,1). The short Co—Co bond length of 2.372(2) Å is consistent with this view. Boag *et al.* (86) have recently reported that $Pt_2(CO)_2Cp_2^*$ reacts with $Co_2(CO)_8$ to afford $PtCo(CO)_5Cp^*$ and the 60-CVE tetrahedral cluster $Co_3Pt(\mu\text{-}CO)_3(CO)_6Cp^*$ (**16**).

Electrochemical studies have been carried out on clusters **106–109** (164,165). Both **106** and **107** show irreversible one-electron reduction waves, with the reduction potential of **106** about 0.45 V more negative than that of **107** due to the higher electron density arising from the diphosphine ligand. The first stage is believed to involve generation of the anion radicals **106⁻**

and **107**⁻, which then undergo loss of [Co(CO)₄]⁻. For cluster **107** it is speculated that an unobserved radical intermediate [CoPt(CO)₄(PPh₃)] · is formed, which dimerizes very rapidly to give cluster **108** (*165*). In contrast, **109** undergoes a reversible one-electron reduction, indicating a more stable radical anion (*164*).

2. High-Nuclearity Clusters

Few high-nuclearity Co–Pt clusters are known. The reaction of *cis*-PtCl₂-(PEt₃)₂ with [Co(CO)₄]⁻ yields a mixture of clusters, including the Co–Pt species Co₂Pt₂(μ-CO)₃(CO)₅(PEt₃)₂, the PEt₃ analogue of **108**, and the pentanuclear Co₂Pt₃(μ-CO)₅(CO)₄(PEt₃)₃ (**111**) (*53*). The trigonal–bipyramidal skeleton (Fig. 24) and 72-CVE count in **111** are consistent with SEPT. The reaction of the methylidyne species Co₃(μ₃-CH)(CO)₉ with Pt(C₂H₄)₂(PR₃) (R = Cy, *i*Pr) affords the carbido clusters Co₃Pt₂(μ-H)(μ₅-C)(μ-CO)₂(CO)₇-(PR₃)₂ (**112**) and Co₃Pt₃(μ-H)(μ₆-C)(μ-CO)₄(CO)₅(P*i*Pr₃)₃ (**113**) (*117*). Similar reactivity involving formation of carbido clusters is observed for Ru–methylene (*116*) (Section V,B,2) and Os–methylidyne species (*42*) (Section V,C,3).

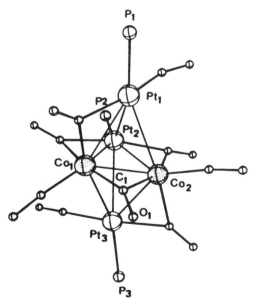

FIG. 24.   Structure of Co₂Pt₃(μ-CO)₅(CO)₄(PEt₃)₃ (**111**). [Reproduced with permission from Barbier *et al.* (*53*).]

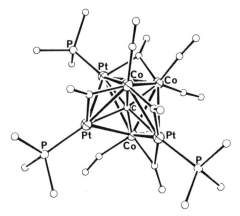

FIG. 25.   Structure of $Co_3Pt_3(\mu\text{-}H)(\mu_6\text{-}C)(\mu\text{-}CO)_4(CO)_5(PiPr_3)_3$ (**113**).

The structure of **112** was determined spectroscopically, and the crystal structure of **113** (Fig. 25) shows an unusual distorted "octahedral" metal core with one long $Pt\cdots Pt$ separation of 3.260(1) Å. The hydride is thought to bridge the bonding $Pt–Pt$ vector.

E. Rhodium–Platinum Clusters

1. Clusters Derived from the 32-Electron Dinuclear Complexes $MRh(\mu\text{-}CO)_2Cp_2^*$ (M = Co, Rh, Ir)

The 32-electron dinuclear species $MRh(\mu\text{-}CO)_2Cp_2^*$ [**114**, M = Co (*166*); **115**, M = Rh (*166,167*); and **116**, M = Ir (*110,168*)] contain short M–Rh separations, consistent with the presence of a formal double bond. Theoretical studies on $Rh_2(\mu\text{-}CO)_2Cp_2$ (*112*) reveal that it possesses an acceptor $2b_2$ and a donor $3a_1$ similar in character to the π and π^* orbitals of ethene. Complexes **114–116** react readily with a variety of ML_n fragments that are isolabal with CH_2, to afford trinuclear clusters (*27,110,166*), and several Rh–Pt

114 M=Co   **115** M=Rh   **118** M=Rh   **119** M=Ir

116 M=Ir   **117** M=Co and M–Rh–Pt (M = Co, Ir) clusters have been synthesized in this fashion $(45,48,49,110,166)$. Syntheses of Rh–Pt clusters using the homonuclear Rh_2 species **115** are outlined in Scheme 5. Treatment of **116** with $Pt_3(\mu\text{-}t\text{BuNC})_3$-$(t\text{BuNC})_3$ affords the trimetallic $IrRhPt(\mu_3\text{-CO})_2(t\text{BuNC})_2Cp_2^*$ (**30**) (110). Reaction of **114**–**116** with $Pt(COD)_2$ results in displacement of one COD ligand, affording high yields of $MRhPt(\mu_3\text{-CO})_2(COD)Cp_2^*$ [**28**, M = Co (166); **3**, M = Rh $(45,49)$; and **29**, M = Ir (110)]. However, treatment of **114**–**116** with $Pt(C_2H_4)_3$ results in displacement of all olefin ligands, giving the novel "bow-tie" clusters $M_2Rh_2Pt(\mu_3\text{-CO})_4Cp_4^*$ [**117**, M = Co (110); **118**, M = Rh $(48,50)$; and **119**, M = Ir (110)], which have all been structurally characterized. The two MRhPt triangles are essentially orthogonal, resulting in a pseudo-D_{2d} core structure. The presence of two ^{13}C NMR signals for the carbonyls in the unsymmetrical clusters **117** and **119** (110) implies a high barrier to mutual rotation of the MRhPt triangles in solution via a pseudo-D_{2h} transition state. The clusters $Rh_2Pt(\mu_3\text{-CO})_2(L)(L')Cp_2^*$ [**3**, (L)(L') = COD; **4**, (L)(L') = $(CO)_2$; or **5**, (L)(L') = $(CO)(PPh_3)$] are easily protonated along a Pt—Rh bond to give the cationic hydrido species **25**–**27** $(48,109)$, in which the $PtHL_2$ unit undergoes dynamic behavior (see Section IV,A).

Dickson *et al.* (169) have reported the syntheses of the closely related clusters $Rh_4Pt(\mu_3\text{-}(\eta^2\text{-}\|)\text{-CF}_3C\equiv CCF_3)_2(\mu\text{-CO})_2Cp_4$ (**120**) and $Rh_2Pt(\mu_3\text{-}(\eta^2\text{-}\|)\text{-CF}_3C\equiv CCF_3)(\mu_3\text{-CO})(COD)Cp_2$ (**121**) from reaction of $Rh_2(\mu_2\text{-}(\eta^2\text{-}\bot)\text{-CF}_3C\equiv CF_3)(\mu\text{-CO})Cp_2$ with $Pt(COD)_2$. Cluster **120** has a "bow-tie"

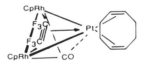

121

Rh_4Pt core (Fig. 26) similar to that found in **118**, with a dihedral angle of 70° between the two Rh_2Pt triangles. The 78-CVE count is consistent with SEPT adapted for condensed polyhedra.

2. High-Nuclearity Rh–Pt Clusters

Italian workers have described the syntheses and structures of a number of high-nuclearity Rh–Pt carbonylate anions, and the Pt atoms in these show a preference for the sites of higher metal–metal connectivity. Several are derived from the octahedral anion $[Rh_5Pt(\mu_3\text{-CO})_4(CO)_{11}]^-$ (**42**) $(56,81)$ (see Scheme 6), which is best prepared by reductive carbonylation of a 1:5 mixture of $PtCl_6^{2-}$ and $RhCl_3 \cdot xH_2O$, although it is also formed from the metal exchange reaction between $[Pt_{12}(CO)_{24}]^{2-}$ and $[Rh_{12}(CO)_{30}]^{2-}$ under a CO

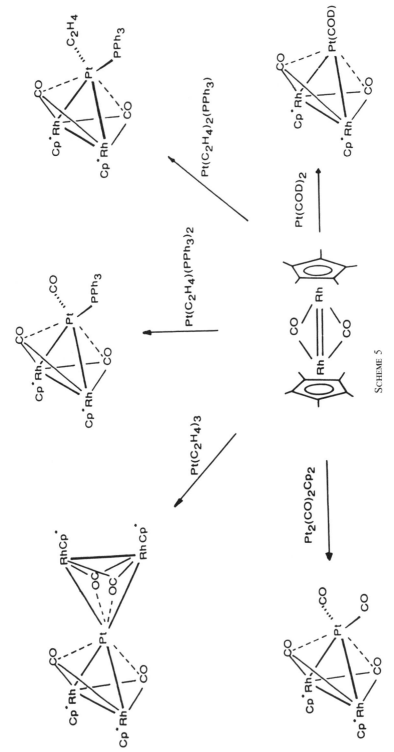

Scheme 5

368

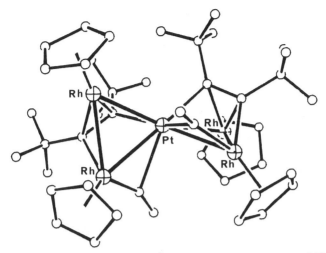

FIG. 26. Structure of $Rh_4Pt(\mu_3-(\eta^2-\|\|)-CF_3C\equiv CCF_3)_2(\mu-CO)_2Cp_4$ (**120**).

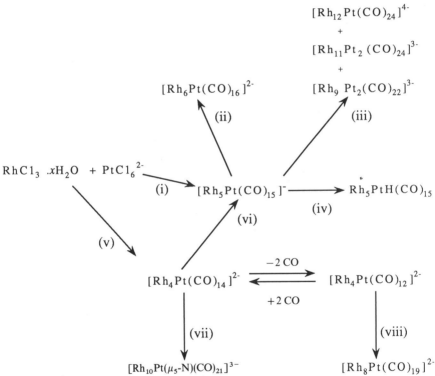

$$[Rh_{12}Pt(CO)_{24}]^{4-}$$
$$+$$
$$[Rh_{11}Pt_2(CO)_{24}]^{3-}$$
$$+$$
$$[Rh_9Pt_2(CO)_{22}]^{3-}$$

$[Rh_6Pt(CO)_{16}]^{2-}$

(ii)   (iii)

$RhCl_3 \cdot xH_2O + PtCl_6^{2-}$ $\xrightarrow{\text{(i)}}$ $[Rh_5Pt(CO)_{15}]^-$ $\longrightarrow$ $Rh_5PtH(CO)_{15}$

(vi)   (iv)

(v)

$[Rh_4Pt(CO)_{14}]^{2-}$ $\underset{+2\,CO}{\overset{-2\,CO}{\rightleftarrows}}$ $[Rh_4Pt(CO)_{12}]^{2-}$

(vii)   (viii)

$[Rh_{10}Pt(\mu_5-N)(CO)_{21}]^{3-}$ $[Rh_8Pt(CO)_{19}]^{2-}$

SCHEME 6. (i) 5:1 molar ratio, CO atmosphere (*81*); (ii) $[Rh(CO)_4]^-$ (*174*); (iii) Δ, MeOH (*8,170*); (iv) aqueous HCl (*81*); (v) 4:1 molar ratio, CO atmosphere (*81*); (vi) $[Rh(CO)_2(NCMe)_2]^+$ (*81*); (vii) $[Rh_6(\mu_6-N)(CO)_{15}]^-$ (*176*); (viii) $Rh_4(CO)_{12}$ (*55*).

atmosphere. Protonation of **42** with aqueous HCl affords the unstable hydride $Rh_5Pt(H)(CO)_{15}$ (*81*). The trigonal–pyramidal anions $[Rh_4Pt(\mu\text{-}CO)_5$-$(CO)_9]^{2-}$ (**43a**) and $[Rh_4Pt(\mu\text{-}CO)_6(CO)_6]^{2-}$ (**11a**) are easily interconverted by addition or loss of CO (*81*). This interconversion results in a skeletal rearrangement, because in the 76-CVE anion **43a** the Pt atom occupies a basal position but occupies an apical one in the 72-CVE anion **11a** (see Fig. 7). A rationale for the existence of both 72- and 76-CVE trigonal–bipyramidal clusters has been discussed in Section III, and their fluxional behavior has been discussed in Section IV,C. Both clusters react with $[Rh(CO)_2(NCCH_3)_2]^+$ under a CO atmosphere to give **42**. Pyrolysis of **42** affords three structurally characterized species, $[Rh_9Pt_2(\mu\text{-}CO)_{11}(CO)_{11}]^{3-}$ (**122**), $[Rh_{11}Pt_2(\mu\text{-}CO)_{12}$-$(CO)_{12}]^{3-}$ (**123**), and $[Rh_{12}Pt(\mu\text{-}CO)_{12}(CO)_{12}]^{4-}$ (**124**) (*8,170*). Anions **123** and **124** (*8*) have the same anticuboctahedral metal core (Fig. 27) found in the isoelectronic series $[Rh_{13}H_{5-n}(CO)_{24}]^{n-}$ (*171,172*), though the differing carbonyl polytope results in overall idealized C_{3v} symmetry rather than the C_s symmetry observed for the Rh_{13} system. A central Pt atom is found in both species, and in **123** the second Pt atom is disordered over the surface sites. The anion **122** (*170*) contains three octahedral units condensed face to face

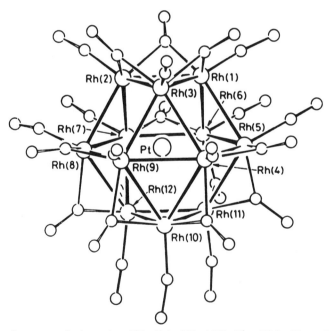

FIG. 27.   Structure of the anion $[Rh_{12}Pt(\mu\text{-}CO)_{12}(CO)_{12}]^{4-}$ (**124**). [Reproduced with permission from Fumagalli *et al.* (*8*).]

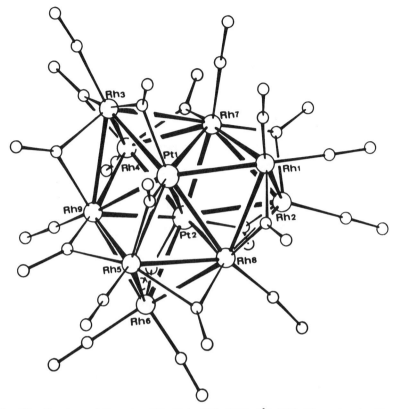

FIG. 28. Structure of the anion $[Rh_9Pt_2(\mu\text{-}CO)_{11}(CO)_{11}]^{3-}$ (122). [Reproduced with permission from Fumagalli *et al.* (*170*).]

along a common Pt–Pt edge (Fig. 28) and has the same idealized D_{3h} skeletal symmetry found in the isoelectronic $[Rh_{11}(CO)_{23}]^{3-}$ (*173*). Again the carbonyl polytopes are different, with the result of no overall symmetry in the case of 122. The 148-CVE count is incompatible with SEPT adapted for condensed structures.

Treatment of 42 with $[Rh(CO)_4]^-$ results in high yields of the heptanuclear $[Rh_6Pt(\mu\text{-}CO)_5(\mu_3\text{-}CO)_3(CO)_8]^{2-}$ (125) (*174*). The Pt caps one face of a Rh_6 octahedron (Fig. 29) in an asymmetric fashion, with Pt—Rh distances of 2.698(2)–3.167(2) Å. The 98-CVE count of 125 is that expected from SEPT for a capped octahedron. A redox condensation reaction of $Rh_4(CO)_{12}$ with 11a affords $[Rh_8Pt(\mu\text{-}CO)_9(\mu_3\text{-}CO)_3(CO)_7]^{2-}$ (41) (*55*), which contains two octahedra fused on one face (Fig. 29). It is isostructural with $[Rh_9(CO)_{19}]^{3-}$

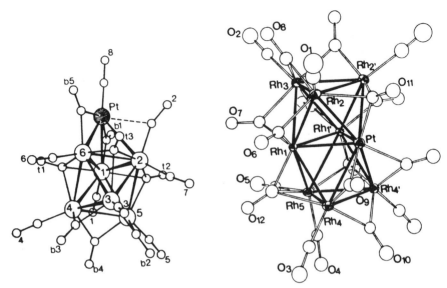

FIG. 29. Structures of the anions $[Rh_6Pt(\mu\text{-}CO)_5(\mu_3\text{-}CO)_3(CO)_8]^{2-}$ (125) and $[Rh_8Pt(\mu\text{-}CO)_9(\mu_3\text{-}CO)_3(CO)_7]^{2-}$ (41). [Reproduced with permission from Fumagalli et al. (174) and (55), respectively.]

(175) and the 122-CVE count is consistent with SEPT adapted for condensed structures. The condensation reaction of 43a with $[Rh_6(\mu_6\text{-}N)(CO)_{15}]^-$ leads to the only example of a hetero-Pt cluster with an interstital nitride $[Rh_{10}Pt\text{-}(\mu_5\text{-}N)(\mu\text{-}CO)_{10}(CO)_{11}]^{3-}$ (126) (176). The unusual core structure shown in Fig. 30 cannot be rationalized by SEPT and has four electrons in excess of that calculated by Ciani and Sironi (177) for this skeletal geometry. The anion $[Rh_{18}Pt_4(\mu\text{-}CO)_{21}(CO)_{14}]^{4-}$ has been briefly reported (177a) as a minor product from the reaction of 11a with 42. The metal skeleton, shown in Fig. 30, contains a distorted tetrahedral Pt_4 unit and consists of a fragment of a body-centered cubic lattice.

F. Iridium–Platinum Clusters

Remarkably few Ir–Pt clusters exist, perhaps due to the unavailability of high-yield syntheses of suitable Ir precursors. The reaction of $Ir(CO)_2Cp^*$ with $Pt(C_2H_4)_3$ affords quantitative yields of the raft cluster $Ir_3Pt_3(\mu\text{-}CO)_3\text{-}(CO)_3Cp_3^*$ (39) (34). The structure of 39 (Fig. 6) closely resembles that found in $[Fe_3Pt_3(CO)_{15}]^{n-}$ (20, $n = 2, 1, 0$) (5,97) in that the raft is not exactly planar, but has one Ir atom lying above the Pt_3 plane and the other two below. The

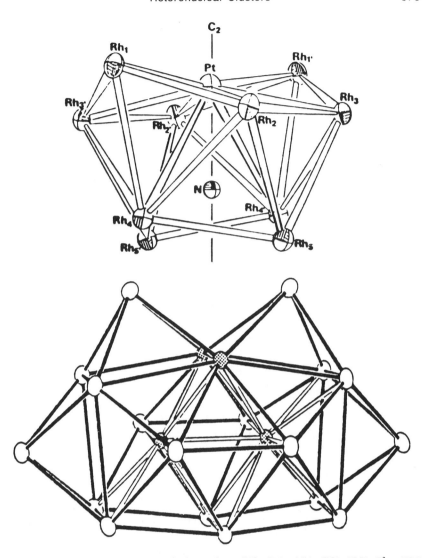

FIG. 30. Metal core structure of the anions $[Rh_{10}Pt(\mu_5\text{-}N)(\mu\text{-}CO)_{10}(CO)_{11}]^{3-}$ (126) [Reproduced with permission from Martinengo *et al.* (176)] and $[Rh_{18}Pt_4(\mu\text{-}CO)_{21}(CO)_{14}]^{4-}$ (Pt atoms shaded).

CVE count of 84 indicates that the a_2', Pt—Pt antibonding orbital (94,95) is unoccupied. The average Pt—Pt distance of 2.704(3) Å is somewhat longer than found in the neutral 84-CVE Fe–Pt raft, **9c** [2.590(2) Å] (97), but is probably a reflection of the differing groups that bridge the Pt–Pt vectors. The indenyl complex $Ir(CO)_2(\eta-C_9H_7)$ behaves similarly to the RhCp* analogue (45) in that reaction with $Pt(C_2H_4)_2(PCy_3)$ affords a trinuclear cluster $Ir_2Pt(CO)_3(PCy_3)(\eta-C_9H_7)_2$ (**31**) (111). Cluster **31** is easily protonated along

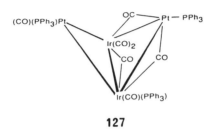

127

an Ir–Pt edge to give the hydrido cation $[Ir_2Pt(\mu-H)(CO)_3(PCy_3)(\eta-C_9H_7)_2]^+$ (111). The tetranuclear butterfly cluster $Ir_2Pt_2(\mu-CO)_3(CO)_4(PPh_3)_3$ (**127**), synthesized from reaction of $[Pt_{12}(CO)_{24}]^{2-}$ with $IrCl(CO)(PPh_3)_2$, has a structure analogous to that of $Co_2Pt_2(\mu-CO)_3(CO)_5(PPh_3)_2$ (**108**), with a nonbonding Pt···Pt separation of 2.976(1) Å (54).

The 76- and 72-CVE trigonal bipyramidal anionic clusters $[Ir_4Pt(\mu-CO)_5-(CO)_9]^{2-}$ (**43b**) and $[Ir_4Pt(\mu-CO)_3(CO)_9]^{2-}$ (**11b**) (57) are closely similar to their Rh cogenors, except for the differing carbonyl distribution in **11b** (see Figs. 7 and 8). Like the Rh analogues, they are easily interconverted by addition or loss of CO. Their fluxionality is discussed in Section IV,C.

G. Nickel–Platinum Clusters

Several series of high-nuclearity Ni–Pt clusters have been described by Longoni and co-workers (9,120,178). The structures of two of the anions in the series, $[Ni_{38}Pt_6H_{6-n}(CO)_{48}]^{n-}$ (**128**, $n = 3-6$), have been reported in detail (9). They contain an octahedral Pt_6 core encapsulated by a v_3 octahedron of 38 Ni atoms (Fig. 31). The mean M–M distances are Pt–Pt = 2.719 Å, Ni–Pt = 2.630 Å, and Ni–Ni = 2.580 Å. There are other series of related "cherry" crystallite clusters with the general formula $[Ni_{44-x}Pt_xH_{6-n}(CO)_{48}]^{n-}$ ($x = 8, 10; n = 3-6$) (178). The physicochemical properties of clusters such as **128** have attracted considerable interest in view of possible quantum-size effects, because the radius of the metal core (10.9 Å) approaches that found in small metal crystallites. Cluster **128** ($n = 5$) has been the subject of analytical electron microscopy (179), while low-temperature magnetic susceptibility (180) and high-field magnetization (181) measurements on

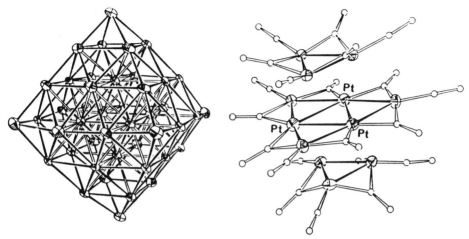

FIG. 31. Structure of the metal core in the anion $[Ni_{38}Pt_6H_3(CO)_{48}]^{3-}$ (**128**) and the anion $[Ni_9Pt_3H(\mu\text{-CO})_{12}(CO)_9]^{3-}$ (**129**). [Reproduced with permission from Ceriotti *et al.* (9) and (*120*).]

128 ($n = 4$) show a small μ_{eff} of $\sim 2\text{-}4.8\ \mu_B$ per cluster. No ^{195}Pt or ^{1}H NMR signals were observed for the Pt nuclei or hydrides in **128** ($n = 3\text{-}5$) (*9,182*) and only broad ^{195}Pt resonances have been observed for the large homo-Pt clusters $[Pt_{38}H_2(CO)_{44}]^{2-}$ and $[Pt_{26}H_2(CO)_{32}]^{2-}$ (*183*). These results have been attributed to the presence of residual paramagnetism in even-electron high-nuclearity clusters (*180,184–189*), which is though to arise from quantum-size effects in this "metametallic" state (*186,190*). However, recently obtained results on single crystal samples of $Os_{10}(\mu_6\text{-C})H_2(CO)_{24}^*$ fail to show (*191*) the previously reported (*190*) EPR signal found in the polycrystalline material and cast some doubt on the validity of the experimental measurements of residual paramagnetism. It is now believed (*191*) that minor surface oxidation occurring during grinding may be responsible for the magnetic behavior of the OS_{10} cluster. The lack of detectable ^{195}Pt NMR signals for **128** may be due to excessively long T_1 values, as has been observed for other cluster-encapsulated nuclei (*182*).

A second series of clusters with the general formula $[Ni_{12-x}Pt_x(CO)_{21}\text{-}H_{4-n}]^{n-}$ (**129**, $x = 2\text{-}7$, $n = 2\text{-}4$) have been reported (*178*), with a crystal structure (Fig. 31) for **129** ($x = 3$, $n = 3$) (*120*). The mean M–M distances are Ni–Ni$_{inlayer}$ = 2.451 Å, Ni–Ni$_{interlayer}$ = 2.918 Å, Ni–Pt$_{inlayer}$ = 2.541 Å, Ni–Pt$_{interlayer}$ = 2.933 Å, and Pt–Pt = 2.845 Å. The structure is closely related to the homo-Ni system $[Ni_{12}(CO)_{21}H_{4-n}]^{n-}$ for which neutron diffraction studies (*192*) have shown that the hydrides occupy trigonal–prismatic interstitial cavities. ^{1}H and ^{195}Pt NMR data suggest a similar hydride site in **129** ($x = 3$, $n = 2, 3$). A preliminary report has appeared (*178*) on the structure of

$[Ni_{36}Pt_4(CO)_y]^{6-}$, which shows a central Pt_4 tetrahedron encapsulated by a v_5-truncated tetrahedron of Ni atoms. A mixture of $[Ni_6(\mu\text{-}CO)_6(CO)_6]^{2-}$ and $[Pt_6(\mu\text{-}CO)_6(CO)_6]^{2-}$ shows a ^{195}Pt NMR signal at δ 20.7, which is attributed to the mixed anion $[Ni_3Pt_3(\mu\text{-}CO)_6(CO)_6]^{2-}$; the simplicity of the signal indicates that no randomization of the metal atoms occurs between triangles (102a).

H. Palladium–Platinum Clusters

Only two bimetallic Pd–Pt clusters have been reported, though there are several M–Pd–Pt species. Treatment of the η^2-phosphaalkyne complex $Pt(PPh_3)_2(\eta^2\text{-}P\equiv CtBu)$ with $Pd(PPh_3)_4$ affords the pentanuclear $Pd_2Pt_3(\mu_4\text{-}\eta^2\text{-}P\equiv CtBu)_3(PPh_3)_5$ (130) (193). The unusual tristellated metal skeleton in 130 (Fig. 32), similar to that observed in $W_3Sb_2(CO)_{15}$ (194), may be rationalized within SEPT, assuming the P atoms are used to define a hexagonal–bipyramidal core. The Pd–Pd distance is 2.679(2) Å. The 44-CVE triangular cluster $[Pd_2Pt(\mu\text{-}Cl)(\mu\text{-}PPh_2)(PPh_3)_3]^+$ (131) is prepared by pyrolysis of a mixture of $[PdCl(PPh_3)_3]BF_4$ and $[PtCl(PPh_3)_3]BF_4$ (195). Pd–Pt and Pd–Pd distances are essentially the same, in the range 2.878(2)–2.908(1) Å, but the X-ray analysis is complicated by the presence of small amounts of the homo-Pd analogue in the crystal.

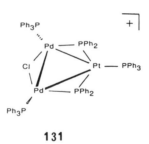

131

I. Clusters Containing More Than Two Different Metals

Several clusters containing platinum and two or more different metal atoms (including at least one from Groups 8–10) have been reported, and almost all fall into three categories. The first structural type, MPdPt-triangular or M_2PdPt-spiked–triangular clusters, has been primarily prepared by Braunstein and co-workers, and most contain bridging bis(diphenylphosphino)-methane ligands. The second major class, prepared from Group 6 alkylidyne

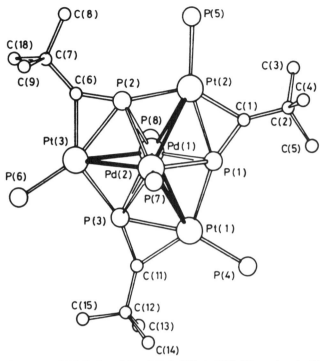

FIG. 32. Structure of $Pd_2Pt_3(\mu_4-\eta^2-P\equiv CtBu)_3(PPh_3)_5$ (**130**). [Reproduced with permission from Al-Resayes *et al.* (*193*).]

precursors, does not generally contain polyhedral arrays of metal atoms, but is included here because of the unusual and interesting structures. A final group is composed of clusters prepared from the 32-electron bimetallic species $MRh(\mu-CO)_2(\eta-C_5Me_5)_2$ (M = Co, Ir). Due to their close relationship to the homo-Rh_2 species, these latter species are discussed in Section V,E,1.

1. MPdPt Tri- and Tetranuclear Clusters

Reaction of the bis(diphenylphosphino)methane (dppm) complex PdPt-$(\mu$-dppm$)_2Cl_2$ (**132**) with carbonylate anions affords several tri- and tetranuclear clusters. Reactions are outlined in Scheme 7. Treatment of **132** with 2 mol equivalents of $[Fe(CO)_3(NO)]^-$ (*196,197*), $[Co(CO)_4]^-$ (*196,198*), or $[Mn(CO)_5]^-$ (*196,197*) results in the tetranuclear spiked triangular clusters **133**, **134**, or **135**. Crystal structures on the dipalladium analogues of **134** and **135** have been reported (*197,199*). Grossel *et al.* (*136*) and Braunstein *et al.* (*200*) have independently reported that reaction of **132** with 1 mol equivalent of $[Fe(CO)_4]^{2-}$ at ambient temperatures leads to an inseparable 1:1 mixture

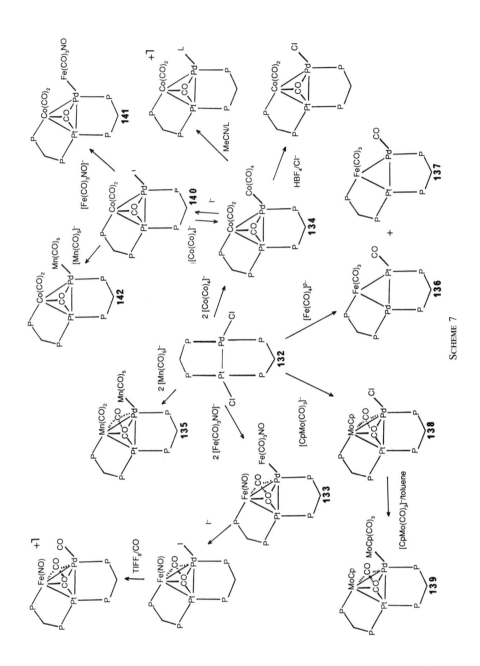

SCHEME 7

of isomers **136** and **137**. At −78°C essentially pure **137** may be isolated, but this reverts to the equilibrium mixture on warming. A mechanism for the isomerization involving a μ-Fe(CO)$_4$ moiety is proposed (*200*). Apart from clusters **136** and **137**, all other species discussed in this section exist solely as the isomer with two phosphine ligands bonded to the Pt atom. The same mixture of **136** and **137** is obtained on reaction of **132** with either [FeH(CO)$_4$]$^-$ or [FeCrH(CO)$_9$]$^-$ (*201*). Tetranuclear species are not formed, however, on treatment of **132** with 2 mol equivalents of [CpMo(CO)$_3$]$^-$, due to the nucleophilicity of the liberated halide (*202*). Instead, the more stable trinuclear cluster **138** is isolated, though the tetranuclear **139** may be isolated by further reaction with [CpMo(CO)$_3$]$^-$ and extraction into toluene (*202*). The dipalladium analogue of **139** has been structurally characterized (*202*). The exocyclic Pd—M or Pd—X bonds in all these clusters are weak and easily cleaved. Thus the Co(CO)$_4$ group in **134** may easily be replaced by Lewis bases such as Cl$^-$ or I$^-$, donor solvents, CO, PPh$_3$, or AsPh$_3$ (*198*), and the iodo derivative **140** has been reacted further with [Fe(CO)$_3$(NO)]$^-$ or [Mn(CO)$_5$]$^-$, forming the clusters **141** and **142**, respectively (*196*), which have four different transition metals. Similar lability has been noted in **133** (*197*) and **138** (*202*), with the position of equilibrium strongly dependent on solvent polarity.

Electrochemical studies on **134** and [CoPdPt(μ-dppm)$_2$(CO)$_3$X]$^+$ (X = I$^-$, PPh$_3$) in DMSO show electrochemically reversible, but chemically irreversible, two-electron reductions attributed to the cationic species [CoPdPt(μ-dppm)$_2$(CO)$_3$(L)]$^+$ (L = DMSO and PPh$_3$) and several ill-defined oxidation steps (*203*). For X = CO an irreversible reduction is observed, indicating a differing electrochemical mechanism (*204*).

Werner and Thometzek (*205*) have described the synthesis of closely related species that contain bridging allylic rather than dppm ligands. The clusters MPdPt(μ-2-MeC$_3$H$_4$)(CO)$_3$(P-iPr$_3$)$_2$Cp are prepared by reaction of PdPt-(μ-2-MeC$_3$H$_4$)(μ-Br)(PiPr$_3$)$_2$ (**143**) with [M(CO)$_3$Cp]$^-$ (M = Cr, Mo, W). Only one ^{13}C NMR signal is observed for the chemically distinct CO ligands in these complexes, indicating CO scrambling, probably by rotation of the M(CO)$_3$ unit about the symmetry axis.

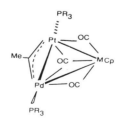

143

2. Clusters Containing Bridging Alkylidyne Ligands

The alkylidyne complexes $M(\equiv CR)(CO)_2Cp$ (M = Mo, W) have been used extensively by Stone (12) as precursors to the directed synthesis of heteronuclear clusters, and there are several examples of relevance to this review. The products from the reaction of $Fe_2(CO)_9$ with $WPt(\mu\text{-}CR')(CO)_2)(PR_3)_2Cp$ ($R' = C_6H_4Me\text{-}4$) depend on the phosphine used (83,83a). With PMe_2Ph the 48-CVE cluster $WFePt(\mu_3\text{-}CR)(\mu\text{-}CO)(CO)_5(PMe_2Ph)_2Cp$ (13) is formed, but with $PMePh_2$, PMe_3, and PEt_3 the 46-CVE species $WFePt(\mu_3\text{-}CR)(CO)_6\text{-}(PR_3)Cp$ (14) result. The reaction using $PMePh_2$ also affords a small amount of $WFePt(\mu_3\text{-}CR)(CO)_5(PMePh_2)_2Cp$.

Numerous examples of open-chain polymetallic species have been reported (15a,15b,206–210) and are prepared by this methodology. Reaction of $W(\equiv CR)(CO)_2L$ ($R = C_6H_4Me\text{-}4$, L = Cp or Cp*) with $Ni(COD)_2$ affords $NiW_2(\mu\text{-}CR)_2(\mu\text{-}CO)_2(CO)_2L_2$, which then further reacts with $Pt(COD)_2$ to afford the tetrametal chain $NiW_2Pt(\mu\text{-}CR)(\mu_3\text{-}CR)(CO)_4(COD)L_2$ (144a, L = Cp; or 144b, L = Cp*). Cluster 144b exists as two diastereomers in solution according to ^{195}Pt and ^{13}C NMR data, and 144a occurs as a single isomer (206). The heptametal chain $NiW_4Pt_2(\mu\text{-}CR)_2(\mu_3\text{-}CR)_2(CO)_8L_4$ (146, R = Me, L = Cp*) is produced by reaction of $W_2Pt(\mu\text{-}CR)_2(\mu\text{-}CO)_2(CO)_2L_2$ (145) with $Ni(COD)_2$ in a 2:1 ratio (207). Treatment of 145 (R = Ph or $C_6H_4Me\text{-}4$, L = Cp) with excess $Ni(COD)_2$ affords an octametal chain that

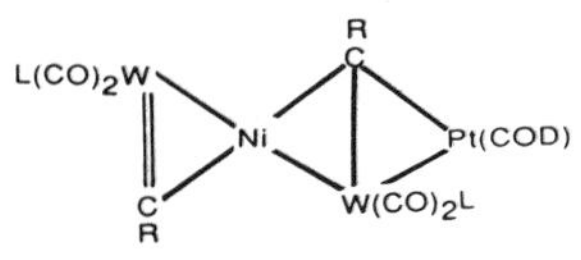

144a L=Cp   144b L=Cp*

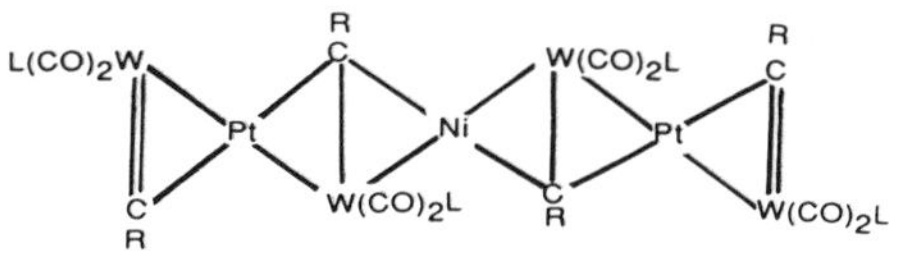

146

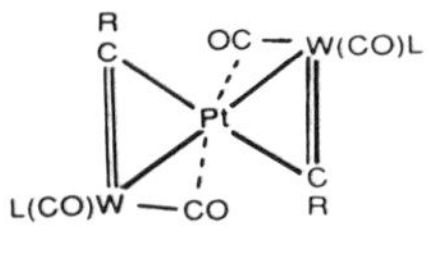

145

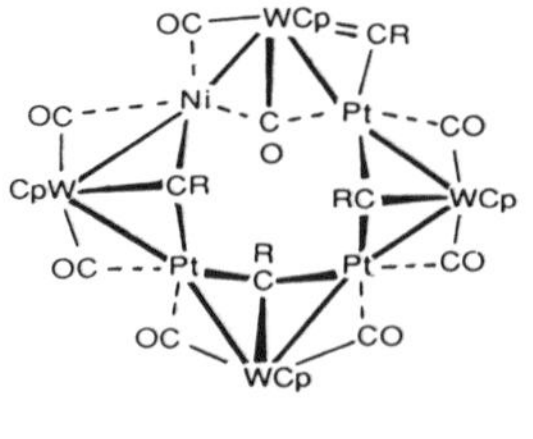

149

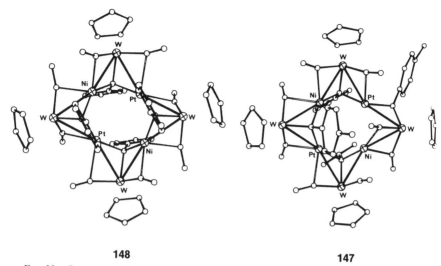

148   **147**

FIG. 33. Structures of the asymmetric "star" cluster $Ni_2W_4Pt_2(\mu\text{-CR})(\mu_3\text{-CR})_3(\mu\text{-CO})$-$(CO)_7Cp_4$ (**147**) and the symmetric "star" cluster $Ni_2W_4Pt_2(\mu_3\text{-CR})_4(CO)_8Cp_4$ (**148**).

ring closes to give the "star" clusters $Ni_2W_4Pt_2(\mu\text{-CR})(\mu_3\text{-CR})_3(\mu\text{-CO})$-$(CO)_7Cp_4$ (**147**) (*3,208*). The structure of **147** (R = $C_6H_4Me\text{-}4$) is shown in Fig. 33. The Ni and Pt sites are disordered in the crystal, and these species exist as two isomers in solution in approximately equal proportions, corresponding to isomers with either a $W(\mu\text{-CR})Pt$ or $W(\mu\text{-CR})Ni$ unit (*208*). These "asymmetric-star" clusters in turn isomerize on reflux in THF to give the "symmetric" species $Ni_2W_4Pt_2(\mu_3\text{-CR})_4(CO)_8Cp_4$ (**148**), which have only $\mu_3\text{-CR}$ groups (*208*). The structure of **148** (R = Ph) is also shown in Fig. 33. There are no direct Pt—Pt or Pt—Ni bonds in these systems. The related· "asymmetric-star" cluster $NiW_4Pt_3(\mu\text{-CR})(\mu_3\text{-CR})_3(\mu\text{-CO})(CO)_7Cp_4$ (**149**, R = $C_6H_4Me\text{-}4$) may be synthesized in a logical fashion. Complex **145** (R = $C_6H_4Me\text{-}4$, L = Cp) reacts with 2 mol of $Pt(COD)_2$ to form the pentametal chain $W_2Pt_3(\mu_3\text{-CR})_2(CO)_4(COD)_2Cp_2$ (**150**) (*206*), and the two COD ligands may be replaced by $W(\equiv CR)(CO)_2$ Cp moieties to afford (*207*) the heptametal chain $W_4Pt_3(\mu\text{-CR})_2(\mu_3\text{-CR})_2(CO)_8Cp_4$ (**151**). Further addition of $Ni(COD)_2$ to **151** results in ring closure, affording **149** (*208*). Cluster **149** exists as two isomers in solution, and on the basis of the two-bond and transannular $J(^{195}Pt-^{195}Pt)$ couplings, the major isomer is assigned to the species containing a $W(\mu\text{-CR})Pt$ rather than $W(\mu\text{-CR})Ni$ unit. Reflux of **149** in THF results in an insoluble black material that is probably a "symmetric" isomer similar to **148**.

Essentially similar chemistry may be carried out using the reagent $Mo(\equiv CC_6H_4Me\text{-}4)(CO)_2Cp$ in the place of the W derivative, allowing

150

151

152

154

153

155

synthesis of the chain clusters $Mo_2NiPt(\mu\text{-}CC_6H_4Me\text{-}4)(\mu_3\text{-}CC_6H_4Me\text{-}4)\text{-}(CO)_4(COD)Cp_2$ (**152**), $Mo_2NiPt_2(\mu_3\text{-}CC_6H_4Me\text{-}4)_2(CO)_4(COD)_2Cp_2$ (**153**), $Mo_2W_2NiPt_2(\mu\text{-}CPh)_2(\mu_3\text{-}CC_6H_4Me\text{-}4)_2(CO)_8Cp_4$ (**154**), and $Mo_2W_2Ni\text{-}Pt_2(\mu\text{-}CMe)_2(\mu_3\text{-}CC_6H_4Me\text{-}4)_2Cp_2^*Cp_2$ (**155**) (*15a*). Ring closure of the heptametal chains affords the "star" clusters $Mo_4Ni_2Pt_2(\mu\text{-}CC_6H_4Me\text{-}4)_4(CO)_8Cp_4$ (both symmetric and asymmetric isomers) and $Mo_2W_2Ni_2\text{-}Pt_2(\mu_3\text{-}CR)_2(\mu_3\text{-}CC_6H_4Me\text{-}4)_2(CO)_8(L)_2Cp_2$ (R = Ph, L = Cp, or R = Me, L = Cp*)(*15a*). The use of Mo reagents has recently also allowed the synthesis of chains up to 11 atoms in length (*15b,209*). Some WRuPt species have been prepared via a similar route (*210*). Reaction of **145** (R = $C_6H_4Me\text{-}4$) with $Ru(CO)_4(C_2H_4)$ results in the tetrametal cluster $W_2RuPt(\mu\text{-}CR)(\mu_3\text{-}CR)\text{-}(CO)_6(L)Cp_2$ (**156**, L = CO), and treatment of this with PMe_2Ph affords **157** (L = PMe_2Ph). The crystal structure of the latter (Fig. 34) shows a WRuPt triangle with an exoligated W atom [W—Pt = 2.733(1) Å] (*210*). The Fe analogue of **156** has been prepared in a similar fashion using $Fe_2(CO)_9$ (*211*). Reaction of the tetrametal chain $W_2Pt_2(\mu\text{-}CR)(\mu_3\text{-}CR)(CO)_4(COD)Cp_2$ with $Ru(CO)_4(C_2H_4)$ results in the hexanuclear cluster $W_2Ru_2Pt_2(\mu_3\text{-}CR)_2(\mu\text{-}CO)_2(CO)_9Cp_2$ (**158**). The unusual core structure, shown in Fig. 34, may be described as a WRu_2Pt butterfly with one W—Ru edge bridged by a Pt atom, and an exoligated W center attached via a short W—Pt bond of 2.663(3) Å.

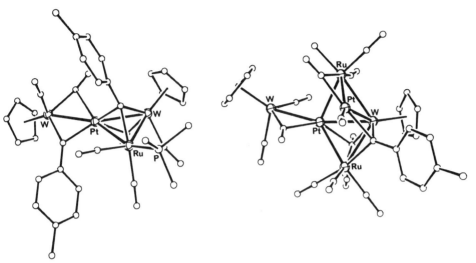

FIG. 34.   Structures of $W_2RuPt(\mu\text{-}CR)(\mu_3\text{-}CR)(CO)_6(PMe_2Ph)Cp_2$ (**157**) and $W_2Ru_2Pt_2(\mu_3\text{-}CR)_{21}(\mu\text{-}CO)_2(CO)_9Cp_2$ (**158**).

VI

CATALYSIS BY HETERO-PT CLUSTERS

Despite the general interest in bimetallic catalysts, there have been relatively few reports on catalysis by hetero-Pt clusters, either in the homogeneous phase or as supported catalysts. The polymer-supported clusters Fe_2Pt-$(CO)_8(Ph_2P\text{-}Poly)_2$ (*212*), $RuPt_2(CO)_5(PPh_2P\text{-}Poly)_3$ (*212*), and Co_2Pt_2-$(CO)_8(Ph_2P\text{-}Poly)_2$ (*213*) [*Poly* = poly(styrene-divinylbenzene)] have been shown to be ethylene hydrogenation catalysts and to retain their integrity through catalysis (*212*), whereas the supported cluster $Os_3Pt(\mu\text{-}H)_2(CO)_{10}$-$(Ph_2P\text{-}Poly)_2$ shows negligible activity (*213*). Chinese workers report that $Co_2Pt_2(\mu\text{-}CO)_3(CO)_5(PPh_3)_2$ is an active catalyst in the hydroformylation of 1-hexene (*214*) and in the hydroformylation, cyclooligomerization, hydrogenation, and isomerization of 1,3-butadiene (*215*). The decarbonylation of this cluster on oxide supports has been studied (*216*). This cluster, together with other Co_2Pt species, has been reported to be active as an alkyne hydrogenation catalyst (*217*). Ichikawa et al. (*218*) report that the anionic clusters $[Fe_3Pt_3(CO)_{15}]^{2-}$ and $[Fe_4Pt(CO)_{16}]^{2-}$, supported on silica, were used to prepare Fe–Pt CO hydrogenation catalysts, and that the former was 100% selective toward formation of CH_3OH. Mössbauer studies on this system show the presence of both (Fe(0) and Fe(III) sites (*218*). The anions [Rh$_5$Pt-(CO)$_{15}$]$^-$ and [Rh$_4$Pt(CO)$_{14}$]$^{2-}$ were examined as homogeneous CO hydrogenation catalysts and were found to give mainly ethylene glycol (*219*). PES studies on [Rh$_5$Pt(CO)$_{15}$]$^-$ supported on amberlite exchange resin (*220*) show degradation of the cluster and surface enrichment of Pt in the resulting alloy. This material is an aromatic-ring hydrogenation catalyst. Ir$_2$Pt$_2$-(μ-CO)$_3$(CO)$_4$(PPh$_3$)$_3$ has been shown to be an active homogeneous catalyst for olefin hydrogenation (*54*).

VII

ADDENDUM

Articles describing the following reactions have appeared since the preparation of this manuscript. The reaction of Ru$_3$(CO)$_{11}$(PPh$_2$H) with Pt(C$_2$H$_4$)(PPh$_3$)$_2$ affords the Pt–Ru clusters **48b** and Ru$_2$Pt(μ-H)(μ-PPh$_2$)-(CO)$_7$(PPh$_3$). The related reaction using Pt(C$_2$H$_4$)$_2$(PCy$_3$) results in Ru$_2$Pt-(μ-H)(μ-PPh$_2$)(CO)$_7$(PCy$_3$) and Ru$_3$Pt(μ-H)(μ-PPh$_2$)(μ-CO)$_3$(CO)$_6$(PCy$_3$), both of which have been crystallographically characterized. The former is a 46 CVE cluster containing an Ru$_2$Pt triangle [Pt-Ru = 2.8755(5) and 2.7248(5) Å]; the latter is a 58 CVE cluster containing an Ru$_3$Pt tetrahedron [Pt–Ru = 2.895(2), 2.846(2) and 2.764(2) Å]. Small amounts of the spectroscopically characterized species RuPt$_2$(μ-CO)$_3$(CO)$_3$(PCy$_3$)$_2$ and Ru$_3$Pt$_2$(μ-H)(μ-PPh$_2$)(CO)$_9$(PCy$_3$)$_2$ are also formed [J. Powell, J. C. Brewer, G. Gulia and J. F. Sawyer, *Inorg. Chem.* **28**, 4470 (1989)].

Full spectroscopic and preparative details for clusters **141** and **142** are given, and the syntheses of positional isomers of MoCoPdPtCp(CO)$_6$(dppm)$_2$ are described. These noninterconverting isomers, which differ in the nature of the exoligated metal atom, are made either by reaction of cluster **138** with [Co-(CO)$_4$]$^-$ or by reaction of cluster **140** with [CpMo(CO)$_3$]$^-$. [P. Braunstein, C. de Meric de Bellefon and M. Ries *Inorg. Chem.* **29**, 1181 (1990)].

ACKNOWLEDGMENTS

Professors F. G. A. Stone, R. D. Adams, J. B. Keister, and A. Fumagalli are kindly thanked for providing preprints of their work prior to publication. Our own research described herein was supported by the SERC and the University of Glasgow.

REFERENCES

1. W. L. Gladfelter and G. L. Geoffroy, *Adv. Organomet. Chem.* **18**, 207 (1980).
2. D. A. Roberts and G. L. Geoffroy, *in* "Comprehensive Organometallic Chemistry" (G. Wilkinson, F. G. A. Stone, and E. Abel, eds.), Vol. 6, Chapter 40. Pergamon, Oxford, 1982.

3. G. P. Elliot, J. A. K. Howard, T. Mise. C. M. Nunn, and F. G. A. Stone, *Angew. Chem., Int. Ed. Engl.* **25**, 190 (1986).
4. R. D. Adams and W. A. Herrmann, *Polyhedron* **7**, 2255–2463 (1988).
5. G. Longoni, M. Manassero, and M. Sansoni, *J. Am. Chem. Soc.* **102**, 7973 (1980).
6. C. Couture and D. H. Farrar, *J. Chem. Soc., Chem. Commun..* 197 (1985).
7. C. Couture and D. H. Farrar, *J. Chem. Soc., Dalton Trans..* 1395 (1986).
8. A. Fumagalli, S. Martinengo, and G. Ciani, *J. Chem. Soc.. Chem. Commun..* 1381 (1983).
9. A. Ceriotti, F. Demartin, G. Longoni, M. Manassero, M. Marchionna. G. Piva. and M. Sansoni, *Angew. Chem., Int. Ed. Engl.* **24**, 697 (1985).
10. M. I. Bruce, G. Shaw, and F. G. A. Stone, *J. Chem. Soc.. Chem. Commun..* 1288 (1971).
11. F. G. A. Stone, *Pure Appl. Chem.* **58**, 529 (1986).
12. F. G. A. Stone, *ACS Symp. Ser.* **211**, 383 (1983).
13. R. Bender, P. Braunstein, J.-M. Jud, and Y. Dusausoy. *Inorg. Chem.* **23**, 4489 (1984); T. Blum, P. Braunstein, A. Tiripicchio, and M. Tiripicchio Camellini. *Organometallics* **8**, 2504 (1989).
14. P. Braunstein, D. Matt, O. Bars, and D. Grandjean, *Angew. Chem.. Int. Ed. Engl.* **18**, 797 (1979); P. Braunstein, D. Matt, O. Bars, M. Louer, D. Grandjean, J. Fischer, and A. Mitschler, *J. Organomet. Chem.* **213**, 79 (1981); E. Kunz, J. Müller, and U. Schubert, *ibid.* **320**, C11 (1987).
15. T. J. Henly, J. R. Shapley, A. L. Rheingold, and S. J. Geib, *Organometallics* **7**, 441 (1988); M. A. Urbancic, S. R. Wilson, and J. R. Shapley, *Inorg. Chem.* **23**, 2954 (1984).
15a. S. J. Davies and F. G. A. Stone, *J. Chem. Soc., Dalton Trans.*, 785 (1989).
15b. S. J. Davies, J. A. K. Howard, R. J. Musgrove, and F. G. A. Stone, *J. Chem. Soc., Dalton Trans.*, 2269 (1989).
16. A. J. Poë, *in* "Metal Clusters" (M Moskovits, ed.), Chapter 4. Wiley, New York.
17. F. R. Hartley "The Chemistry of Platinum and Palladium." Applied Science, London.
18. F. R. Hartley, *in* "Comprehensive Organometallic Chemistry" (G. Wilkinson, F. G. A. Stone, and E. Abel, eds.), Vol. 6 Chapter 39. Pergamon, Oxford, 1982.
19. D. M. P. Mingos and R. W. M. Wardle, *Transition Met. Chem. (N.Y.)* **10**, 441 (1985).
20. N. K. Eremenko, E. G. Mednikov, and S. S. Kurasov, *Russ. Chem. Rev. (Engl. Transl.)* **54**, 394 (1985).
21. R. B. King, *Inorg. Chim. Acta* **116**, 119 (1986).
22. M. D. Vargas and J. N. Nicholls, *Adv. Inorg. Chem. Radiochem.* **30**, 123 (1986).
23. P. Lemoine, *Coord. Chem. Rev.* **83**, 169 (1988).
24. E. Sappa, A. Tiripicchio, A. J. Carty, and G. E. Toogood, *Prog. Inorg. Chem.* **35**, 437 (1987).
25. R. D. Adams, *Polyhedron* **4**, 2003 (1985).
26. A. J. Deeming, *Adv. Organomet. Chem.* **26**, 1 (1986).
27. F. G. A. Stone, *Angew. Chem., Int. Ed. Engl.* **23**, 89 (1984).
28. F. G. A. Stone, *Acc. Chem. Res.* **14**, 318 (1981).
29. F. G. A. Stone, *Inorg. Chim. Acta* **50**, 33 (1981).
30. J. Powell, M. R. Gregg, and J. F. Sawyer, *Inorg. Chem.* **27**, 4526 (1988).
31. L. J. Farrugia, J. A. K. Howard, P. Mitrprachachon, F. G. A. Stone, and P. Woodward, *J. Chem. Soc., Dalton Trans.*, 1134 (1981).
32. M. Green, K. A. Mead, R. M. Mills, I. D. Salter, F. G. A. Stone, and P. Woodward, *J. Chem. Soc., Chem. Commun.*, 51 (1982).
33. R. D. Adams and T. S. A. Hor, *Inorg. Chem.* **23**, 4723 (1984).
34. M. J. Freeman, A. D. Miles, M. Murray, A G. Orpen, and F. G. A. Stone, *Polyhedron* **3**, 1093 (1984).
35. P. D. Macklin, C. A. Mirkin, N. Viswanathan, G. D. Williams, G. L. Geoffroy, and A. L. Rheingold, *J. Organomet. Chem.* **334**, 117 (1987).
36. M. I. Bruce, G. Shaw, and F. G. A. Stone, *J. Chem. Soc., Dalton Trans.*, 1082 (1972).

37. M. I. Bruce, G. Shaw, and F. G. A. Stone, *J. Chem. Soc., Dalton Trans.*, 1781 (1972).

38. R. D. Adams, T. S. A. Hor, and P. Mathur, *Organometallics* **3**, 634 (1984).

39. R. D. Adams and S. Wang, *Inorg. Chem.* **24**, 4447 (1985).

39a. R. D. Adams, I. T. Horváth, and S. Wang, *Inorg. Chem.* **25**, 1617 (1986).

40. L. J. Farrugia, J. A. K. Howard, P. Mitrprachachon, J. L. Spencer, F. G. A. Stone, and P. Woodward, *J. Chem. Soc., Chem. Commun.*, 260 (1978).

40a. L. J. Farrugia, J. A. K. Howard, P. Mitrprachachon, F. G. A. Stone, and P. Woodward, *J. Chem. Soc., Dalton Trans.*, 155 (1981).

41. L. J. Farrugia, J. A. K. Howard, P. Mitrprachachon, F. G. A. Stone, and P. Woodward, *J. Chem. Soc., Dalton Trans.*, 1274 (1981).

42. L. J. Farrugia, A. D. Miles, and F. G. A. Stone, *J. Chem. Soc., Dalton Trans.*, 2437 (1985).

43. R. D. Adams, J. E. Babin, R. Mahtab, and S. Wang, *Inorg. Chem.* **25**, 4 (1986).

43a. R. D. Adams, J. E. Babin, R. Mathab, and S. Wang, *Inorg. Chem.* **25**, 1623 (1986).

44. R. D. Adams, T. S. A. Hor, and I. T. Horváth, *Inorg. Chem.* **23**, 4733 (1984).

45. M. Green, R. M. Mills, G. N. Pain, F. G. A. Stone, and P. Woodward, *J. Chem. Soc., Dalton Trans.*, 1309 (1982).

46. R. Hoffmann, *Angew. Chem., Int. Ed. Engl.* **21**, 711 (1982).

47. L. J. Farrugia, J. A. K. Howard, P. Mitrprachachon, F. G. A. Stone, and P. Woodward, *J. Chem. Soc., Dalton Trans.*, 162 (1981).

48. M. Green, J. A. K. Howard, R. M. Mills, G. N. Pain, F. G. A. Stone, and P. Woodward, *J. Chem. Soc., Chem. Commun.*, 869 (1981).

49. N. M. Boag, M. Green, R. M. Mills, G. N. Pain, F. G. A. Stone, and P. Woodward, *J. Chem. Soc., Chem. Commun.*, 1171 (1980).

50. M. Green, J. A. K. Howard, G. N. Pain, and F. G. A. Stone, *J. Chem. Soc., Dalton Trans.*, 1327 (1982).

51. G. Longoni, M. Manassero, and M. Sansoni, *J. Am. Chem. Soc.* **102**, 3242 (1980).

52. J. Fischer, A. Mitschler, R. Weiss, J. Dehand, and J. F. Nennig, *J. Organomet. Chem.* **91**, C37 (1975).

52a. P. Braunstein, J. Dehand, and J. F. Nennig, *J. Organomet. Chem.* **92**, 117 (1975).

53. J.-P. Barbier, P. Braunstein, J. Fischer, and L. Ricard, *Inorg. Chim. Acta* **31**, L361 (1978).

54. S. Bhaduri, K. R. Sharma, W. Clegg, G. M. Sheldrick, and D. Stalke, *J. Chem. Soc., Dalton Trans.*, 2851 (1984).

55. A. Fumagalli, S. Martinengo, G. Ciani, and G. Marturano, *Inorg. Chem.* **25**, 592 (1986).

56. A. Fumagalli, S. Martinengo, P. Chini, A. Albinati, S. Bruckner, and B. T. Heaton, *J. Chem. Soc., Chem. Commun.*, 195 (1978).

57. A. Fumagalli, R. Della Pergola, F. Bonacina, L. Garlaschelli, M. Moret, and A. Sironi, *J. Am. Chem. Soc.* **111**, 165 (1989).

58. S. M. Owen, *Polyhedron* **7**, 253 (1988).

59. K. Wade, *Adv. Inorg. Chem. Radiochem.* **18**, 1 (1976).

60. D. M. P. Mingos, *Nature (London), Phys. Sci.* **236**, 99 (1972).

61. D. M. P. Mingos, *Acc. Chem. Res.* **17**, 311 (1984).

62. D. M. P. Mingos, *J. Chem. Soc., Chem. Commun.*, 706 (1983).

63. R. L. Johnston and D. M. P. Mingos, *J. Organomet. Chem.* **280**, 419 (1985).

64. B. K. Teo, *Inorg. Chem.* **23**, 1251 (1984).

65. B. K. Teo, G. Longoni, and F. R. K. Chung, *Inorg. Chem.* **23**, 1257 (1984).

66. B. K. Teo, *Inorg. Chem.* **24**, 1627 (1985).

67. D. M. P. Mingos, *Inorg. Chem.* **24**, 114 (1985).

68. B. K. Teo, *Inorg. Chem.* **24**, 115 (1985).

69. A. J. Stone, *Inorg. Chem.* **20**, 563 (1981).

70. A. J. Stone and M. J. Alderton, *Inorg. Chem.* **21**, 2297 (1982).
71. A. J. Stone, *Polyhedron* **3**, 2051 (1984).
72. R. L. Johnston and D. M. P. Mingos, *J. Organomet. Chem.* **280**, 407 (1985).
73. R. L. Johnston and D. M. P. Mingos, *J. Chem. Soc., Dalton Trans.*, 1445 (1987).
74. L. Zhenyang and D. M. P. Mingos, *J. Organomet. Chem.* **339**, 367 (1988).
75. D. G. Evans and D. M. P. Mingos, *J. Organomet. Chem.* **232**, 171 (1982).
76. D. G. Evans and D. M. P. Mingos, *J. Organomet. Chem.* **240**, 321 (1982).
77. D. I. Gilmour and D. M. P. Mingos, *J. Organomet. Chem.* **302**, 127 (1986).
77a. D. G. Evans, *J. Organomet. Chem.* **352**, 397 (1988).
78. K. P. Hall and D. M. P. Mingos, *Prog. Inorg. Chem.* **32**, 237 (1984).
79. D. M. P. Mingos and D. G. Evans, *J. Organomet. Chem.* **251**, C13 (1983).
80. P. Ewing and L. J. Farrugia, *J. Organomet. Chem.* **347**, C31 (1988).
81. A. Fumagalli, S. Martinengo, P. Chini, D. Galli, B. T. Heaton, and R. Della Pergola, *Inorg. Chem.* **23**, 2947 (1984).
82. V. E. Lopatin, Yu, L. Slovokhotov, and Yu. T. Struchkov, *Koord. Khim.* **14**, 116 (1988).
83. M. J. Chetcuti, M. Green, J. A. K. Howard, J. C. Jeffery, R. M. Mills, G. N. Pain, S. J. Porter, F. G. A. Stone, A. A. Wilson, and P. Woodward, *J. Chem. Soc., Chem. Commun.*, 1057 (1980).
83a. M. J. Chetcuti, J. A. K. Howard, R. M. Mills, F. G. A. Stone, and P. Woodward, *J. Chem. Soc., Dalton Trans.*, 1757 (1982).
84. L. J. Farrugia, *Acta Crystallogr., Sect. C* **C44**, 818 (1988).
85. C. Couture and D. H. Farrar, *J. Chem. Soc., Dalton Trans.*, 2253 (1987).
86. N. M. Boag, K. M. Rao, and N. J. Terril, unpublished results.
87. N. M. Boag, *J. Chem. Soc., Chem. Commun.*, 617 (1988).
88. P. Ewing and L. J. Farrugia, *New J. Chem.* **12**, 409 (1988).
89. M. Green, D. R. Hankey, M. Murray, A. G. Orpen, and F. G. A. Stone, *J. Chem. Soc., Chem. Commun.*, 689 (1981).
89a. L. J. Farrugia, M. Green, D. R. Hankey, M. Murray, A. G. Orpen, and F. G. A. Stone, *J. Chem. Soc., Dalton Trans.*, 177 (1985).
90. P. Ewing and L. J. Farrugia, *Organometallics* **8**, 1665 (1989).
90a. L. J. Farrugia, M. Green, D. R. Hankey, A. G. Orpen, and F. G. A. Stone, *J. Chem. Soc., Chem. Commun.*, 310 (1983).
91. L. J. Farrugia, *Acta Crystallogr., Sect. C* **C44**, 1307 (1988).
92. D. M. P. Mingos, M. I. Forsyth, and A. J. Welch, *J. Chem. Soc., Dalton Trans.*, 1363 (1978); M. J. Calhorda, D. M. P. Mingos, and A. J. Welch, *J. Organomet. Chem.* **228**, 309 (1982).
93. C. Mealli, *J. Am. Chem. Soc.* **107**, 2245 (1985).
94. D. J. Underwood, R. Hoffmann, K. Tatsumi, A. Nakamura, and Y. Yamamoto, *J. Am. Chem. Soc.* **107**, 5968 (1985).
95. D. G. Evans and D. M. P. Mingos, *Organometallics* **2**, 435 (1983).
96. G. Longoni and F. Morazzoni, *J. Chem. Soc., Dalton Trans.*, 1735 (1981).
97. R. D. Adams, G. Chen, and J.-G. Wang, *Polyhedron* **8**, 2521 (1989).
98. B. E. R. Schilling and R. Hoffmann, *J. Am. Chem. Soc.* **101**, 3456 (1979).
99. R. L. Johnston and D. M. P. Mingos, *J. Chem. Soc., Dalton Trans.*, 647 (1987).
100. B. F. G. Johnson and R. E. Benfield, *in* "Transition Metal Clusters" (B. F. G. Johnson, ed.), Chapter 7. Wiley, New York, 1980.
101. E. Band and E. L. Muetterties, *Chem. Rev.* **78**, 639 (1978).
102. V. J. Johnston, F. W. B. Einstein, and R. K. Pomeroy, *Organometallics* **7**, 1867 (1988).
102a. C. Brown, B. T. Heaton, A. D. C. Towl, P. Chini, A. Fumagalli, and G. Longoni, *J. Organomet. Chem.* **181**, 233 (1979).
103. M. Grassi, B. E. Mann, B. T. Pickup, and C. M. Spencer, *J. Magn. Reson.* **69**, 92 (1986).

104. R. Willem, *Prog. NMR Spectrosc.* **20**, 1 (1988).
105. E. W. Abel, T. P. J. Coston, K. G. Orrell, V. Šik, and D. Stephenson, *J. Magn. Reson.* **70**, 34 (1986).
106. L. J. Farrugia, unpublished results.
107. L. J. Farrugia, *Organometallics* **8**, 2410 (1989).
108. P. Ewing, L. J. Farrugia, and D. S. Rycroft, *Organometallics* **7**, 859 (1988).
109. M. Green, R. M. Mills, G. N. Pain, F. G. A. Stone, and P. Woodward, *J. Chem. Soc., Dalton Trans.*, 1321 (1982).
110. A. C. Bray and F. G. A. Stone, unpublished results; A. C. Bray, Ph.D. Thesis, University of Bristol, 1984.
111. J. A. Abad, *Inorg. Chim. Acta* **121**, 213 (1986).
112. A. R. Pinhas, T. A. Albright, P. Hofmann, and R. Hoffmann, *Helv. Chim. Acta* **63**, 29 (1980).
113. P. Ewing and L. J. Farrugia, *Organometallics* **8**, 1246 (1989).
114. E. Rosenberg, C. B. Thorsen, L. Milone, and S. Aime, *Inorg. Chem.* **24**, 231 (1985).
115. G. E. Hawkes, L. Y. Lian, E. W. Randall, and K. D. Sales, *J. Magn. Reson.* **65**, 173 (1985).
116. D. L. Davies, J. C. Jeffery, D. Miguel, P. Sherwood, and F. G. A. Stone, *J. Chem. Soc., Chem. Commun.*, 454 (1987); *J. Organomet. Chem.* **383**, 463 (1990).
117. J. C. Jeffery, M. J. Parrott, and F. G. A. Stone, *J. Organomet. Chem.* **382**, 225 (1990).
118. P. Ewing and L. J. Farrugia, *Organometallics* **7**, 871 (1988).
119. L. R. Nevinger and J. B. Keister, *Organometallics* (to be published).
120. A. Ceriotti, F. Demartin, G. Longoni, M. Manassero, G. Piva, G. Piro, and M. Sansoni, *J. Organomet. Chem.* **301**, C5 (1986).
121. D. J. Wales and D. M. P. Mingos, *Polyhedron* **8**, 1933 (1989).
122. D. J. Wales, D. M. P. Mingos, and L. Zhenyang, *Inorg. Chem.* **28**, 2754 (1989).
123. A. Rodger and B. F. G. Johnson, *Polyhedron* **7**, 1107 (1988).
124. B. F. G. Johnson, *J. Chem. Soc., Chem. Commun.*, 27 (1986).
125. P. Ewing and L. J. Farrugia, *J. Organomet. Chem.* **373**, 259 (1989).
126. A. Moor, P. S. Pregosin, L. Venanzi, and A. J. Welch, *Inorg. Chim. Acta* **85**, 103 (1984).
127. C. Couture and D. H. Farrar, *J. Chem. Soc., Dalton Trans.*, 2245 (1987).
128. L. J. Farrugia, N. Macdonald, and R. D. Peacock, unpublished results.
129. V. G. Albano, G. Ciani, M. I. Bruce, G. Shaw, and F. G. A. Stone, *J. Organomet. Chem.* **42**, C99 (1972); V. G. Albano and G. Ciani, *ibid.* **66**, 311 (1974).
130. A. Modinos and P. Woodward, *J. Chem. Soc., Dalton Trans.*, 1534 (1975).
131. M. I. Bruce, J. G. Matisons, B. W. Skelton, and A. H. White, *Aust. J. Chem.* **35**, 687 (1982).
132. M. I. Bruce, M. R. Snow, and E. R. T. Tiekink, *Aust. J. Chem.* **39**, 2145 (1986).
133. R. Mason, J. Zubieta, A. T. T. Hsieh, J. Knight, and M. J. Mays, *J. Chem. Soc., Chem. Commun.*, 200 (1972); R. Mason and J. A. Zubieta, *J. Organomet. Chem.* **66**, 289 (1974).
134. R. Bender, P. Braunstein, D. Bayeul, and Y. Dusausoy, *Inorg. Chem.* **28**, 2381 (1989).
135. P. Braunstein, N. Guarino, C. de Méric de Bellefon, and J-L. Richert, *Angew. Chem., Int. Ed. Engl.* **26**, 88 (1987).
136. M. C. Grossel, R. P. Moulding, and K. R. Seddon, *J. Organomet. Chem.* **253**, C50 (1983).
137. A. A. Ioganson, O. N. Chimkova, N. A. Deikhina, A. B. Antonova, A. I. Rubailo, and V. P. Selina, *Izv. Akad. Nauk SSSR, Ser. Khim.* **4**, 938 (1983).
138. A. A. Ioganson, G. V. Burmakina, V. A. Trukhacheva, A. I. Rubailo, N. G. Maksimov, and N. A. Deikhina, *Izv. Akad. Nauk SSSR, Ser. Khim.* **6**, 1296 (1987).
139. B. M. Peake, B. H. Robinson, J. Simpson, and D. J. Watson, *J. Chem. Soc., Chem. Commun.*, 945 (1974).
140. S. I. Al-Resayes, P. B. Hitchcock, M. F. Meidine, and J. F. Nixon, *J. Chem. Soc., Chem. Commun.*, 1080 (1984).

140a. S. I. Al-Resayes, P. B. Hitchcock, M. F. Meidine, and J. F. Nixon, *J. Organomet. Chem.* **341**, 457 (1988).
141. J. F. Blount, L. F. Dahl, C. Hoogzand, and W. Hübel, *J. Am. Chem. Soc.* **88**, 292 (1966).
142. L. J. Farrugia, unpublished results.
143. S. V. Kovalenko, A. B. Antonova, N. A. Deikhina, A. A. Ioganson, E. D. Korniets, A. G. Ginzburg, A. I. Yanovskii, Yu. L. Slovokhotov, and Yu. T. Struchkov, *Izv. Akad. Nauk SSSR, Ser. Khim.* **12**, 2864 (1987).
144. G. Longoni, A. Ceriotti, R. Della Pergola, M. Manassero, M. Perego, G. Piro, and M. Sansoni, *Philos. Trans. R. Soc. London, Ser. A* **308**, 47 (1982).
144a. R. P. Brint, M. P. Collins, T. R. Spalding, F. T. Deeney, G. Longoni, and R. Della Pergola, *J. Organomet. Chem.* **319**, 219 (1987).
145. D. A. Lesch and T. B. Rauchfuss, *J. Organomet. Chem.* **199**, C6 (1980).
145a. V. W. Day, D. A. Lesch, and T. B. Rauchfuss, *J. Am. Chem. Soc.* **104**, 1290 (1982).
146. D. Seyferth, R. K. Henderson, and M. K. Gallagher, *J. Organomet. Chem.* **193**, C75 (1980).
147. D. Seyferth, R. S. Henderson, and L.-C. Song, *Organometallics* **1**, 125 (1982).
148. L. E. Bogan, Jr., D. A. Lesch, and T. B. Rauchfuss, *J. Organomet. Chem.* **250**, 429 (1983).
149. P. Mathur and I. J. Mavunkal, *Inorg. Chim. Acta* **126**, L9 (1987); *J. Organomet. Chem.* **350**, 251 (1988).
150. P. Mathur, I. P. Mavunkal, and V. Rugmini, *J. Organomet. Chem.* **367**, 243 (1989).
151. A. M. Mazany, J. P. Fackler, Jr., M. K. Gallagher, and D. Seyferth, *Inorg. Chem.* **22**, 2593 (1983).
152. L. J. Farrugia, *Organometallics* **9**, 105 (1990).
153. P. Sundberg, *J. Chem. Soc., Chem. Commun.*, 1307 (1987).
154. P. Ewing, Ph.D. Thesis, University of Glasgow, 1989.
155. P. Ewing and L. J. Farrugia, *J. Organomet. Chem.* **320**, C47 (1987).
156. B. Norén and P. Sundberg, *J. Chem. Soc., Dalton Trans.*, 3103 (1987).
157. G. D. Williams, M.-C. Lieszkovszky, C. A. Mirkin, G. L. Geoffroy, and A. L. Rheingold, *Organometallics* **5**, 2228 (1986).
158. C. Couture, D. H. Farrar, and R. J. Goudsmit, *Inorg. Chim. Acta* **89**, L29 (1984).
159. L. J. Farrugia, unpublished results using HYDEX [A. G. Orpen, *J. Chem. Soc., Dalton Trans.*, 2509 (1980)].
160. R. D. Adams and T. S. A. Hor, *Organometallics* **3**, 1915 (1984).
161. B. F. G. Johnson, D. A. Kaner, J. Lewis, P. R. Raithby, and M. J. Taylor, *Polyhedron* **1**, 105 (1982).
162. J. Dehand and J. F. Nennig, *Inorg. Nucl. Chem. Lett.* **10**, 875 (1974).
163. R. Bender, P. Braunstein, J. Fischer, L. Ricard, and A. Mitschler, *Nouv. J. Chim.* **5**, 81 (1981).
164. R. Bender, P. Braunstein, B. Metz, and P. Lemoine, *Organometallics* **3**, 381 (1984).
165. P. Lemoine, A. Giraudeau, M. Gross, R. Bender, and P. Braunstein, *J. Chem. Soc., Dalton Trans.*, 2059 (1981).
166. M. Green, D. R. Hankey, J. A. K. Howard, P. Louca, and F. G. A. Stone, *J. Chem. Soc., Chem. Commun.*, 757 (1983).
167. A. Nutton and P. M. Maitlis, *J. Organomet. Chem.* **166**, C21 (1979).
168. A. C. Bray, M. Green, D. R. Hankey, J. A. K. Howard, O. Johnson, and F. G. A. Stone, *J. Organomet. Chem.* **281**, C12 (1985).
169. R. S. Dickson, G. D. Fallon, M. J. Liddell, B. W. Skelton, and A. H. White, *J. Organomet. Chem.* **327**, C51 (1987).
170. A. Fumagalli, S. Martinengo, and G. Ciani, *J. Organomet. Chem.* **273**, C46 (1984).
171. V. G. Albano, G. Ciani, S. Martinengo, and A. Sironi, *J. Chem. Soc., Dalton Trans.*, 978 (1979).

172. G. Ciani, A. Fumagalli, and S. Martinengo, *J. Chem. Soc., Dalton Trans.*, 519 (1981).
173. A. Fumagalli, S. Martinengo, G. Ciani, and A. Sironi, *J. Chem. Soc., Chem. Commun.*, 453 (1983).
174. A. Fumagalli, S. Martinengo, D. Galli, A. Albinati, and F. Ganazzoli, *Inorg. Chem.* **28**, 2476 (1989).
175. S. Martinengo, A. Fumagalli, R. Bonfichi, G. Ciani, and A. Sironi, *J. Chem. Soc., Chem. Commun.*, 825 (1982).
176. S. Martinengo, G. Ciani, and A. Sironi, *J. Am. Chem. Soc.* **104**, 328 (1982).
177. G. Ciani and A. Sironi, *J. Organomet. Chem.* **197**, 233 (1980).
177a. A. Fumagalli, S. Martinengo, G. Ciani, N. Masciocchi, and A. Sironi, *Natl. Congr. Inorg. Chem., 18th*, Como, Italy, *1985*, Abstr. No. A30 (1985).
178. A. Ceriotti, F. Demartin, P. Ingallina, G. Longoni, M. Manassero, M, Marchionna, N. Masciocchi, and M. Sansoni, *Int. Conf. Chem. Platinum Met., 3rd*, Sheffield, *1987*, Abstr. O-30 (1987).
179. B. T. Heaton, P. Ingallina, R. Devenish, C. J. Humphreys, A. Ceriotti, G. Longoni, and M. Marchionna, *J. Chem. Soc., Chem. Commun.*, 765 (1987).
180. B. J. Pronk, H. B. Brom, L. J. De Jongh, G. Longoni, and A. Ceriotti, *Solid State Commun.* **59**, 349 (1986).
181. L. J. De Jongh, H. B. Brom, G. Longoni, B. Pronk, G. Schmid, and M. P. J. van Staveren, *J. Chem. Res., Synop.*, 150 (1987).
182. B. T. Heaton, personal communication.
183. B. J. Pronk, H. B. Brom, A. Ceriotti, and G. Longoni, *Solid State Commun.* **64**, 7 (1987).
184. J. A. O. De Aguiar, A. Mees, J. Darriet. L. J. De Jongh, S. Drake, P. P. Edwards, B. F. G. Johnson, and J. Lewis, *Solid State Commun.* **66**, 913 (1988).
185. L. J. De Jongh, J. A. O. De Aguiar, H. B. Brom, G. Longoni, J. M. van Ruitenbeek, G. Schmid, H. H. A. Smit, M. P. J. van Staveren, and R. C. Thiel, *Z. Phys. D* **12**, 445 (1989).
186. S. R. Drake, P. P. Edwards, B. F. G. Johnson, J. Lewis, E. A. Marseglia, S. D. Obertelli, and N. C. Pyper, *Chem. Phys. Lett.* **139**, 336 (1987).
187. R. E. Benfield, *Z. Phys. D* **12**, 453 (1989).
188. R. E. Benfield, P. P. Edwards, A. M. Stacy, *J. Chem. Soc., Chem. Commun.*, 525 (1982).
189. B. K. Teo, F. J. DiSalvo, J. V. Waszczak, G. Longoni, and A. Ceriotti, *Inorg. Chem.* **25**, 2262 (1986).
190. R. E. Benfield, *J. Phys. Chem.* **91**, 2712 (1987).
191. B. F. G. Johnson, personal communication.
192. W. Broach, L. F. Dahl, G. Longoni, P. Chini, A. J. Schultz, and J. M. Williams, *Adv. Chem. Ser.* **167**, 93 (1978).
193. S. I. Al-Resayes, P. B. Hitchcock, J. F. Nixon, and D. M. P. Mingos, *J. Chem. Soc., Chem. Commun.*, 365 (1985).
194. G. Huttner, U. Weber, B. Sigwarth, and O. Scheidsteger, *Angew. Chem., Int. Ed. Engl.* **21**, 215 (1982).
195. D. E. Berry, G. W. Bushnell, K. R. Dixon, P. M. Moroney, and C. Wan, *Inorg. Chem.* **24**, 2625 (1985).
196. P. Braunstein, C. de Méric de Bellefon, and M. Ries, *J. Organomet. Chem.* **262**, C14 (1984).
197. P. Braunstein, C. de Méric de Bellefon, M. Ries, and J. Fischer, *Organometallics* **7**, 332 (1988).
198. P. Braunstein, C. de Méric de Bellefon, and M. Ries, *Inorg. Chem.* **27**, 1338 (1988).
199. P. Braunstein, C. de Méric de Bellefon, M. Ries, J. Fischer, S.-E. Bouaoud, and D. Grandjean, *Inorg. Chem.* **27**, 1327 (1988).
200. P. Braunstein, J. Kervennal, and J.-L. Richert, *Angew. Chem., Int. Ed. Engl.* **24**, 768 (1985).
201. P. Braunstein and B. Oswald, *J. Organomet. Chem.* **328**, 229 (1987).

202. P. Braunstein, M. Ries, C. de Méric de Bellefon, Y. Dusausoy, and J.-M. Mangeot, *J. Organomet. Chem.* **355**, 533 (1988).

203. G. Nemra, P. Lemoine, P. Braunstein, C. de Méric de Bellefon, and M. Ries, *J. Organomet. Chem.* **304**, 245 (1986).

204. G. Nemra, P. Lemoine, M. Gross, P. Braunstein, C. de Méric de Bellefon, and M. Ries, *Electrochim. Acta* **31**, 1205 (1986).

205. P. Thometzek and H. Werner, *J. Organomet. Chem.* **252**, C29 (1983); *Organometallics* **6**, 1169 (1987).

206. G. P. Elliot, J. A. K. Howard, T. Mise, I. Moore, C. M. Nunn, and F. G. A. Stone, *J. Chem. Soc., Dalton Trans.*, 2091 (1986).

207. S. J. Davies, G. P. Elliot, J. A. K. Howard, C. M. Nunn, and F. G. A. Stone, *J. Chem. Soc., Dalton Trans.*, 2177 (1987).

208. G. P. Elliot, J. A. K, Howard, T. Mise, C. M. Nunn, and F. G. A. Stone, *J. Chem. Soc., Dalton Trans.*, 2189 (1987).

209. S. J. Davies, J. A. K. Howard, R. J. Musgrove, and F. G. A. Stone, *Angew. Chem., Int. Ed. Engl.* **28**, 624 (1989).

210. S. J. Davies, J. A. K. Howard, M. U. Pilotti, and F. G. A. Stone, *J. Chem. Soc., Dalton Trans.*, 2289 (1989).

211. M. J. Chetcuti, Ph.D. Thesis, Bristol University, 1980.

212. R. Pieantozzi, K. J. McQuade, B. C. Gates, M. Wolf, H. Knözinger, and W. Ruhmann, *J. Am. Chem. Soc.* **101**, 5436 (1979).

213. R. Pierantozzi, K. J. McQuade, and B. C. Gates, *Stud. Surf. Sci. Catal.* **7B**, 941 (1981).

214. W. Zhai, D. Li, Y. Ma, and Z. Zhao, *Youji Huaxue*, 180 (1983); *Chem. Abstr.* **99**, 114918e.

215. D. Li, W. Weixu, Z. Chen, Y. Sun, X. Zhao, and Z. Wang, *Huaxue Xuebao* **44**, 990 (1986); *Chem. Abstr.* **107**, 28819a.

216. X. Tang, H. Xu, L. Huang, and W. Zhai, *Fenzi Cuihua* **2**, 95 (1988); *Chem. Abstr.* **110**, 45653w.

217. C. U. Pittman, W. Honnick, M. Absi-Halabi, M. G. Richmond, R. Bender, and P. Braunstein, *J. Mol. Catal.* **32**, 177 (1985).

218. A. Fukuoda, T. Kimura, and M. Ichikawa, *J. Chem. Soc., Chem. Commun.*, 428 (1988).

219. M. Roper, M. Schieren, and A. Fumagalli, *J. Mol. Catal.* **34**, 173 (1986).

220. S. Bhaduri and K. R. Sharma, *J. Chem. Soc., Dalton Trans.*, 2309 (1984).

Index

A

Alkyl and aryl metal amide
  alkaline earth element, 270–271
    chelating ligand, 271
    dimerization, 271
    substituent effect on structure, 271
  aluminum, gallium, indium, and thallium, 271–276
    boat-chair isomerization, 272
    bond length, 272
    bulky substituent effect, 272–273
    cyclic and "chelated", 273–276
    dimeric, 273–275
    formation, 271
    macrocycle, 276
    monomeric, 273
    polycyclic oligomer, 272
Alkyl and aryl metal compound
  amide substitution, 270
  Grignard reagent, 269
  solvent-induced equilibrium, 269
  stabilization, 269–270
Alkylidyne carbaborane anionic salt
  dynamic behavior, 57–58
  protonation, 84–87
    acid dependent, 85
    hydrochloric acid, 86–87
    mechanism, 85–86
    tetrafluoroboric acid, 85–86
  reaction
    general, 58
    platinum complex, 60–61
    rhodium complex, 61–63
    ruthenium complex, 63–64
    trimetal compound, 60–61
    triphenylphosphinegold(I) chloride, 59, 61
  structure, 57
  synthesis, 56
Alkylidyne carbaborane complex
  dynamic behavior, 59
  exopolyhedral σ-boron-transition metal bond
    iridium, 76
    iron, 71–74
    mechanism, 71-72
    platinum, 74–75

ruthenium, 70
substituent effect, 70
exopolyhedral bonding
  B-H-iron, 74
  B-H-metal, 63–70
  B-H-rhodium, 62
  di-B-H-metal bond, 68–69
  isomerism, 67–68
  removal, 66
isomerism, 59–60, 62
Alkylidyne ligand
  alkyne comparison, 54
  asymmetric spanning, 59
  bridging, 62
  bonding scheme, 54
  carbaborane isolobal mapping, 55
  complexing agent, 53
  synthesis
    carbametallaborane, 55
    dimetal, 55
    polynuclear, 54
  transfer to carbaborane cage, 77–84
    iron complex, 82
    isomer equilibrium, 81
    mechanism, 77, 79–80
    molybdenum complex, 77
    platinum complex, 77–81
    rhenium, 83–84
    rhodium, 82–83
  tungsten–platinum species, 53–54
Aluminum
  alkyl and aryl metal amide, 271-273
  cyclic and "chelated" amide, 273–276
Antimony and bismuth transition metal complex, di-element
  bond length, 136
  bonding trimetal species, 135
  chromium, molybdenum, and tungsten, 135–137
  cobalt, rhodium, and indium anionic species
    paramagnetic, 151
    structure, 152
    synthesis mechanism, 153
  four-electron donor, 136
  iron, ruthenium, and osmium cluster
    anionic species, 145–146
    mutually bonded element, 141–143

393

Antimony and bismuth transition metal
   complex, di-element (*continued*)
     mutually nonbonded element, 143–147
     synthesis, 143–144
     synthetic interaction, 146–147
     thermodynamic control of isomer, 144
     trigonal bipyramidal geometry, 143
     Zintl metal carbonylate, 145
   manganese and rhenium, 137–139
   stabilization, 135
Antimony and bismuth transition metal
   complex, mono-element
   bond length, 131–133
   chromium, molybdenum, and tungsten
     anionic, 130–133
     mixed metal, 134
     trigonal planar coordination, 133
     trimetal species, 134
   cobalt, rhodium, and iridium
     anionic species, 150–151
     bridging carbonyl, 149
     cationic species, 151
     paramagnetism, 151
     phosphine derivative, 148
     steric consideration, 149
     trimetal species, 148
   iron, ruthenium, and osmium cluster
     hydride, 139–140
     pentametal species, 141
     tetrahedral, 139
     tetrametal species, 140
   manganese and rhenium, 137
Antimony and bismuth transition metal
   complex, tetra-element
   cobalt, 153–154
   cubic cluster, 153

B

Bismuth transition metal complex *see*
   Antimony and bismuth transition metal
   complex, mono, di, and tetra

C

Carbaborane ligand
   B-H-metal bridge tungsten complex,
     64–70
   cobalt, 69
   iridium, 67
   molybdenum, 64–65
   ruthenium, 63–64, 68

   tungsten–metal unsaturated bonding, 65
   nonspectator ligand, 62–84
   spectator ligand, 59–63, 69–70
Carbonylmetallate anion
   chromium, molybdenum, and tungsten,
     39–46
   cobalt, rhodium, and iridium, 31–38
   manganese and rhenium, 2–15
   "super-reduced" species prospect, 47
   vanadium, niobium, and tantalum, 15–31
Chromium complex
   hexacarbonyl, 15, 21
   main group
     antimony and bismuth, 130–137
     germanium, tin, and lead, 108–113
     tellurium, 155–157
   pentacarbonyl(2-) anion, 19, 21
   perflurobenzene, 199
   tetracarbonyl(4-) anion, 39–43
Cluster
   iron and ruthenium, main group, 101–106
   platinum heteronuclear, 301–391
   with more than two different metals,
     376–383
     tri- and tetra-nuclear, 377–379
     bridging alkylidyne ligand, 380–383
Cobalt complex
   alkylidyne carbaborane, 69
   μ-hexafluorocyclooctatrieneyne, 255–260
   main group
     antimony and bismuth, 148–154
     gallium, indium, and thallium, 106–107
     germanium, tin, and lead, 122–126
     tellurium, 167–168, 172–173, 175
     octafluorocycloocta-1,3,5,7-tetraene
       derivative, 217–239
     pentamethylcyclopentadienyl, 225–229
     perfluorobicyclo[3.3.0]octadienediyl,
       230–233, 221–225
   perfluorocyclooctatrieneyne, 255–260
   tricarbonylmetallate(3-) anion, 31–36
Cobalt–platinum cluster
   crystallographically characterized, 337
   high nuclearity
     carbido species, 365–366
     trigonal bipyramidal skeleton, 365
   synthesis, 303–305
   tri- and tetra-nuclear
     electrochemical study, 364–365
     phosphido-bridged species, 364
     preparation, 363–364

Cyclic and "chelated" organometal amide
  aluminum, gallium, indium, and thallium
    bonding, 276
    chelating macrocycle, 276
    equilibrium, 274, 276
    intermolecular chelate, 274–275
    isomeric form, 276
    mono-metal centered ring, 274–275
    ring dimer, 273
Cyclooctatetraene, metallation
  η^1-cyclooctatetraenyl ligand, 245
  isodynamical process in substituted system
    bond shift, 246
    NMR study, 247
    ring inversion, 246

D

Difluoromethylidene complex
  bridging, 190
  terminal
    cationic species, 189
    geometry, 189
    reactivity, 189–190

E

Electrochemical study
  cobalt complex, 226–228
    ESR of radical anion, 227–228
    rearrangement rate of radical anion, 228
  cycloocta-1,3,5,7-tetraene complex,
    226–227
  octafluorocycloocta-1,3,5,7-tetraene
    complex, 227

F

Fluorine
  ligand with α-fluorine, 200–201
    coupling, 201
    fluoroacetylene complex, 201
  perfluorinated ligand, 183–267
    definition, 184–185
Fluorocarbon
  hydrocarbon comparison, 183–184
  octafluorocycloocta-1,3,5,7-tetraene
    cyclooctatetraene comparison, 202
    isomerism, 202–203
    reduction, 204
    synthesis, 202
    unsuccessful complex preparation, 204

perfluorobicyclo[4.2.0]octa-2,4,7-triene,
    cobalt, and rhodium complex,
    217–221
perfluorinated ligand, 183–267
perfluorobenzocyclobutene, 222
perfluoroolefin complex, 191–194
perfluorotricyclo[4.2.0.0]octa-3,7-diene
    manganese complex, 210–211
    iron complex, 214
transition metal compound, 183–267
unsaturated
    metallation reaction, 200
    perfluorocyclohexa-1,3-diene cobalt
      complex, 200
Fluoromethylidyne complex
  bridging, 190
  difluoroacetylene, 191

G

Gallium, indium, and thallium transition
    metal complex
  alkyl and aryl metal amide, 271–273
  cobalt, 106–107
    anionic, 106
    heterolytic bond dissociation, 106
    preparation, 106
    structure, 106
  cyclic and "chelated" amide, 273–276
  iron and ruthenium anionic cluster, 101–106
    bonding, 103
    bridging thallium, 105
    formation of larger cluster, 102–103
    preparation, 101, 105
    reaction bidentate nitrogenous base, 104
    structure, 101–102
  manganese and rhenium
    cluster formation, 98
    isolobal bonding, 102
    mixed metal species, 99–101
    polar bonding, 97
    preparation, 96–97, 98–100
    reaction pentacarbonyl species, 97
  molybdenum and tungsten
    aluminum comparison, 95–96
    bond length, 93–94
    ionic metal–metal bonding, 95
    preparation, 92, 95
    structure, 92–93
  platinum
    cationic cluster, 107
    structure and bonding, 108

Germanium, tin, and lead transition metal
    complex
    chromium, molybdenum and tungsten
        bond length, 111
        cyclopentadienyl ligand, 108–109
        isocarbonyl bridging, 110
        mixed metal species, 113
        substituted cyclopentadienyl ligand,
            109–110
        three transition metal, 110
    cobalt
        synthesis divalent germanium species,
            122–123
        high nuclearity cluster, 125–126
        oxidation divalent species, 123
        spirobicyclic, 124
        tetracobalt cluster, 123
        tin and lead preparation, 124
    iron
        carbonyl bridging, 120
        carbonyl cluster, 117, 120–122
        mixed metal, 119
        spirocyclic tetra-iron species,
            118–120
        tetrahedral tetra-iron, 121
        tetra-iron species, 117
        tri-iron species, 117
    manganese
        conformer, 115
        linear two-coordinate geometry,
            115–117
        multiple bonding, 113–116
        triple bond character, 115
    mixed transition metal
        cobalt–iron cluster, 129–129
        cobalt–molybdenum, 129
        cobalt–nickel, 129
        iron–rhenium cluster, 128–129
    nickel cluster
        carbonyl, 127
        icosahedral, 128
        pentagonal antiprismatic, 127
    spectroscopic study, 129–130
Gold complex
    alkylidene carbaborane
        iron, 72–73
        molybdenum, 59
    manganese, 14–15
    tantalum, 29–30
    vanadium, 24–26

I

Indium complex, transition metal, *see*
    Gallium, indium, and thallium transition
        metal complex
Iridium
    alkylidene carbaborane complex, 67,76
    antimony and bismuth complex, 148–154
    platinum cluster
        crystallographically characterized, 340
        fluxional behavior, 319, 321–323
        raft structure, 372–373
        synthesis, 303–305, 372
        trinuclear, 366–367, 374
    tricarbonylmetallate(3-) anion, *see*
        tricarbonylmetallate(3-) anion of
        rhodium and iridium
Iron complex
    alkylidene carbaborane, 71–74
    η^1-cyclooctatetraenyl, 245–255
    η^1-heptafluorocyclooctatetraenyl
        synthesis, 247–248
        crystallographic study, 248–250
        ligand substitution, 250–251
    main group
        antimony and bismuth, 139–147
        gallium, indium, and thallium, 101–106
        germanium, tin, and lead, 117–122
        indium mixed metal, 99–101
        tellurium, 165–167, 169–174
    octafluorocycloocta-1,3,5,7-tetraene,
        211–213
    perfluoro-1,3-diene, 197–198
    perfluorotricyclo[4.2.0.0]octa-3,7-diene,
        214–217
Iron–platinum cluster, 328–383
    chalcogenide
        structure, 346
        synthesis, 346
    crystallographically characterized,
        330–331
    fluxional behavior, 322
    high nuclearity
        "bow-tie" anion, 344
        ^{57}Fe Mössbauer spectroscopy, 344–345
        planar species, 345
        structure, 343–344
    synthesis, 303–305
    tetranuclear
        diplatinum species, 342–343
        structure, 342

vinylidene, 343
trinuclear, 329–342
  bonding mode, 342
  polarographic study, 341
  preparation, 341–342

L

Lead, transition metal complex, *see*
  Germanium, tin, and lead transition
  metal complex

M

Magnesium, alkyl, and aryl metal amide,
  270–271
Manganese complex
  main group,
    bismuth, 137–139
    gallium, indium, and thallium, 96–101
    germanium, tin, and lead, 113–117
    tellurium, 157–165, 172
    octafluorocycloocta-1,3,5,7-tetraene,
      205–211
    tetracarbonyl(3-) anion *see*
      tetracarbonylmetallate(3-) of
      manganese and rhenium
Molybdenum
  alkylidene carbaborane complex, 64–65
    gold, 59
  main group complex
    gallium, indium, and thallium, 92–96
    germanium, tin, and lead, 108–113
    antimony and bismuth, 130–137
    tellurium, 155–157, 169–171, 175
    platinum heteronuclear cluster, 380–382
    tetracarbonylmetallate(4-) anion, 39–45

N

Nickel complex
  main group
    germanium, tin, and lead, 127–128
    tellurium, 168–169, 173
    1,2,5,6-η^2-octafluorocycloocta-1,3,5,7-
      tetraene, 241–242
    perfluorobicyclo[3.3.0]octadienyl,
      242–245
Nickel–platinum cluster
  high nuclearity
    "cherry" crystallite, 374
    hydride, interstitial site, 375–376

octahedral core, 374
physicochemical property, 374–375
molybdenum species, 381–382
tungsten species, 380
Niobium, pentacarbonylmetallate anion, *see*
  pentacarbonylmetallate(3-) of niobium
  and tantalum

O

Octafluorocycloocta-1,3,5,7-tetraene and
  valence isomer metallation
  cyclooctatetraene, 245–247
  dimetallation, iron complex, 251–252
  heptafluorobicyclo[4.2.0]octatrienyl,
    247–255
  η^1-heptafluorocyclooctatetraenyl complex
    chromatography, 251
    crystallographic study, iron, 248–250
    hydrocarbon comparison, 254–255
    isodynamic process, 254–255
    ligand substitution, 250–251
    NMR dynamic study, 254
    synthesis, 247–248
  perfluorotricyclo[4.2.0.0]octa-3,7-diene
    dimetallation, 252–253
    structure, 252
    synthesis, 252
Octafluorocycloocta-1,3,5,7-tetraene
    bridging indenyl rhodium complex
  intermediate structure, 236–239
    cis coordination of two rhodium,
      236–238
    mechanism of bridge formation, 238
  selective C–F bond activation
    by chromatography, 239
    F–H exchange, 239
  stereochemical nonrigidity of bridging
      ligand
    activation energy, 236
    ^{19}F NMR spectrum, 235–236
    mechanism interconversion, 235–236
  synthesis and structure, 233–235
    bonding, 233–235
    hydrocarbon analogue, 235
Octafluorocycloocta-1,3,5,7-tetraene
    complex, pentamethylcyclopentadienyl
    cobalt and rhodium
  formation of
    perfluorobicyclo[3.3.0]octadienyl,
      230–233

Octafluorocycloocta-1,3,5,7-tetraene
     complex, pentamethylcyclopentadienyl
     cobalt and rhodium (*continued*)
     cyclopentadienyl comparison,
          231–232
     indenyl comparison, 232–233
     mechanism, 231
     pyramidalization, 233
     structure, 230–231
     t-butylisocyanide ligand, 230
     formation of stable
          perfluorocyclooctatrienediyl,
          229–230
     trimethylphosphine ligand, 229
     η^3–η^1 transformation, 229–230
Octafluorocycloocta-1,3,5,7-tetraene
     derivative cobalt complex
     cyclopentadienyl
          structure, 221
          synthesis, 222
          transannular ring closure, 222
     electrochemical comparative study,
          226–228
     cyclic voltammetry, 227
     cycloocta-1,3,5,7-tetraene, 226–227
     reduction, 227
     electron spin resonance study, 227–228
     pentamethylcyclopentadienyl, 225–233
          electrochemistry, 226–228
          isomer, 225–226
          synthesis, 225
     radical anion
          cyclic voltammetry, 227
          preparation, 226–227
     relative rearrangement rate
          isomerization slowing, 228
          hydrocarbon comparison, 228
Octafluorocycloocta-1,3,5,7-tetraene
     derivative nickel complex
     ligand effect on bonding mode, 244–245
     1,2,5,6-η^2-octafluorocyclooctatetraene,
          241–242
     perfluorobicyclo[3.3.0]octadienediyl,
          242–245
     five coordinate species, 243
     mechanism, 244
     phosphine ligand dissociation, 244
     phosphine reaction, 242
     t-butylisocyanide preparation, 242
Octafluorocycloocta-1,3,5,7-tetraene
     derivative palladium complex,

perfluorobicyclo[3.3.0]octadienediyl,
          241
Octafluorocycloocta-1,3,5,7-tetraene
     derivative platinum complex
     perfluorobicyclo-[3.3.0]octadienediyl
          synthesis, 240
          mechanism, 240–241
          transannular ring closure, 244
Octafluorocycloocta-1,3,5,7-tetraene
     derivative rhodium complex
     cyclopentadienyl
          bonding, 217–218
          synthesis, 217
     indenyl
          bonding, 223
          dynamic NMR study, 225
          NMR study, 224
          slip-folding, 223–224
          structure, 223–224
          synthesis, 218
     pentamethylcyclopentadienyl
          synthesis, 225
          isomer, 225–226
     perfluorobicyclo[4.2.0]octatriene
          ring closure control, 221
          structure and bonding, 220
          synthesis, 218–219
          valence isomer, 219
Octafluorocycloocta-1,3,5,7-tetraene iron
     complex
     bonding mode, 212
     hydrocarbon comparison, 211–212
     perfluorocyclooctadienyl formation,
          215–217
          mechanism of nucleophilic attack, 215
          bonding, 217
          hard and soft nucleophile, 215–217
     synthesis, 211–212
     X-ray structural study, 212–214
Octafluorocycloocta-1,3,5,7-tetraene
     manganese complex
     η^6-hydrocarbon comparison
          dynamic NMR study, 209–210
          structural, 207–209
     synthesis and physical property of η^6-
          species, 205–207
          carbonyl replacement, 205
          hydrocarbon comparison, 205–207
          tricyclic route, 210
Octafluorocycloocta-1,3,5,7-tetraene, vicinyl
     defluorination, 255–260

cyclooctatrieneyne structure, 255–256
dicobalt complex, 256–260
   ^{31}P NMR, 259
   bonding, 258–259
   crystallographic study, 257–259
   inversion barrier comparison, 260
   isodynamical process, 259–260
   ^{31}P NMR, 259
   synthesis, 256–257
Osmium complex
main group
   antimony and bismuth, 139–147
   tellurium, 165–167, 172
Osmium–platinum cluster
crystallographically characterized,
      333–336
exchange spectroscopy, 315–317
fluxional behavior, 317–320
high nuclearity, 358–363
   bicapped octahedral species, 358
   bicapped tetrahedron species, 358–360
   carbido species, 362–363
   diplatinum bridged species, 360–361
   preparation from sulphido cluster,
         361–362
   skeletal rarrangement, 360
   skeletal structure, 361–362
isomerization, 327–328
synthesis, 303–305
tetranuclear, 351–358
   butterfly metal core, 351–358
   carbido species, 362–363
   isomerism phosphine adduct, 352–353
   ligand loss, 353
   preparation, 351, 355, 357
   reaction, 354, 356
   spiked triangular core, 355, 357–358
   thermal behavior, 352
   two-electron donor ligand reaction, 352
trinuclear, 350–351
   bridging carbonyl, 350–351
   diplatinum, 351
   preparation, 350

P

Palladium complex
main group, tellurium, 173
perfluorobicyclo[3.3.0]octadienyl, 241
platinum cluster
   bimetallic, 376

different metal, 377–379
Pentacarbonylmetallate(3-) of niobium and
   tantalum
amine derivative
   π-acceptor ligand reaction, 31
   preparation, 30
   structure, 30
characterization
   elemental, 19
   infrared spectra, 18
   tricesium salt, 19–20
hydride derivative
   ^{1}H, ^{93}Nb, and ^{13}C NMR spectra, 27–29
   hydride transfer agent, 29
   metal carbonyl reaction, 29
   preparation, 27
   triphenylphosphine gold complex, 30
isocyanide derivative, 31
stability, 18–19
synthesis
   reduction hexacarbonyl(1-) anion, 18
   tricesium salt, 19–20
   trisodium salt, 20
Pentacarbonylmetallate(3-) of vanadium
aminepentacarbonylvanadate(1-) anion
   reaction, 25–26
   synthesis, 25–26
characterization
   trirubidium, cesium, and potassium salt,
         16
   ^{51}V, ^{13}C, and ^{1}H NMR study, 20–23
protonation
   isolation and identification, 20
   ^{1}H and ^{51}V NMR study, 20–23
salt bonding, 17
synthesis, 15–18
thermal stability, 16–17
with main group electrophile
   reaction, 24
   mechanism, 24
with triphenylphosphine gold(I) chloride
   eight-coordinate complex, 24
   structure, 24, 26
   vanadium carbonyl comparison, 24
Perfluoroalkyl complex
alkyl comparison
   α or β elimination, 188
   bond length, 186
   bond strength, 185
   carbonyl insertion, 187–188
   Fenske–Hall calculation, 186–187

Perfluoroolefin complex (*continued*)
  reactivity, 187–188
  thermodynamic stability, 185
Perfluoroallylic complex
  anionic species, 196
  η^3-pentafluoroallyl compound, 194–196
  isomerization, 196
  MO calculation, 196–197
Perfluorobenzene complex
  η^2-bonded, 199
  η^6-bonded, 199
  σ-bonded, 199
  platinum Dewar valence isomer, 199–200
Perfluorobicyclo[3.3.0]octadienediyl ligand
  cobalt, 221–222
  nickel, 242–245
  palladium, 241
Perfluoro-1,3-diene complex
  bonding, 197
  η^5-pentafluorocyclopentadienyl, 198
  hydrocarbon comparison, 197
  oxidative addition, 198
Perfluoroolefin complex
  cyclopropene ring opening, 193–194
    irreversible, 194
    stereoselectivity, 194
  dynamic behavior
    bonding, 193
    cyclization, 193
    NMR study, 192–193
    rotation activation barrier, 192–193
  structure and bonding, 191–192
    hydrocarbon comparison, 191–192
    metallacyclopentane, 192
    pyramidalization of ligated carbon, 192
Platinum cluster with more than two different
    metals, 376–383
  bridging alkylidyne, 380–383
    heptametal chain, 382
    iron and tungsten, 380
    isomerization, 381
    nickel and tungsten "star", 381
    open chain polymetallic species,
        380–381
  tri- and tetra-nuclear palladium species
    bridging allylic ligand, 379
    electrochemical study, 379
    isomerization, 379
    preparation, 377–379
Platinum complex
  alkyliyne carbaborane, 60–61, 77–81
  hexafluorobenzene, 199–200

main group
  tellurium, 168–169, 173
  thallium, 108
  perfluorobicyclo[3.3.0]octadienyl,
      240–241
Platinum heteronuclear cluster
  bimetallic, 301
  bonding and electron counting, 306–312
    anionic raft, 310
    dianion hypothetical, 311
    frontier orbital, 306–307
    high nuclearity, 309
    molecular orbital study, 310–312
    nonplanar, 309
    nonutilization of tangential orbital,
        306–307
    skeletal electron pair theory, 306–308
    tetrahedral, 309
    topological electron counting theory,
        306–307
    triangular, 308
    trigonal bipyramidal, 312
    utilization of tangential orbital, 310
    valence electron, 307–308, 312
  site-specific reactivity, 302
  structural and reactivity aspect, 328–383
    cobalt platinum, *see* cobalt–platinum
        cluster
    iron platinum, *see* iron–platinum cluster
    nickel platinum, *see* nickel–platinum
        cluster
    osmium platinum, *see* osmium–platinum
        cluster
    palladium platinum, 376
    rhodium platinum, *see* rhodium–
        platinum cluster
    ruthenium platinum, *see* ruthenium–
        platinum cluster
Platinum heteronuclear cluster, fluxional
    behavior, 312–327
  exchange spectroscopy, 315–317
  intermetallic ligand migration, 315,
      322–326
    carbonyl exchange, 322–323
    carbonyl migration in anionic complex,
        322–323
    free energy of activation, 325
    hydride exchange, 323–326
  localized exchange in ML group
    iron and cobalt triad, 320
    "rocking", 321
    tripodal rotation, 320–321

metal framework rearrangement, 315,
    326–328
    butterfly core geometry, 327
    carbonyl scrambling, 327
    time-averaged mirror plane, 326
metal-localized ligand scrambling, 315
^{195}Pt NMR, 312–315
rotation and localized exchange in PtL$_2$ and
    PtL$_3$ group, 317–320
    ^{187}Os NMR coupling, 317
    MO calculation, 319–320
    olefinic and phosphine environment
        exchange, 320
    phosphine substitution effect, 317–318
    triangular comparison, 319
Platinum heteronuclear cluster isomerization
metal core rearrangement, 327–328
    effect of ligand, 328
    isotopic labeling, 328
    kinetic study, 328
Platinum heteronuclear cluster synthesis
carbonylate anion with halide complex
    high nuclearity, 305
    metal exchange, 305
    redox condensation, 305
carbonylation of metal halide, 305
saturated carbonyl complex
    cobalt triad, 303–304
    iron triad, 303
    phosphine, 304
unsaturated carbonyl complex
    isolobal analogy, 304
    osmium, 304
    rhodium, 304–305

R

Rhenium
alkylidyne carbaborane complex, 83–84
main group complex
    bismuth, 137
    gallium, indium, and thallium, 96–101
    tellurium, 157–165
tetracarbonyl(3-) anion, *see*
    tetracarbonyl(3-) anion of manganese
    and rhenium
Rhodium complex
alkylidyne carbaborane, 61–63, 82–84
fluoroalkene, 188
main group
    antimony and bismuth, 148–154
    tellurium, 167–168, 172–173

Octafluorocyclooncta-1,3,5,7-tetraene
    derivative, 217–239
    cyclopentadienyl and indenyl, 217–218
    dinuclear indenyl, 233–239
    pentamethylcyclopentadienyl, 225–229
    perfluorobicyclo[3.3.0]octadienyl,
        230–233
    perfluorobicyclo[4.2.0]octatriene,
        218–221
perfluorobenzene, 199
perfluorocyclooctatrienediyl, 229–230
tricarbonylmetallate(3-) anion, *see*
    tricarbonylmetallate (3-) anion of
    rhodium and iridium
Rhodium–platinum cluster
crystallographically characterized,
    338–339
dinuclear complex derivative
    protonation, 367
    synthesis, 367–368
    trinuclear, 366–367
fluxional behavior, 322–323
high nuclearity
    anticuboctahedral metal core, 370
    heptanuclear, 371–372
    octahedral anion synthesis, 367, 369
synthesis, 304–305
Ruthenium complex
alkylidene carbaborane, 63–64, 68
main group
    antimony and bismuth, 139–147
    tellurium, 165–167, 172
    thallium, 105
Ruthenium–platinum cluster
crystallographically characterized, 332
fluxional behavior, 320, 323–325
high nuclearity, 348–350
tetranuclear
    preparation, 348
    reaction, 348–350
trinuclear, 346–348
    axially bound phosphine, 347
    preparation, 346
    steric factor, 347
tungsten species, 382–383

S

Silazane, intramolecular bond fluctuation
chelate
    bond breaking and rotation, 295
    enantiomeric form, 295–296

Silazane, intramolecular bond fluctuation
  chelate (*continued*)
  electron rich and deficient center, 294
  "ligand exchange"
    conformer, 297
    enantiomer equilibrium, 296–297
    steric effect, 296
  magnesium amide, 297–298
    ¹H NMR spectrum, 297
    "rapid-rocking", 298
  solid state energy minimum, 295
Silazane, organometallic derivative
  structure, 280–284
    aluminum polycycle, 283
    dimetallasilazane, 281–282
    indium dimer, 282–283
    magnesium cube, 283
    monomeric and cyclic molecule,
      280–281
    secco norcubane-like polycycle, 284
  synthesis, 276–280
    from amide, 276
    from oxoamine, 277
    from cyclic "cis" amine, 277
    metallacycle formation, 277–279
      condition, 278
Silazane, synthetic use
  alkylating agent, silicon, and tin halide,
    285
  base addition, 285
  polymetal, 286–294
    anion, 294
    lithium and aluminum, 286
    lithium and indium, 286
    from alkyl metal silazane, 286–287
    property, 287
    structure, 287–290
      bond distance, 290
      bond reactivity, 290–291
      dimagnesium with differing organic
        ligand, 291
      X-ray diffraction study, 291–292
      two-coordinate sodium, 293
      Grignard reagent, 293–294

                        T

Tantalum, pentacarbonylmetallate anion, *see*
  pentacarbonylmetallate(3-) anion of
  niobium and tantalum

Tellurium mixed transition metal complex
  cobalt and molybdenum, 175
  iron
    cobalt and rhodium, 172–173
    manganese, 172
    molybdenum, 169–171
    nickel, palladium, and platinum, 173
    ruthenium and osmium, 172
    structure, 169–170
    thermolysis, 170
Tellurium transition metal complex
  bond length, 160–161, 164
  chromium, molybdenum, and tungsten,
    155–157
    antimony comparison, 156
    ditellurium moiety, 156
    geometry, 155
    reactivity, 155
    Zintl-type, 157
  cobalt and rhodium, 167–168
    cluster, 167
    resonance structure, 168
  iron, ruthenium, and osmium, 165–167
    cubic cluster, 167
    dimetalla-ditelluride, 165
    ditellurium ligand, 165
    heterometallic species, 166
    hydride cluster, 166–167
    triply-bridged species, 166
  manganese and rhenium, 157–165
    bonding, 158–159, 162
    cationic species, 164
    dimetalla–ditelluride, 163
    semiconductor material, 164
    six-electron donor ligand, 162
    trimetal, 159
  nickel and platinum, 168–169
    dicationic species, 169
    phosphine, 168
  spectroscopic study, 175–176
    ¹²⁵Te Mössbauer, 176
    ¹²⁵Te NMR, 175–176
  zirconium and vanadium, 154–155
Tetracarbonylmetallate(4-) anion of
  chromium, molybdenum, and tungsten
  characterization, 40–42
    infrared spectrum, 40–41
  derivative, 42–45
    carbonylhydrido cluster, 44–45
    diamine, 42

dianion, 44
  hydride, 42–43
  tiphenyltin, 42
  synthesis, 39–40
Tetracarbonylmetallate(3-) anion of
    manganese and rhenium
  acylation, 13
  ^{13}C NMR study, 4
  characterization, infrared spectrum, 5–8
    first row transition metal carbonyl
        comparison, 7
    hexamethylphosphoric triamide solution,
        5–6
    ruthenium complex comparison, 5–6
    sodium–manganese interaction,
        evidence, 8
  conductivity study, 8
  protonation evidence, 8–10
    conversion to dihydride, 9
    ^{1}H NMR spectrum, 8–9
    infrared spectrum, 8–9
    structure of dihydride, 10
    trihydride, 10
  synthesis, 2–5
  thermal stability, 4
  with alkylating agent
    1,4-butaneditosylate, 12
    complex stability, 11
    ethylating, 12
    infrared spectrum, 11–12
    methylating, 10–11
  with main group electrophile, 10–13
    dianion, 10–11
    octahedral, 10
    seven-coordinate, 10
  with transition metal electophile, 14–15
    complex structure, 14–15
    triphenylphosphinegold(I) chloride, 14
Thallium, transition metal complex, see
    Gallium, indium, and thallium transition
    metal complex
Tin, transition metal complex, see
    Germanium, tin, and lead transition
    metal complex
Titanium, cyclopentadienyltetracarbonyl(1-)
    anion, 19

Tricarbonylmetallate(3-) anion of cobalt
  characterization
    infrared spectrum, 33–34
    purification of trisodium salt, 33
  germanium, tin, and lead derivative, 34–35
  hydride derivative, 35
  isocyanide derivative, 35–36
  synthesis
    reduction tetracarbonyl(1-) anion, 31–33
    tripotassium salt, 32–34
    trisodium salt, 32
Tricarbonylmetallate(3-) anion of rhodium
    and iridium, 36–39
  synthesis, 36–37
  characterization, 37
  protonation, 37–38
  germanium and tin derivative, 38
Tricarbonyltungsten(6-) anion
  infrared spectral evidence, 46
  speculative preparation, 45–46
Tungsten
  alkylidyne carbaborane complex, 53–87
  main group complex
    antimony and bismuth, 130–137
    gallium, indium, and thallium, 92–96
    germanium, tin, and lead, 108–113
    tellurium, 155–157

V

Vanadium
  main group complex, tellurium, 154–155
  pentacarbonylmetallate anion, see
      pentacarbonylmetallate(3-) of
      vanadium

Z

Zirconium complex
  fluoroalkyl, 188
  main group, tellurium, 154–155
  platinum heteronuclear cluster, 380–383
  tetracarbonylmetallate(4-) anion, 39–45
  tricarbonylmetallate(6-) anion, 45–46

Cumulative List of Contributors

Abel, E. W., **5**, 1; **8**, 117
Aguilo, A., **5**, 321
Albano, V. G., **14**, 285
Alper, H., **19**, 183
Anderson, G. K., **20**, 39
Angelici, R. J., **27**, 51
Aradi, A. A., **30**, 189
Armitage, D. A., **5**, 1
Armor, J. N., **19**, 1
Ash, C. E., **27**, 1
Ashe III, A. J., **30**, 77
Atwell, W. H., **4**, 1
Baines, K. M., **25**, 1
Barone, R., **26**, 165
Bassner, S. L., **28**, 1
Behrens, H., **18**, 1
Bennett, M. A., **4**, 353
Birmingham, J., **2**, 365
Blinka, T. A., **23**, 193
Bogdanović, B., **17**, 105
Bottomley, F., **28**, 339
Bradley, J. S., **22**, 1
Brinckman, F. E., **20**, 313
Brook, A. G., **7**, 95; **25**, 1
Brown, H. C., **11**, 1
Brown, T. L., **3**, 365
Bruce, M. I., **6**, 273; **10**, 273; **11**, 447; **12**, 379; **22**, 59
Brunner, H., **18**, 151
Buhro, W. E., **27**, 311
Cais, M., **8**, 211
Calderon, N., **17**, 449
Callahan, K. P., **14**, 145
Cartledge, F. K., **4**, 1
Chalk, A. J., **6**, 119
Chanon, M., **26**, 165
Chatt, J., **12**, 1
Chini, P., **14**, 285
Chisholm, M. H., **26**, 97; **27**, 311
Chiusoli, G. P., **17**, 195
Chojnowski, J., **30**, 243
Churchill, M. R., **5**, 93
Coates, G. E., **9**, 195
Collman, J. P., **7**, 53
Compton, N. A., **31**, 91
Connelly, N. G., **23**, 1; **24**, 87
Connolly, J. W., **19**, 123

Corey, J. Y., **13**, 139
Corriu, R. J. P., **20**, 265
Courtney, A., **16**, 241
Coutts, R. S. P., **9**, 135
Coyle, T. D., **10**, 237
Crabtree, R. H., **28**, 299
Craig, P. J., **11**, 331
Csuk, R., **28**, 85
Cullen, W. R., **4**, 145
Cundy, C. S., **11**, 253
Curtis, M. D., **19**, 213
Darensbourg, D. J., **21**, 113; **22**, 129
Darensbourg, M. Y., **27**, 1
Davies, S. G., **30**, 1
Deacon, G. B., **25**, 237
de Boer, E., **2**, 115
Deeming, A. J., **26**, 1
Dessy, R. E., **4**, 267
Dickson, R. S., **12**, 323
Dixneuf, P. H., **29**, 163
Eisch, J. J., **16**, 67
Ellis, J. E., **31**, 1
Emerson, G. F., **1**, 1
Epstein, P. S., **19**, 213
Erker, G., **24**, 1
Ernst, C. R., **10**, 79
Errington, R. J., **31**, 91
Evans, J., **16**, 319
Evans, W. J., **24**, 131
Faller, J. W., **16**, 211
Farrugia, L. J., **31**, 301
Faulks, S. J., **25**, 237
Fehlner, T. P., **21**, 57; **30**, 189
Fessenden, J. S., **18**, 275
Fessenden, R. J., **18**, 275
Fischer, E. O., **14**, 1
Ford, P. C., **28**, 139
Forniés, J., **28**, 219
Forster, D., **17**, 255
Fraser, P. J., **12**, 323
Fritz, H. P., **1**, 239
Fürstner, A., **28**, 85
Furukawa, J., **12**, 83
Fuson, R. C., **1**, 221
Gallop, M. A., **25**, 121
Garrou, P. E., **23**, 95
Geiger, W. E., **23**, 1; **24**, 87

Geoffroy, G. L., **18**, 207; **24**, 249; **28**, 1
Gilman, H., **1**, 89; **4**, 1; **7**, 1
Gladfelter, W. L., **18**, 207; **24**, 41
Gladysz, J. A., **20**, 1
Glänzer, B. I., **28**, 85
Green, M. L. H., **2**, 325
Griffith, W. P., **7**, 211
Grovenstein, Jr., E., **16**, 167
Gubin, S. P., **10**, 347
Guerin, C., **20**, 265
Gysling, H., **9**, 361
Haiduc, I., **15**, 113
Halasa, A. F., **18**, 55
Hamilton, D. G., **28**, 299
Harrod, J. F., **6**, 119
Hart, W. P., **21**, 1
Hartley, F. H., **15**, 189
Hawthorne, M. F., **14**, 145
Heck, R. F., **4**, 243
Heimbach, P., **8**, 29
Helmer, B. J., **23**, 193
Henry, P. M., **13**, 363
Heppert, J. A., **26**, 97
Herberich, G. E., **25**, 199
Herrmann, W. A., **20**, 159
Hieber, W., **8**, 1
Hill, E. A., **16**, 131
Hoff, C., **19**, 123
Horwitz, C. P., **23**, 219
Hosmane, N. S., **30**, 99
Housecroft, C. E., **21**, 57
Huang, Yaozeng (Huang, Y. Z.), **20**, 115
Hughes, R. P., **31**, 183
Ibers, J. A., **14**, 33
Ishikawa, M., **19**, 51
Ittel, S. D., **14**, 33
Jain, L., **27**, 113
Jain, V. K., **27**, 113
James, B. R., **17**, 319
Jolly, P. W., **8**, 29; **19**, 257
Jonas, K., **19**, 97
Jones, M. D., **27**, 279
Jones, P. R., **15**, 273
Jukes, A. E., **12**, 215
Jutzi, P., **26**, 217
Kaesz, H. D., **3**, 1
Kaminsky, W., **18**, 99
Katz, T. J., **16**, 283
Kawabata, N., **12**, 83
Kemmitt, R. D. W., **27**, 279
Kettle, S. F. A., **10**, 199

Kilner, M., **10**, 115
Kim, H. P., **27**, 51
King, R. B., **2**, 157
Kingston, B. M., **11**, 253
Kitching, W., **4**, 267
Köster, R., **2**, 257
Kreiter, C. G., **26**, 297
Krüger, G., **24**, 1
Kudaroski, R. A., **22**, 129
Kühlein, K., **7**, 241
Kuivila, H. G., **1**, 47
Kumada, M., **6**, 19; **19**, 51
Lappert, M. F., **5**, 225; **9**, 397; **11**, 253; **14**, 345
Lawrence, J. P., **17**, 449
Le Bozec, H., **29**, 163
Lednor, P. W., **14**, 345
Longoni, G., **14**, 285
Luijten, J. G. A., **3**, 397
Lukehart, C. M., **25**, 45
Lupin, M. S., **8**, 211
McKillop, A., **11**, 147
McNally, J. P., **30**, 1
Macomber, D. W., **21**, 1; **25**, 317
Maddox, M. L., **3**, 1
Maguire, J. A., **30**, 99
Maitlis, P. M., **4**, 95
Mann, B. E., **12**, 135; **28**, 397
Manuel, T. A., **3**, 181
Mason, R., **5**, 93
Masters, C., **17**, 61
Matsumura, Y., **14**, 187
Mingos, D. M. P., **15**, 1
Mochel, V. D., **18**, 55
Moedritzer, K., **6**, 171
Morgan, G. L., **9**, 195
Mrowca, J. J., **7**, 157
Müller, G., **24**, 1
Mynott, R., **19**, 257
Nagy, P. L. I., **2**, 325
Nakamura, A., **14**, 245
Nesmeyanov, A. N., **10**, 1
Neumann, W. P., **7**, 241
Norman, N. C., **31**, 91
Ofstead, E. A., **17**, 449
Ohst, H., **25**, 199
Okawara, R., **5**, 137; **14**, 187
Oliver, J. P., **8**, 167; **15**, 235; **16**, 111
Onak, T., **3**, 263
Oosthuizen, H. E., **22**, 209
Otsuka, S., **14**, 245

Pain, G. N., **25**, 237
Parshall, G. W., **7**, 157
Paul, I., **10**, 199
Petrosyan, W. S., **14**, 63
Pettit, R., **1**, 1
Pez, G. P., **19**, 1
Poland, J. S., **9**, 397
Poliakoff, M., **25**, 277
Popa, V., **15**, 113
Pourreau, D. B., **24**, 249
Powell, P., **26**, 125
Pratt, J. M., **11**, 331
Prokai, B., **5**, 225
Pruett, R. L., **17**, 1
Rao, G. S., **27**, 113
Rausch, M. D., **21**, 1; **25**, 317
Reetz, M. T., **16**, 33
Reutov, O. A., **14**, 63
Rijkens, F., **3**, 397
Ritter, J. J., **10**, 237
Rochow, E. G., **9**, 1
Rokicki, A., **28**, 139
Roper, W. R., **7**, 53; **25**, 121
Roundhill, D. M., **13**, 273
Rubezhov, A. Z., **10**, 347
Salerno, G., **17**, 195
Salter, I. D., **29**, 249
Satgé, J., **21**, 241
Schade, C., **27**, 169
Schmidbaur, H., **9**, 259; **14**, 205
Schrauzer, G. N., **2**, 1
Schubert, U., **30**, 151
Schulz, D. N., **18**, 55
Schwebke, G. L., **1**, 89
Setzer, W. N., **24**, 353
Seyferth, D., **14**, 97
Shen, Yanchang (Shen, Y. C.), **20**, 115
Shriver, D. F., **23**, 219
Siebert, W., **18**, 301
Sikora, D. J., **25**, 317
Silverthorn, W. E., **13**, 47
Singleton, E., **22**, 209
Sinn, H., **18**, 99
Skinner, H. A., **2**, 49
Slocum, D. W., **10**, 79
Smallridge, A. J., **30**, 1

Smith, J. D., **13**, 453
Speier, J. L., **17**, 407
Stafford, S. L., **3**, 1
Stańczyk, W., **30**, 243
Stone, F. G. A., **1**, 143; **31**, 53
Su, A. C. L., **17**, 269
Suslick, K. M., **25**, 73
Sutin, L., **28**, 339
Swincer, A. G., **22**, 59
Tamao, K., **6**, 19
Tate, D. P., **18**, 55
Taylor, E. C., **11**, 147
Templeton, J. L., **29**, 1
Thayer, J. S., **5**, 169; **13**, 1; **20**, 313
Theodosiou, I., **26**, 165
Timms, P. L., **15**, 53
Todd, L. J., **8**, 87
Touchard, D., **29**, 163
Treichel, P. M., **1**, 143; **11**, 21
Tsuji, J., **17**, 141
Tsutsui, M., **9**, 361; **16**, 241
Turney, T. W., **15**, 53
Tyfield, S. P., **8**, 117
Usón, R., **28**, 219
Vahrenkamp, H., **22**, 169
van der Kerk, G. J. M., **3**, 397
van Koten, G., **21**, 151
Veith, M., **31**, 269
Vezey, P. N., **15**, 189
von Ragué Schleyer, P., **24**, 353; **27**, 169
Vreize, K., **21**, 151
Wada, M., **5**, 137
Walton, D. R. M., **13**, 453
Wailes, P. C., **9**, 135
Webster, D. E., **15**, 147
Weitz, E., **25**, 277
West, R., **5**, 169; **16**, 1; **23**, 193
Werner, H., **19**, 155
Wiberg, N., **23**, 131; **24**, 179
Wiles, D. R., **11**, 207
Wilke, G., **8**, 29
Winter, M. J., **29**, 101
Wojcicki, A., **11**, 87; **12**, 31
Yashina, N. S., **14**, 63
Ziegler, K., **6**, 1
Zuckerman, J. J., **9**, 21